Aquatic Biology and Ecology

Aquatic Biology and Ecology

Editor: Jim Williams

CALLISTO
REFERENCE

www.callistoreference.com

Callisto Reference,
118-35 Queens Blvd., Suite 400,
Forest Hills, NY 11375, USA

Visit us on the World Wide Web at:
www.callistoreference.com

ISBN: 978-1-64116-137-4 (Hardback)

Cataloging-in-Publication Data

Aquatic biology and ecology / edited by Jim Williams.
    p. cm.
Includes bibliographical references and index.
ISBN 978-1-64116-137-4
1. Aquatic biology. 2. Aquatic ecology. 3. Aquatic sciences. I. Williams, Jim.
QH90 .A68 2019
577.6--dc23

Table of Contents

Preface

In my initial years as a student, I used to run to the library at every possible instance to grab a book and learn something new. Books were my primary source of knowledge and I would not have come such a long way without all that I learnt from them. Thus, when I was approached to edit this book; I became understandably nostalgic. It was an absolute honor to be considered worthy of guiding the current generation as well as those to come. I put all my knowledge and hard work into making this book most beneficial for its readers.

The study of aquatic ecosystems and the organisms that inhabit them are studied in the two fields of aquatic ecology and biology. An aquatic ecosystem is a combination of various biotic communities and abiotic factors. Abiotic factors include parameters of water depth, nutrient level, salinity, temperature, etc. Maintenance of normal values of these variables is essential for the continued sustainability of the diverse flora and fauna of such ecosystems. Aquatic ecosystems are primarily grouped into marine and freshwater ecosystems, along with lentic, pond, lotic and wetlands forming other smaller classes. Aquatic ecosystems are crucial to the efficient recycling of nutrients, purification of water, ground water replenishment and in the provision of habitats to aquatic life. This book provides significant information on these disciplines to help develop a good understanding of aquatic biology and ecology and their related fields. It also includes some of the vital pieces of work being conducted across the world. This book is a vital tool for all researching or studying aquatic sciences as it gives incredible insights into the emerging trends and concepts.

I wish to thank my publisher for supporting me at every step. I would also like to thank all the authors who have contributed their researches in this book. I hope this book will be a valuable contribution to the progress of the field.

Editor

Maternal influence on timing of parturition, fecundity, and larval quality in three shelf rockfishes (*Sebastes* spp.)

David M. Stafford[1,2,*], Susan M. Sogard[2], Steven A. Berkeley[3]

[1]Moss Landing Marine Laboratories, Moss Landing, California 95039, USA
[2]Fisheries Ecology Division, Southwest Fisheries Science Center, National Marine Fisheries Service, NOAA, Santa Cruz, California 95060, USA
[3]University of California Santa Cruz, Santa Cruz, California 95060, USA

ABSTRACT: Differences in maternal investment and reproductive timing can have important consequences for offspring survival. Prior studies on nearshore rockfishes have shown significant effects of maternal age and size on the timing of parturition, fecundity, and larval quality, offering advantages to population persistence by maintaining age diversity in rockfish populations. In this study, reproduction in chilipepper *Sebastes goodei*, widow rockfish *S. entomelas*, and yellowtail rockfish *S. flavidus* was examined to determine whether age- and size-related effects on maternal investment and reproductive timing are exhibited in deeper-dwelling species of this genus. Parturition dates were derived from fine-scale staging of pre-parturition embryos from gravid females. Measurements of embryonic energy reserves (oil globule and yolk), indicators of condition, were used to estimate depletion rates and to test for maternal age and size effects on larval quality. For widow and yellowtail rockfish, larger or older rockfish gave birth earlier in the parturition season than younger, smaller fishes. Maternal factors of weight, length, or age were positively correlated to absolute and relative (weight-specific) fecundity in all species. A trade-off was observed between egg size and fecundity among species, with chilipepper displaying larger egg size and lower fecundity relative to widow and yellowtail rockfish. Embryonic reserves were weakly but significantly related to age only in chilipepper, with embryos from larger, older mothers having larger oil globules. Since the strength of maternal effects varies among *Sebastes* species, information on maternal influence can assist managers in identifying species most likely to benefit from the protection of age structure afforded by marine reserves or other fisheries regulations.

KEY WORDS: *Sebastes* · Maternal effects · Larval quality · Timing of parturition · Fecundity

INTRODUCTION

The commercial, recreational, and ecological importance of rockfishes *Sebastes* spp. merit additional early life history research to understand factors determining their recruitment success or failure. Knowledge of species-specific reproductive phenologies and factors determining both the quality and quantity of offspring is integral to evaluating reproductive success. All nearshore *Sebastes* species researched to date have exhibited some significant effect of maternal size or age on reproductive traits (Berkeley et al. 2004a, Bobko & Berkeley 2004, Fisher et al. 2007, Sogard et al. 2008, Rodgveller et al. 2012). This study extends our understanding of maternal effects to deep-water rockfishes, examining the timing of parturition, fecundity, and larval quality.

*Corresponding author: david.stafford@noaa.gov

Rockfishes are among the longest-lived marine fishes along the Pacific coast (Cailliet et al. 2001) and do not display reproductive senescence (de Bruin et al. 2004). All *Sebastes* spp. are live-bearers and can mate several months prior to egg fertilization. Females have the ability to store sperm, control fertilization, and ultimately determine the timing of parturition (Love et al. 2002). Offspring develop and hatch within the ovaries synchronously. Species north of central California typically extrude 1 brood per year, whereas some southern populations may have an additional brood (MacGregor 1970, Wyllie Echeverria 1987, Love et al. 2002). Gestation period ranges from 1 to 2 mo, with a longer duration in colder temperatures. At birth, larvae are free-swimming and able to begin exogenous feeding in their pelagic environment. High fecundity and typically long reproductive lifespan in rockfishes serve as a bet-hedging strategy to buffer against high variability in recruitment success in the physically dynamic environment of the Northeast Pacific (Leaman & Beamish 1984, Longhurst 2002).

An age/size-related maternal effect on the timing of parturition potentially provides a stabilizing influence on recruitment during periods of poor environmental conditions. In some rockfishes, older females extrude larvae earlier in the season than younger females, effectively extending the parturition season (Nichol & Pikitch 1994, Bobko & Berkeley 2004, Sogard et al. 2008, Rodgveller et al. 2012). This protracted parturition season results in a greater likelihood of larval production during times of peak food availability, as in the match-mismatch hypothesis (Cushing 1990). Sogard et al. (2008) suggest that a progression of earlier parturition dates with increasing female age represents a diversified bet-hedging strategy, effectively spreading annual reproductive effort over time within a maternal lineage despite the constraint of each individual spawning only once each year.

In some nearshore rockfishes, older mothers provide greater nutrient reserves to their offspring compared to younger mothers (Berkeley et al. 2004a, Sogard et al. 2008, Rodgveller et al. 2012). During embryonic development, each larva is provisioned with endogenous energy in the form of yolk containing an oil globule (see Fig. 1) comprised of triacylglycerol lipids (Norton & MacFarlane 1999). At parturition, larvae have absorbed their yolk, but a portion of their oil globule remains, sustaining the larvae after extrusion until they are able to locate prey and feed (Fisher et al. 2007). In laboratory experiments, rockfish larvae from older, larger mothers had larger oil globules and exhibited increased resistance to starvation compared to larvae from younger, smaller mothers (Berkeley et al. 2004a, Sogard et al. 2008).

In many teleosts, absolute fecundity (total number of eggs produced), often used as a proxy for reproductive success, typically increases as maternal size increases (Blaxter 1969, Love et al. 2002). In rockfishes, older, larger females also exhibit increased relative fecundity (number of eggs produced per gram of female body weight), thereby contributing more larvae per unit biomass than younger females (Boehlert et al. 1982, Bobko & Berkeley 2004, Sogard et al. 2008). The strength of this maternal effect on fecundity varies among species of *Sebastes* (Dick 2009). In long-term studies, interannual differences have been found in the fecundities of young (<15 yr old) yellowtail rockfish (Eldridge & Jarvis 1995). Older yellowtail, however, were less affected by environmental influences, thus further demonstrating the role older fishes play in stabilizing populations during poor environmental conditions.

Stock assessment models have assumed constant relative fecundity among mature females such that annual egg production is directly proportional to the population's spawning stock biomass regardless of age or size structure (Berkeley 2006, Dick 2009). For species displaying age- and size-modulated differences in fecundity and larval quality, the assumption of equivalent relative fecundity can result in a misestimation of population productivity. Increasing recognition of the disproportionate contribution of larger, older females to fecundity has resulted in the incorporation of maternal effects in assessments for some but not all *Sebastes* species. Larval quality differences among mothers have received less recognition, although modeling studies suggest that their inclusion may improve estimates of recovery time for overfished populations (Lucero 2008, 2009) and stock productivity (Spencer & Dorn 2013).

This study investigates whether age- and size-related effects in maternal investment and reproductive timing are exhibited in deeper-dwelling species of this genus (chilipepper *Sebastes goodei* and widow rockfish *S. entomelas*). In prior studies (Berkeley et al. 2004a, Sogard et al. 2008), gravid females of nearshore species were held in laboratory tanks until parturition, allowing the precise measurement of larval traits upon completion of development. However, many species that live in deeper habitats are more susceptible to barotrauma-related injuries when brought to the surface, making laboratory-rearing studies impractical. Our approach here was to use fine-scale developmental staging to determine maternal effects on the oil globule depletion rate during embryogenesis and to predict parturition date.

Yellowtail rockfish *S. flavidus* are included in this study to compare methodologies for the prediction of parturition timing, larval quality, and fecundity. Although they are a shelf species, yellowtail rockfish migrate vertically to feed (Pereyra et al. 1969, Pearcy 1992) and, in contrast to most *Sebastes* species, can purge their swim bladders quickly upon ascent (Love et al. 2002), reducing barotrauma-related injuries. Adult females have successfully been held in the laboratory until parturition, allowing the assessment of larval quality, timing of parturition, and fecundity (Eldridge et al. 2002, Fisher et al. 2007, Sogard et al. 2008). Results from laboratory studies of yellowtail rockfish are compared with predictions made from pre-parturition yellowtail rockfish embryos in this study.

MATERIALS AND METHODS

Sampling

Gravid female chilipepper, widow, and yellowtail rockfishes were collected via hook and line fishing at 150 to 400 m depth from Cordell Bank, California (approximate location 38.0666° N, 123.3678° W) from 2005 through 2008 (Table 1). All 3 species give birth in the winter (Table 2, Wyllie Echeverria 1987, Love et al. 2002); thus, sampling began in December and continued bi-weekly, weather permitting, through March or until fish with developing embryos were no longer found. Morphometric data collected from each adult included fork length, total weight, ovary weight, and somatic weight (total weight minus ovary weight).

Sagittal otoliths were extracted from each adult to obtain maternal age data. Age validation has been conducted by marginal increment analysis on all 3 species: yellowtail rockfish (Leaman & Nagtegaal 1987), widow rockfish (Lenarz 1987), and chilipepper (J. Mello pers. comm.). It was therefore assumed that all study species deposit annual growth band pairs, composed of 1 opaque and 1 translucent band (Cailliet et al. 2006), which were counted for age estimates. Otoliths were prepared using the break and burn method (Beamish & Chilton 1982, Laidig et al. 2003), which is routinely used for rockfish stock assessments.

Embryonic staging/timing of parturition

Upon dissection, each fish was assigned a macroscopic maturity stage (Table 3, Gunderson et al.

Table 1. Collection summary reporting sample size and fork length, weight, and age means (±SD) and ranges for study species during collection winters 2005/2006 to 2007/2008 in Cordell Bank, CA. na = not available

| Year/species | n | Fork length (mm) | | Weight (g) | | Age (yr) | |
		Mean ± SD	Range	Mean ± SD	Range	Mean ± SD	Range
2005/2006							
Widow rockfish	91	428 ± 36	324–489	1170 ± 286	463–1775	12 ± 4	5–22
Yellowtail rockfish	144	419 ± 29	320–505	1101 ± 201	475–1782	14 ± 4	6–27
Chilipepper	316	416 ± 30	290–515	1004 ± 215	458–1799	8 ± 3	5–23
2006/2007							
Widow rockfish	na						
Yellowtail rockfish	na						
Chilipepper	53	407 ± 14	363–435	903 ± 100	610–1106	7 ± 1	6–9
2007/2008							
Widow rockfish	66	429 ± 29	382–482	1182 ± 217	859–1627	13 ± 5	7–27
Yellowtail rockfish	na						
Chilipepper	38	435 ± 29	386–502	1109 ± 221	387–1593	12 ± 4	8–26

Table 2. Parturition period, community, and life history traits for rockfishes in this study

Species	Parturition period[a]	Community[b]	Longevity[b,c] (yr)	Length at maturity[b] (mm)	Maximum length[b] (mm)
Widow rockfish	Dec–Apr	Shelf	60	350–380	590
Yellowtail rockfish	Jan–July	Shelf	64	360–450	660
Chilipepper	Nov–Mar	Deep shelf	35	340	590

[a]Wyllie Echeverria (1987); [b]Love et al. (2002); [c]Cailliet et al. (2001)

Table 3. Macroscopic descriptions of stages used to describe female widow rockfish, yellowtail rockfish, and chilipepper ovarian development (Gunderson et al. 1980, Bobko & Berkeley 2004)

Maturity stage	Macroscopic description
1 Immature	Ovary small and translucent or small and yellow.
2 Vitellogenesis	Ovary firm, eggs yellowish and opaque. Widow rockfish eggs are cream in color.
3 Fertilization	Eggs are golden and translucent. Ovary extremely large in relation to body cavity. Ovary wall thin and easily torn.
4 Eyed larvae	Eyes of developing embryos visible, giving ovary an overall greyish color. Ovary fills a large portion of the body cavity.
5 Spent	Ovary flaccid, purplish-red in color. Eyed larvae may still be visible.

Table 4. Fine-scale embryonic staging (Yamada & Kusakari 1991), with converted developmental timing (Eldridge et al. 2002) for yellowtail rockfish. Refer to Fig. 1 for a visual representation of embryonic stages

Yamada/ Kusakari stage	Developmental stage description	Eldridge et al. time to birth (d)
1	Mature unfertilized ovum	
2	Formation of the germ disc	29
3	2-celled ovum	
4	4-celled ovum	
5	8-celled ovum	
6	16-celled ovum	
7	32-celled ovum	
8	64-celled ovum	
9	Morula	26
10	Early blastula	
11	Late blastula	
12	Beginning of epiboly	
13	Early gastrula	24
14	Late gastrula	
15	Embryonic shield	
16	Head fold	22
17	Optic vesicles	
18	Somite formation begins	
19	Finfold	
20	Optic cups and auditory vesicle	18
21	Auditory placodes	
22	Lens formation, motility	
23	Appearance of otoliths	
24	Pectoral fins	
25	Retinal pigmentation	11
26	Blood circulation	
27	Lens transparent	
28	Mouth and anus open	8
29	Pigmentation of the peritoneal wall	
30	Depletion of yolk	
31	Pre-hatching	6
32	Hatching	
33	Hatched, pre-born larva	0

1980, Bobko & Berkeley 2004). Frequency histograms of stage by collection month were created to provide a coarse-scale assessment of differences in the timing of ovarian development during the reproductive season based on maternal size (in 20 mm bins). For finer scale analyses, samples of fertilized eggs and developing larvae were photographed using a Nikon dissection microscope equipped with a high-resolution calibrated camera. Larvae were assigned one of 33 developmental stages, each corresponding to an estimated number of days to birth, according to the classification scheme for kurosoi *Sebastes schlegeli* (Yamada & Kusakari 1991, Table 4, Fig. 1). Gestation time was modified using a conversion formula developed by Eldridge et al. (2002) ($r^2 = 0.94$) for yellowtail rockfish also captured on Cordell Bank:

$$\text{Days to parturition} = -0.022x^2 - 0.097x + 28.9$$

where x = developmental stage following the Yamada & Kusakari (1991) scheme. Under laboratory conditions at 12.0°C, the yellowtail rockfish in Eldridge et al.'s (2002) study had a mean gestation period of 29.2 d (range 27 to 33 d).

Chilipepper and widow rockfish larvae were assumed to follow a similar time course of development under the same temperature regime. Mean sea surface temperatures in the Cordell Bank area ranged from 10.5 to 13.0°C during our collections, suggesting that 12°C provided a reasonable approximation of the temperatures experienced by female rockfish during the well-mixed, pre-upwelling conditions of the winter spawning period.

To calculate parturition date, the estimated number of days to parturition was added to the capture date for each mother. Regression analyses (ordinary least squares, OLS) were used to determine if there were age- or size-related maternal influences on the timing of parturition in each species. Results for yellowtail rockfish were compared with those previously observed for fish held in laboratory tanks until larval release (Sogard et al. 2008).

Fecundity

The gravimetric method (Bagenal & Braum 1978) was used for fecundity analyses. Fecundity estimates

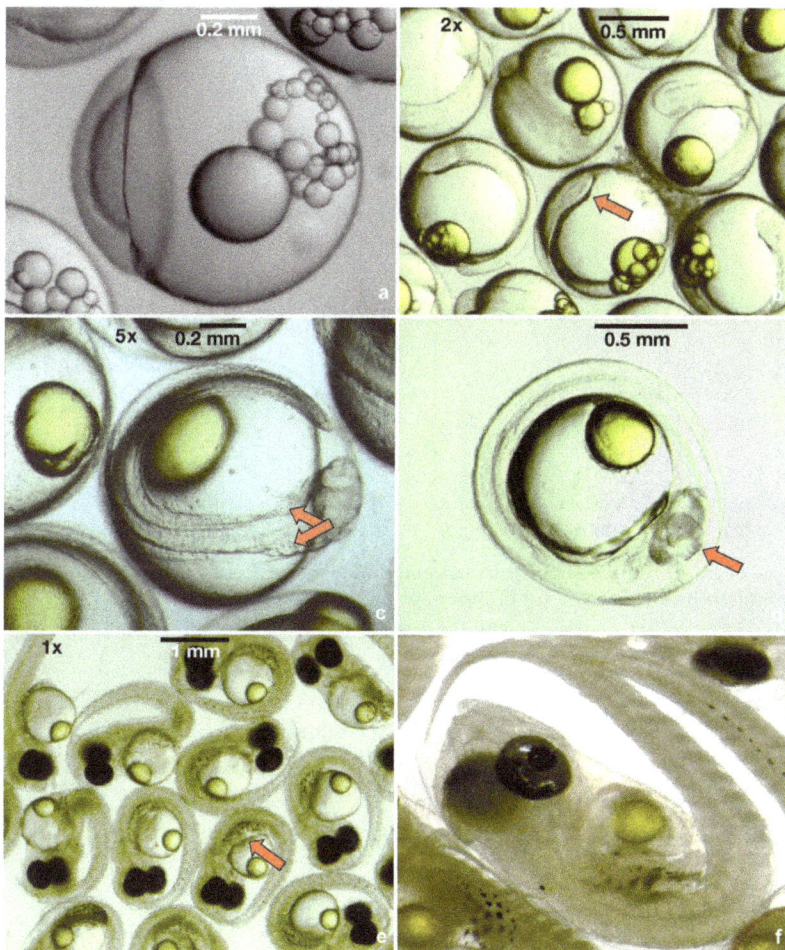

Fig. 1. Examples of embryonic features (highlighted by arrows) used in fine-scale staging of developing chilipepper embryos: (a) late blastula (stage 11), (b) head fold (stage 16), (c) appearance of otoliths (stage 23), (d) pigmentation of retina (stage 25), (e) pigmentation of peritoneal wall (stage 29), and (f) pre-hatching (stage 31). Note the oil globule surrounded by clear yolk in (d)

Egg size comparison among species

A calibrated pen tablet (Wacom Technology) was used to measure the circumference of fertilized eggs in stages <22 (lens formation) from embryos of each species. These stages were chosen to confine measurements to early development, when eggs remain spherical. ANCOVA was used to determine if there were differences in egg size among species, with stage and maternal size (fork length) as covariates. The relationship between egg size and relative fecundity was compared among species at a standardized size of 450 mm fork length.

Larval quality

Mean oil globule volumes from each developing brood of larvae were used to determine the oil utilization rate during gestation for each species. Optimas image analysis software (version 6.51) was used to measure the oil globule volume (mm^3) from 5 to 20 randomly subsampled embryos from each female. The average coefficient of variation for measurements within a female's brood was less than 5%. Since meristics from photographs only offer 2-dimensional viewing, oil globule volumes, assumed to be spherical, were obtained by averaging 2 diameter measurements and calculating the volume for a sphere, $V = (4/3)\pi r^3$. Embryonic yolk utilization rates were also determined using photomicrographs. For the non-spherical yolk sac, a proxy for size was obtained by tracing the circumference of the yolk using a calibrated pen tablet and subtracting the enclosed oil globule circumference.

A regression (OLS) of oil globule volume versus embryonic developmental time to parturition (derived from fine-scale staging as noted above) was used to examine the rate of oil depletion for each species. Models were selected by choosing the regression with the highest coefficient of determination. The residuals of measured oil volumes from this regression were examined to determine if there was a relationship with maternal age and size. Data from previously published laboratory studies on yellowtail

were obtained by counting all of the eggs/larvae in each of 2 weighed subsamples (0.5 to 1 g) per mother and scaling by total ovary weight. The 2 fecundity estimates were averaged to determine total fecundity for each fish.

Relative fecundity was determined by dividing the total fecundity by somatic weight, providing a standardized measure to compare fish of different sizes. Regression analyses (OLS) were used to determine if there were age- or size-related maternal effects on relative fecundity in each species. Data from previous laboratory studies on yellowtail rockfish relative fecundity at parturition (Sogard et al. 2008) were compared with predicted values from pre-parturition samples.

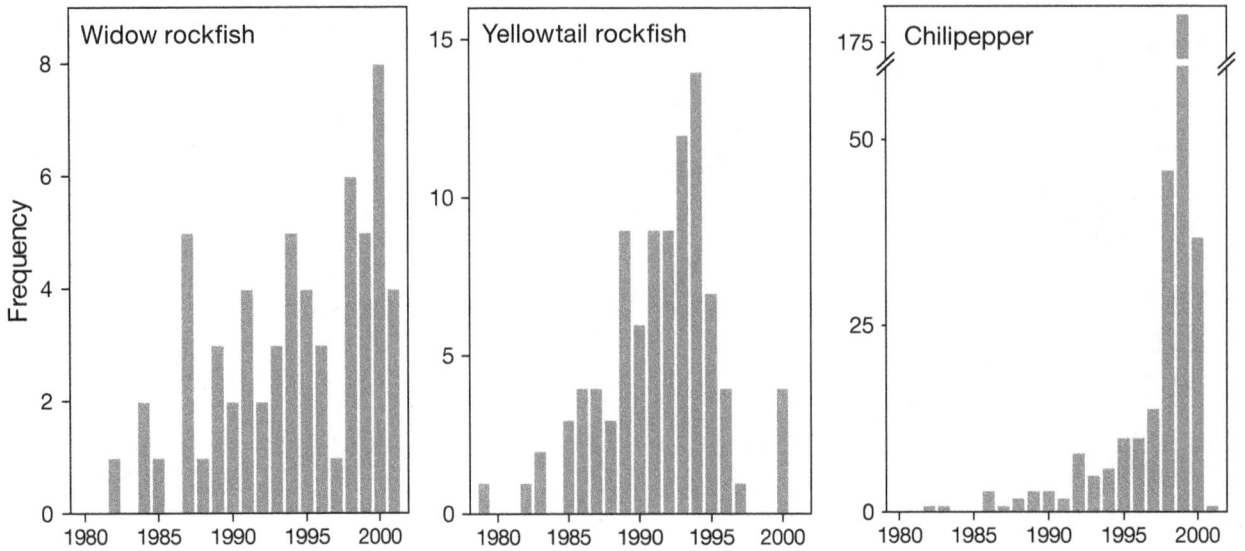

Fig. 2. Year class histogram based on otolith ages of adult female rockfishes collected at Cordell Bank, all collection years combined. Note break in scale for chilipepper

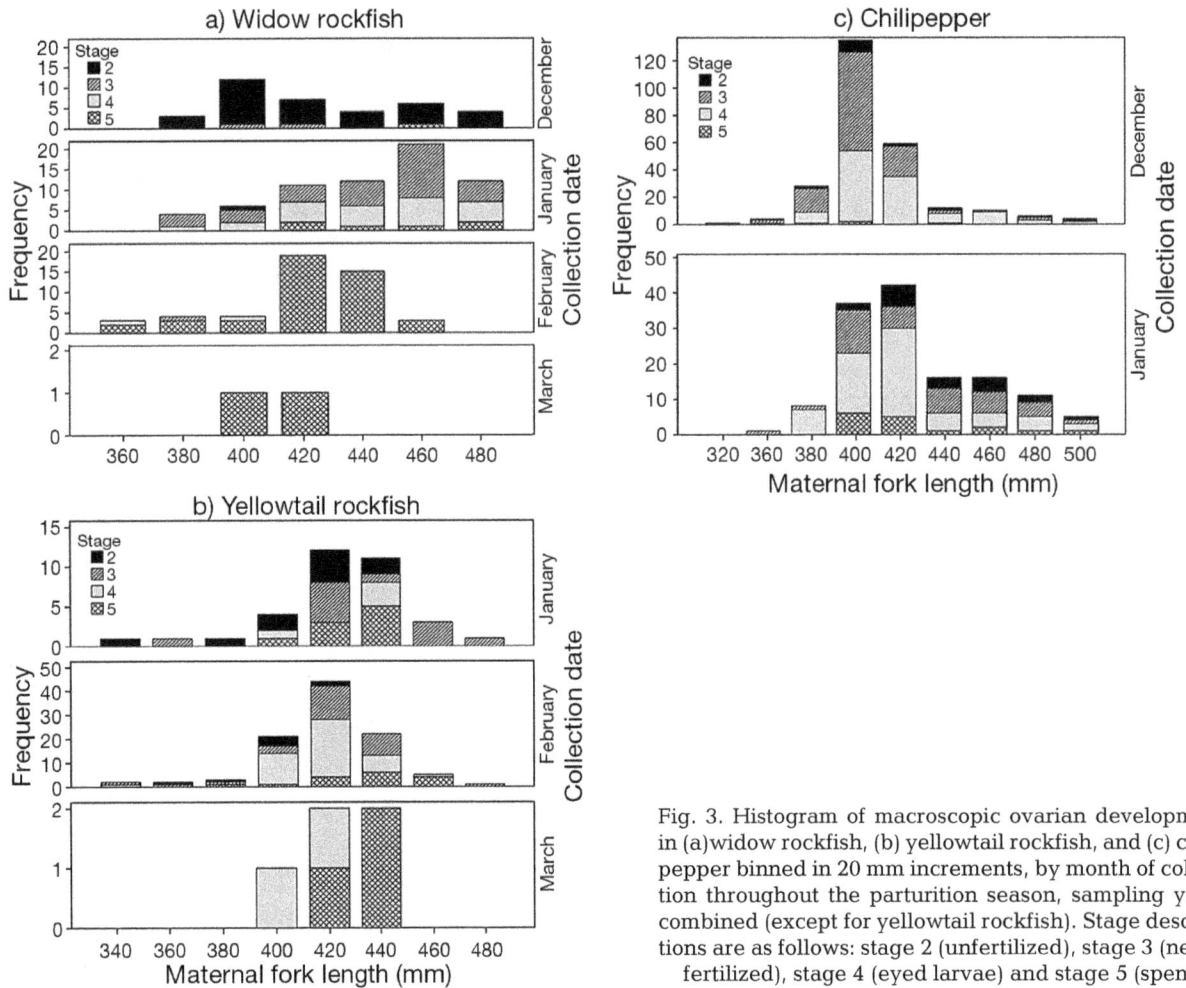

Fig. 3. Histogram of macroscopic ovarian development in (a) widow rockfish, (b) yellowtail rockfish, and (c) chilipepper binned in 20 mm increments, by month of collection throughout the parturition season, sampling years combined (except for yellowtail rockfish). Stage descriptions are as follows: stage 2 (unfertilized), stage 3 (newly fertilized), stage 4 (eyed larvae) and stage 5 (spent)

rockfish oil globule volume at parturition (Sogard et al. 2008) were compared with the observed relationship in this study.

IBM SPSS Statistics (v.19) software was used for all statistical analyses in this study. ANOVAs and linear regressions were performed following verification of the assumptions of normality and homoscedasticity of variance. Preliminary analyses found no interannual differences for any of the response variables; therefore, collection years were pooled.

RESULTS

Collection summary

A broad age and size range of each species was collected (Table 1). Based on female ages derived from otoliths, the proportion of fish in each year class (year of birth) was determined. Chilipepper ranged in age from 5 to 26 yr but were clearly dominated by the 1999 year class (Fig. 2). Yellowtail rockfish ranged from 6 to 27 yr of age. The 1994 year class was most abundant but did not strongly dominate the population, as observed for the 1999 year class of chilipepper. Widow rockfish ranged in age from 5 to 27 yr, with the 2000 year class most abundant but, as with yellowtail, not clearly dominant within the population.

Timing of parturition

Macroscopic staging

Based on macroscopic staging, larger widow rockfish females appeared to fertilize their eggs and initiate embryo development earlier in the parturition season than smaller fish (Fig. 3a). In December, the majority of widow rockfish had unfertilized eggs, whereas a few larger fish (460 mm bin) were already spent. In January, larger fish (420 to 480 mm bins) were further along in ovarian development, with some spent individuals, but none of the smaller size classes (<420 mm) were spent. By February, the largest fish in the collection were spent, but some smaller fish (360 to 400 mm) were still developing larvae.

Macroscopic staging of yellowtail rockfish suggested that larger females extruded larvae earlier in the parturition season (Fig. 3b). In January and February, smaller fish (340 to 440 mm) in the collection had unfertilized eggs or newly fertilized larvae, whereas the majority of the largest fish (>440 mm) were spent.

In chilipepper, macroscopic staging suggested a weak trend of the reverse pattern, with larger fish somewhat delayed in parturition timing compared to smaller fish (Fig. 3c). In December, both small and large chilipepper were represented by all stages. In January, a higher proportion of larger fish (>400 mm) were earlier in development, at stages 2 and 3, compared to smaller fish (<420 mm).

Fine-scale staging

For widow rockfish with developing embryos (macroscopic stages 3 and 4), larger, older females released larvae earlier in the season than smaller, younger fish (Fig. 4). Maternal factors of length, weight, and age were all significantly related to parturition date (linear regression with length: $F_{1,58} = 18.393$, $p < 0.001$, $r^2 = 0.24$, with weight: $F_{1,58} = 21.794$, $p < 0.001$, $r^2 = 0.28$, with age: $F_{1,58} = 6.144$, $p = 0.016$, $r^2 = 0.08$).

Yellowtail rockfish displayed a similar pattern to widow rockfish, with larger, older females giving birth earlier in the parturition season (Fig. 4, linear regression with length: $F_{1,93} = 11.534$, $p = 0.001$, $r^2 = 0.10$, with weight: $F_{1,92} = 22.882$, $p < 0.001$, $r^2 = 0.19$, with age $F_{1,92} = 7.496$, $p = 0.007$, $r^2 = 0.08$). Yellowtail rockfish held in laboratory tanks until parturition (Sogard et al. 2008) showed a similar significant trend, with larger females releasing larvae earlier in the season than smaller females (n = 36, $r^2 = 0.50$).

Chilipepper displayed unusual residual patterns from the linear regression between estimated parturition date and age or size; thus, parametric statistical analyses were not possible. Younger fish released larvae throughout the season, whereas older fish showed a bimodal trend in the timing of parturition (Fig. 4).

Fecundity

Absolute fecundity

Absolute fecundity for widow rockfish ranged between 39 585 and 734 732 larvae per mother and increased with maternal weight, length, and age (Fig. 5a). The highest correlation for the relationship was linear for maternal weight ($F_{1,53} = 45.077$, $p < 0.001$, $r^2 = 0.46$), a power function for maternal length ($F_{1,53} = 55.053$, $p < 0.001$, $r^2 = 0.51$), and logarithmic for maternal age ($F_{1,53} = 18.306$, $p < 0.001$, $r^2 = 0.26$).

In yellowtail rockfish, absolute fecundity ranged between 64 824 and 720 113 larvae per mother (Fig. 5b). The highest correlation for the relationship was lin-

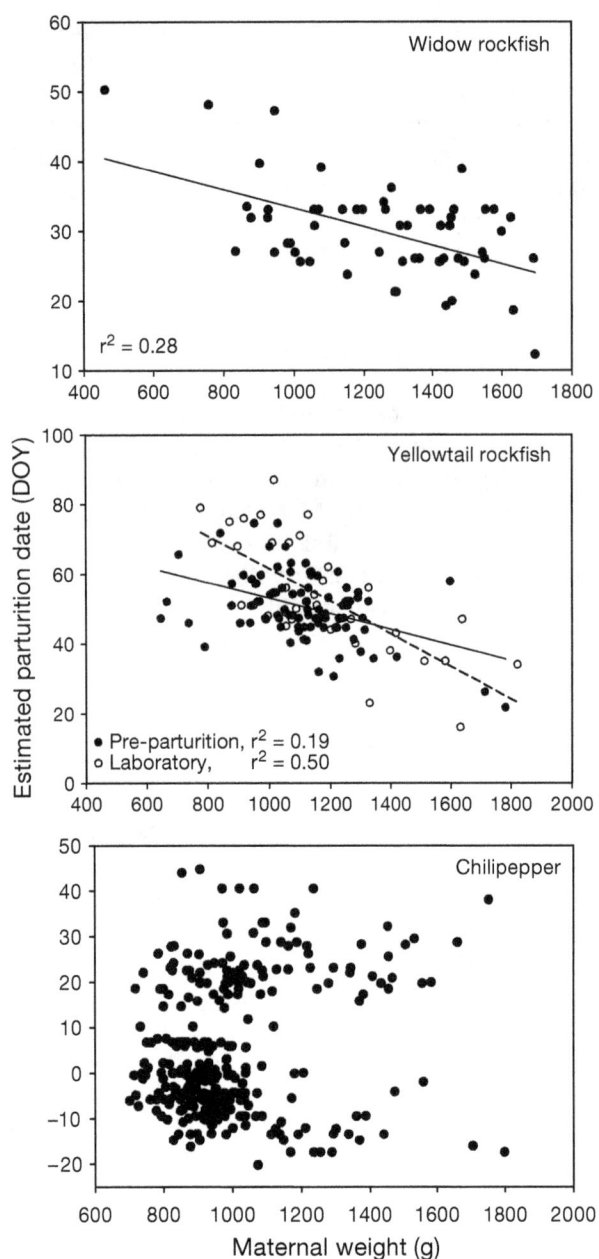

Fig. 4. Maternal influence on parturition date for widow rockfish, yellowtail rockfish, and chilipepper. Each symbol shows the estimated parturition date (day of the year [DOY], 1 = January 1) for the brood of 1 female, based on fine-scale staging of embryos. The maternal factor (age, length, or weight) with the strongest relationship with extrusion date is displayed. Dashed line in (b) presents yellowtail rockfish data from laboratory studies for comparison (Sogard et al. 2008)

ear for maternal weight ($F_{1,91} = 60.474$, p < 0.001, $r^2 = 0.40$), a power function for maternal length ($F_{1,91} = 43.896$, p < 0.001, $r^2 = 0.33$), and logarithmic for maternal age ($F_{1,90} = 11.693$, p = 0.001, $r^2 = 0.12$).

In chilipepper, absolute fecundity increased significantly with maternal size and age and ranged be-

tween 79 382 and 488 119 larvae per mother (Fig. 5c). The highest correlation for the relationship was linear for maternal weight ($F_{1,171} = 354.478$, p < 0.001, $r^2 = 0.74$), a power function for maternal length ($F_{1,171} = 321.679$, p < 0.001, $r^2 = 0.65$), and logarithmic for maternal age ($F_{1,171} = 254.223$, p < 0.001, $r^2 = 0.60$).

Relative fecundity

After adjusting for maternal size by dividing absolute fecundity by female somatic weight, a maternal effect on larval production continued to be evident in widow rockfish (Fig. 5d). Relative fecundity increased with maternal size (linear regression with weight: $F_{1,54} = 5.295$, p = 0.025, $r^2 = 0.13$, power regression with length: $F_{1,54} = 7.757$, p = 0.007, $r^2 = 0.09$) and age (power regression with age: $F_{1,53} = 4.464$, p = 0.039, $r^2 = 0.08$).

There was a positive relationship between maternal size and relative fecundity in yellowtail rockfish (Fig. 5e, linear regression with weight: $F_{1,92} = 11.012$, p = 0.001, $r^2 = 0.11$, linear regression with length: $F_{1,92} = 12.440$, p = 0.001, $r^2 = 0.12$). There was no significant relationship between relative fecundity and age ($F_{1,91} = 0.419$, p = 0.519) in yellowtail rockfish. In comparing yellowtail rockfish pre-parturition relative fecundity with previously published data on fish held in laboratory tanks until parturition (Sogard et al. 2008), both groups displayed a positive relationship between relative fecundity and maternal fork length; however, the coefficient of determination was greater for the laboratory study ($r^2 = 0.34$, Fig. 5e). The slope of the relationship did not differ between the 2 studies (fixed-effects ANCOVA: $F_{1,106} = 0.558$, p = 0.457), suggesting a similar maternal effect, but the intercept was lower in the laboratory study, suggesting a lower overall fecundity estimate compared to that calculated from pre-parturition fish (fixed-effects ANCOVA: $F_{1,107} = 5.616$, p = 0.02).

In chilipepper, relative fecundity increased with maternal size and age (Fig. 5f), with linear relationships providing the highest correlation for weight ($F_{1,171} = 43.314$, p < 0.001, $r^2 = 0.28$), length ($F_{1,171} = 60.582$, p < 0.001, $r^2 = 0.26$), and age ($F_{1,171} = 61.210$, p < 0.001, $r^2 = 0.27$).

Egg size comparison among species

Egg size was significantly different among species (ANCOVA: $F_{2,85} = 66.846$, p < 0.001, maternal length covariate: $F_{1,85} = 3.377$, p = 0.070, stage covariate:

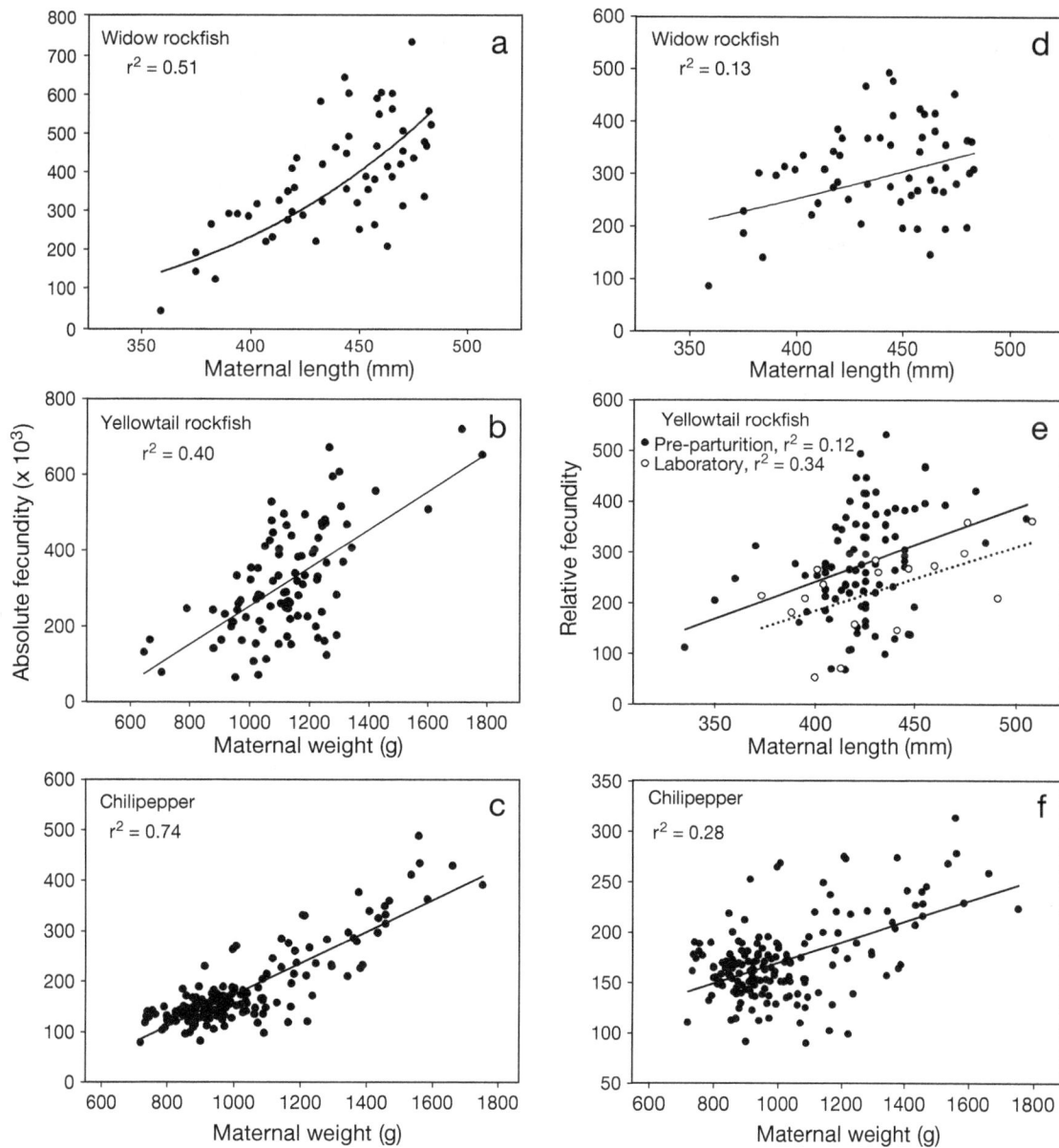

Fig. 5. (a–c) Absolute (no. of larvae) and (d–f) relative (larvae g^{-1} somatic weight) fecundity versus the maternal factor (length or weight, or age) with the highest significant coefficient of determination for (a,d) widow rockfish (power function), (b,e) yellowtail rockfish (linear function), and (c,f) chilipepper (linear function). Dashed line in (e) presents yellowtail rockfish data from laboratory studies for comparison (Sogard et al. 2008)

$F_{1,85}$ = 22.900, p < 0.001). Chilipepper had larger mean size eggs than both yellowtail and widow rockfishes, which did not differ from each other based on Tukey post hoc analyses. When egg size was plotted versus relative fecundity at a common female size of 450 mm, greater egg size was associated with reduced relative fecundity in chilipepper (mean egg circumference 3.30 ± 0.05 mm, mean relative fecundity 194 ± 10.1 larvae g^{-1}) in comparison to that of widow (mean egg circumference 2.60 ± 0.16 mm, mean relative fecundity 314 ± 87.5 larvae g^{-1}) and

yellowtail (mean egg circumference 2.34 ± 0.09 mm, mean relative fecundity 314 ± 50.8 larvae g^{-1}).

Larval quality

Oil and yolk depletion rates

The 2 forms of energy reserves, yolk and oil globule, were utilized differently during embryogenesis (Fig. 6). Early in development, yolk was utilized at a

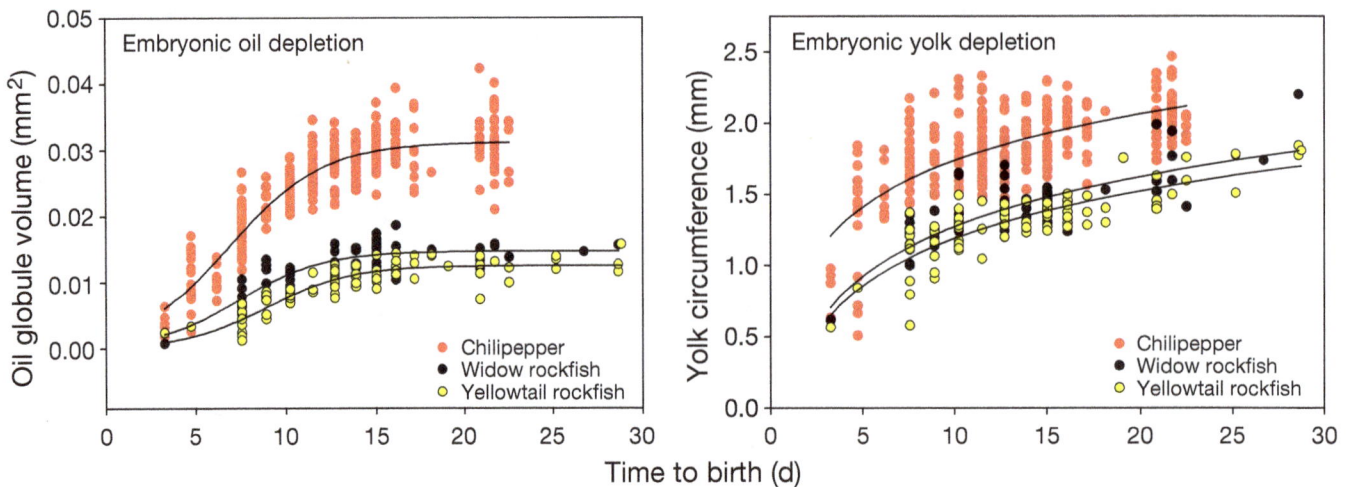

Fig. 6. Energy reserve depletion. Each of the 3 species displayed a sigmoidal trend in the depletion of oil reserves during embryogenesis: chilipepper, $y = 0.0312/[1 + e^{-(x - 6.9512)/2.5956}]$; widow rockfish, $y = 0.0147/[1 + e^{-(x - 7.4240)/2.3289}]$; yellowtail rockfish, $y = 0.0125/[1 + e^{-(x - 8.9894)/2.3041}]$. Yolk reserves depleted in a logarithmic fashion in the 3 species: chilipepper, $y = 0.6489 + 0.4724 \times \ln(x)$; widow rockfish, $y = 0.1084 + 0.5054 \times \ln(x)$; yellowtail rockfish, $y = 0.0813 + 0.4810 \times \ln(x)$

faster rate than oil, followed by the rapid depletion of both resources beginning around 15 d prior to parturition. Just prior to parturition, yolk depletion accelerated whereas oil depletion slowed. In all 3 species, yolk depletion curves followed a logarithmic trend with time, whereas oil depletion displayed a sigmoidal trend. Matching their larger egg sizes, chilipepper embryos were provisioned with more oil and yolk than widow and yellowtail rockfishes.

Relationship between maternal age/size and oil globule volume

Oil globule volume was not related to maternal age, length, or weight in widow or yellowtail rockfishes but was weakly correlated in chilipepper, which had a positive relationship between oil globule volume residuals from the depletion curve and maternal weight (linear regression: $F_{1,304} = 5.523$, p = 0.019, $r^2 = 0.02$, Fig. 7).

DISCUSSION

Timing of parturition

Both methods of analyzing maternal effects on the timing of parturition—macroscopic staging and fine-scale staging—showed analogous patterns for each species, with a clear trend of older, larger fish releasing larvae earlier than younger, smaller fish in widow and yellowtail rockfish and a bimodal pattern for

older chilipepper. Macroscopic staging analysis provided a larger sample size that included stage 2 (unfertilized) and stage 5 (spent) fish and showed the general course of gonadal development for different size classes through time. Fine-scale staging allowed increased precision in estimating the date of parturition but with a reduced sample size. Our estimates of parturition dates based on fine-scale staging relied on the assumption that widow rockfish and chilipepper have similar gestation times to those established for yellowtail rockfish by Eldridge et al. (2002). Few studies have characterized rockfish embryonic development from fertilization through parturition; thus, knowledge of accurate gestation times is limited. Developmental sequence, however, appears to be common in the genus, as observed in studies of yellowtail rockfish (Eldridge et al. 2002), black rockfish *Sebastes melanops* (Boehlert & Yoklavich 1984), and kurosoi *S. schlegeli* (Yamada & Kusakari 1991), although there were variations in the rate of development because of differences in water temperatures. Despite the much greater resource provisioning of chilipepper embryos compared to yellowtail and widow, the 3 species exhibited a strikingly similar pattern of yolk and oil depletion by developmental stage (Fig. 6, where time to birth is a function of stage). Stage definition generally depends on physical features, such as the appearance of otoliths, and is therefore independent of yolk and oil quantities. This similarity in embryo development among species suggests that the assumption of a common, temperature-dependent gestation time is reasonable, but we were not able to test this assumption with our

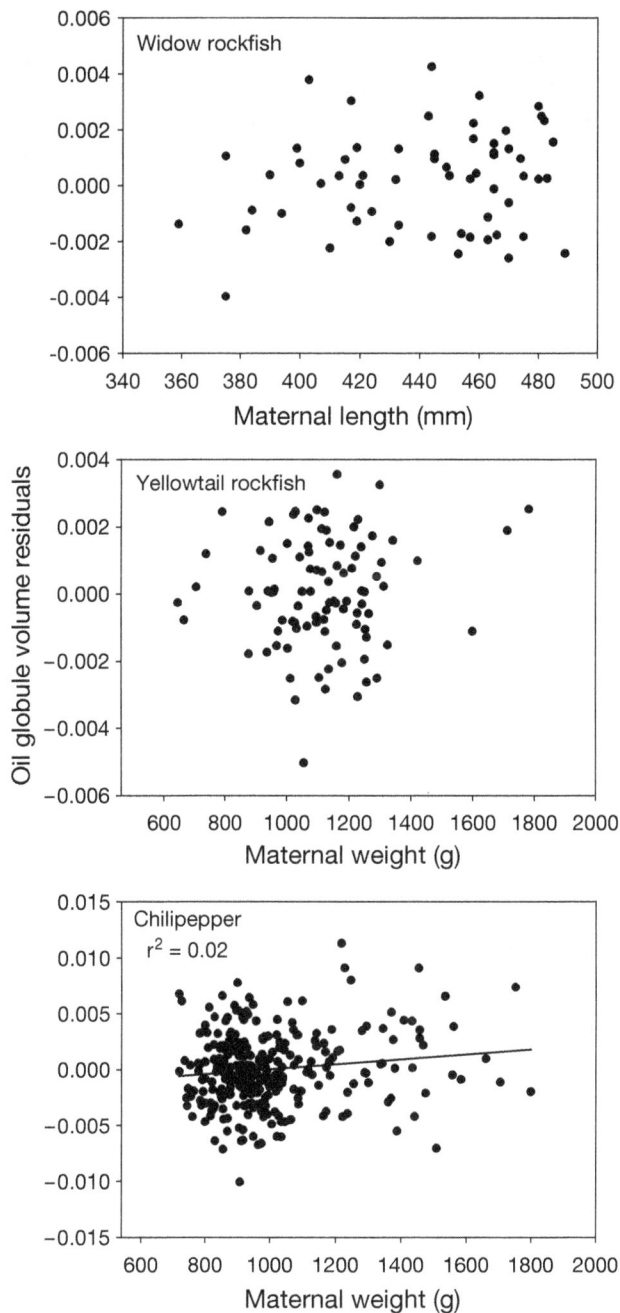

Fig. 7. Maternal influence on oil globule volume. The *y*-axis represents residuals from the sigmoidal depletion curves shown in Fig. 6. The maternal factor with the highest coefficient of determination for each species is shown; the relationship was significant only for chilipepper

can also impact precision in estimates of extrusion date. The extent of individual variability as well as species differences in gestation time need to be addressed in future research.

By giving birth earlier in the season, older, larger yellowtail and widow rockfishes effectively extended parturition at the population level, thus offering an advantage to the population in contending with seasonal environmental variability encountered by newly released larvae in a match-mismatch scenario. At the individual level, this pattern is consistent with the diversified bet-hedging strategy for maternal lineages proposed by Sogard et al. (2008). A pattern of earlier spawning by older, larger females is evident in a phylogenetically diverse range of species (review in Hixon et al. 2013). Age truncation in fish populations, with a consequent shift of reproduction to younger females, has been associated with increased variability in recruitment, potentially a consequence of the restricted period of spawning and increased mismatch with environmental conditions (Hixon et al. 2013).

In contrast to yellowtail and widow rockfish, older chilipepper did not contribute to an extension of the parturition season; the 1999 year class alone spanned the range of observed parturition dates. The bimodal pattern observed in the parturition dates of older chilipepper (Fig. 3) visually suggests multiple parturition within a season; however, we saw no evidence of multiple broods in ovary examination. In a more recent study (Beyer et al. 2014), multiple broods were common in chilipepper collected in a southern population (Santa Barbara Channel) and occurred occasionally in central California populations. Our detection of multiple broods depends on the presence of residual larvae that were not released from the first brood. It is possible that multiple broods occurred in chilipepper during this study but were not detected.

Fecundity

Older, larger fishes produced more offspring in both absolute and relative terms; thus, the assumption that larval production is a function of female biomass regardless of size structure is inappropriate for these species. In general, fecundity estimates derived in this study were comparable to estimates for the 3 species in other studies (Boehlert et al. 1982, Haldorson & Love 1991, Eldridge & Jarvis 1995, Beyer et al. 2014). Likewise, the differences among species, with the reduced absolute fecundity of chilipepper com-

data. If the gestation period is longer or shorter than predicted, the estimated dates of parturition would be comparably expanded or compressed, but any maternal effects would remain. Individual variability in gestation period, documented to range between 27 and 33 d in yellowtail rockfish (Eldridge et al. 2002),

pared to widow or yellowtail rockfish and the much weaker maternal effect on relative fecundity in chilipepper, were consistent with Haldorson & Love's (1991) comparisons of relative fecundity at 50% maturity and at maximum fish size.

Although a significant maternal effect on relative fecundity was observed in all 3 species, there was substantial variability among individual females, suggesting that female age/size only partly determines the allocation of resources to egg production. The success of females in storing energy reserves during the pre-reproductive period potentially plays a major role. For example, Eldridge & Jarvis (1995) found that yellowtail rockfish collected in Washington had greater mesenteric fat reserves and higher relative fecundity than fish collected in central California. Beyer et al. (2014) found several maternal condition indices correlated with relative fecundity in yellowtail, chilipepper, blackgill *Sebastes melanostomus*, and speckled rockfish *S. ovalis*. In particular, liver size, as indexed by the hepato-somatic index, appeared to be influential.

Pre-parturition yellowtail rockfish in this study had higher relative fecundity than laboratory-held yellowtail rockfish collected from the same location in the same year but a similar slope for the maternal effect of increasing relative fecundity with increasing female size (Fig. 5c, Sogard et al. 2008). This difference may reflect the different methods used to calculate fecundity. A gravimetric method was used for pre-parturition fish in this study, whereas fecundity of the laboratory fish was estimated by the collection of all released larvae followed by the use of a plankton splitter to create countable subsamples.

Larval quality

No clear age- or size-related maternal effects were evident on the quantity of lipid provisioning and thus larval quality in the 3 study species. A statistically significant (p = 0.019) effect of increased oil globule volume with increasing maternal weight, observed in chilipepper, was weak ($r^2 = 0.02$) and perhaps biologically insignificant (Fig. 7b).

The lack of a maternal effect on larval quality in yellowtail rockfish in this study was contrary to findings for yellowtail rockfish in laboratory studies, which showed a positive relationship between maternal weight and oil globule size at parturition ($r^2 = 0.23$, Sogard et al. 2008). Laboratory studies have also found maternal effects on larval quality in black, blue (*S. mystinus*), and gopher (*S. carnatus*) rock-

fishes (Berkeley et al. 2004a, Sogard et al. 2008). Variations in methodology between laboratory rearing until parturition versus sampling pre-term gravid females could account for this difference. Measurement of larvae at the time of parturition allows more precise estimates of available energy reserves in the oil globule. Examining pre-parturition larvae allows for greater sample sizes; however, drawbacks include possible misestimation in projected time intervals between fine-scale developmental stages and inaccuracies in depletion rate estimations, including lack of accounting for matrotrophic effects.

Small errors in staging have the potential to greatly influence the oil depletion relationship, particularly during the period of rapid catabolism of oil reserves. During the later stages, depletion rates change over a short time period and are difficult to estimate with small sample sizes. A greater sample size of late-stage embryos (stages 30 to 33, <7 d to birth) would be beneficial to critically analyze yolk and oil utilization in stages just prior to parturition.

Although laboratory-rearing studies cannot duplicate the natural environment completely, they provide standardization by reducing the variability from factors associated with field-caught specimens. These include differences in maternal nutrition, activity levels, and water temperature, which can affect developmental rates of embryos. By allowing females to complete gestation, more precise estimates of parturition timing and larval quality can be obtained.

Species differences

In this study, variability in reproductive strategies among 3 shelf rockfish species was observed, with an apparent trade-off between egg size and fecundity. Chilipepper had the largest egg size and yolk and oil reserves and the lowest absolute and relative fecundity, whereas the reverse pattern was observed for both widow and yellowtail rockfishes. Beyer et al. (2014) also found the mass of chilipepper eggs/larvae to be greater than that of yellowtail at all embryonic stages. Larger larval size at parturition may aid in survivorship because of increased motility for escaping prey and capturing food, while the greater energy reserves in the larger oil globule provide resistance to starvation (Fisher et al. 2007). Several life history traits distinguish chilipepper from widow and yellowtail rockfish (Table 2). Chilipepper tend to be more demersal, their parturition season begins earlier in the winter, they have a shorter lifespan, and they are capable of producing multiple broods. Interestingly, speckled rockfish *Sebastes ovalis* also have low fecundity but large egg size and, like chilipepper, are multiple brooders, begin parturition early, and have a comparable life span of about 34 yr (Beyer et al. 2014). These differences in life history may relate to the importance of quality over quantity for chilipepper larvae.

Conclusions

Widow rockfish, yellowtail rockfish, and chilipepper each exhibited a positive age- or size-related maternal effect on 1 or more reproductive metrics in this study, extending this pattern previously reported in nearshore species (Berkeley et al. 2004a, Bobko & Berkeley 2004, Fisher et al. 2007, Sogard et al. 2008, Rodgveller et al. 2012) to deeper-dwelling rockfishes. The strength and pattern of such maternal effects differ among species, potentially reflecting differences in life histories. These results further support the suggested benefits of preserving age structure in rockfishes as well as accounting for differences in life histories among species. Numerous rockfish species have experienced age/size truncation because of fishing pressure on these slow-growing, late-maturing fishes (Parker et al. 2000, Berkeley et al. 2004b, Harvey et al. 2006). A recent review by Hixon et al. (2013) demonstrates that positive maternal effects on reproductive capacity are common in exploited fish species and that extensive age truncation results in widespread diminished ability to buffer against temporal environmental variability. The effects of continual size/age truncation may be phenotypic and reversible or they may induce evolutionary responses, with less plastic shifts toward reduced growth rate, age at maturity, body size, and productivity (Conover & Munch 2002, Law 2007). Management options that protect older fish, accounting for the phenology of parturition and oceanic regime, can enhance conditions for the successful rebuilding of species in the genus *Sebastes*.

Acknowledgements. We thank G. Cailliet, S. Parker, and L. Allen for their advice and helpful comments. We appreciate R. Powers of the New Sea Angler and the efforts of the volunteer fisherman who assisted in rockfish collection. N. Kashef and N. Parker provided valuable assistance in field work and fish processing. L. Jennings, T. Archer, and S. Brown assisted with fecundity estimation. We thank J. Field for a thorough review of an earlier draft of the paper. We are grateful to D. Pearson for help with ageing. Financial support for this study was provided by California Sea Grant project R/MLPA-03 to S.A.B., and S. Parker, entitled 'Using life history characteristics to determine optimum placement of marine reserves', and by NOAA Fisheries.

LITERATURE CITED

Bagenal TB, Braum E (1978) Eggs and early life history. In: Bagenal T (ed) Methods for assessment of fish production in fresh waters, Blackwell Scientific, Oxford, p 165–201

Beamish RJ, Chilton DE (1982) Preliminary evaluation of a method to determine the age of sablefish (*Anoplopoma fimbria*). Can J Fish Aquat Sci 39:277–287

Berkeley SA (2006) Pacific rockfish management: Are we circling the wagons around the wrong paradigm? Bull Mar Sci 78:655–668

Berkeley SA, Chapman C, Sogard SM (2004a) Maternal age as a determinant of larval growth and survival in a marine fish, *Sebastes melanops*. Ecology 85:1258–1264

Berkeley SA, Hixon MA, Larson RJ, Love MS (2004b) Fisheries sustainability via protection of age structure and spatial distribution of fish populations. Fisheries 29: 23–32

Beyer SG, Sogard SM, Harvey CJ, Field JC (2014) Variability in rockfish (*Sebastes* spp.) fecundity: species contrasts, maternal size effects, and spatial differences. Environ Biol Fishes, doi: 10.1007/s10641-014-0238-7

Blaxter JHS (1969) Development: eggs and larvae. In: Hoar WS, Randall DJ (eds) Fish physiology, Vol 3. Academic Press, New York, NY, p 177–252

Bobko SJ, Berkeley SA (2004) Maturity, ovarian cycle, fecundity, and age-specific parturition of black rockfish (*Sebastes melanops*). Fish Bull 102:418–429

Boehlert GW, Yoklavich MM (1984) Reproduction, embryonic energetics, and the maternal-fetal relationship in the viviparous genus *Sebastes* (Pisces: Scorpaenidae). Biol Bull 167:354–370

Boehlert GW, Barss WH, Lamberson PB (1982) Fecundity of the widow rockfish, *Sebastes entomelas*, off the coast of Oregon. Fish Bull 80:881–884

Cailliet GM, Andrews AH, Burton EJ, Watters DL, Kline DE, Ferry-Graham LA (2001) Age determination and validation studies of marine fishes: Do deep dwellers live longer? Exp Gerontol 36:739–764

Cailliet GM, Smith WD, Mollet HF, Goldman KJ (2006) Age and growth studies of chondrichthyan fishes: the need for consistency in terminology, verification, and growth function fitting. Environ Biol Fishes 77:211–228

Conover DO, Munch SB (2002) Sustaining fisheries yields over evolutionary time scales. Science 297:94–96

Cushing DH (1990) Plankton production and year-class strength in fish populations: an update of the match/mismatch hypothesis. Adv Mar Biol 26:249–293

de Bruin JP, Gosden RG, Finch CE, Leaman BM (2004) Ovarian aging in two species of long-lived rockfish, *Sebastes aleutianus* and *S. alutus*. Biol Reprod 71:1036–1042

Dick EJ (2009) Modeling the reproductive potential of rockfishes (*Sebastes* spp.). PhD thesis, University of California, Santa Cruz

Eldridge MB, Jarvis BM (1995) Temporal and spatial variation in fecundity of yellowtail rockfish. Trans Am Fish Soc 124:16–25

Eldridge MB, Norton EC, Jarvis BM, MacFarlane RB (2002) Energetics of early development in the viviparous yellowtail rockfish. J Fish Biol 61:1122–1134

Fisher R, Sogard SM, Berkeley SA (2007) Trade-offs between size and energy reserves reflect alternative strategies for optimizing larval survival potential in rockfish. Mar Ecol Prog Ser 344:257–270

Gunderson DR, Calahan P, Goiney B (1980) Maturation and

fecundity of four species of *Sebastes*. Mar Fish Rev 42: 74–79

Haldorson L, Love M (1991) Maturity and fecundity in the rockfishes, *Sebastes* spp., a review. Mar Fish Rev 53: 25–31

▶ Harvey CJ, Tolimieri N, Levin PS (2006) Changes in body size, abundance and energy allocation in rockfish assemblages of the northeast Pacific. Ecol Appl 16:1502–1515

▶ Hixon MA, Johnson DW, Sogard SM (2013) BOFFFFs: on the importance of conserving old-growth age structure in fishery populations. ICES J Mar Sci, doi: 10.1093/icesjms/fst200

Laidig TE, Pearson DE, Sinclair LL (2003) Age and growth of blue rockfish (*Sebastes mystinus*) from central and northern California. Fish Bull 101:800–808

▶ Law R (2007) Fisheries-induced evolution: present status and future directions. Mar Ecol Prog Ser 335:271–277

Leaman BM, Beamish RJ (1984) Ecological and management implications of longevity in some northeast Pacific groundfishes. Bulletin INPFC 42:85–97

▶ Leaman BM, Nagtegaal DA (1987) Age validation and revised natural mortality rate for yellowtail rockfish. Trans Am Fish Soc 116:171–175

Lenarz WH (1987) Ageing and growth of widow rockfish. In: Lenarz WH, Gunderson DR (eds) Widow rockfish: proceedings of a workshop, Tiburon, California, December 11–12, 1980. NOAA Tech Rep NMFS 48, p 31–35

▶ Longhurst A (2002) Murphy's law revisited: longevity as a factor in recruitment to fish populations. Fish Res 56: 125–131

Love M, Yoklavich M, Thorsteinson L (2002) The rockfishes of the Northeast Pacific. University of California Press, Berkeley, CA

Lucero Y (2008) Maternal effects and time to recovery. Bull Mar Sci 83:217–234

▶ Lucero Y (2009) A multivariate stock-recruitment function for cohorts with sympatric subclasses: application to

maternal effects in rockfish (genus *Sebastes*). Can J Fish Aquat Sci 66:557–564

MacGregor JS (1970) Fecundity, multiple spawning, and description of the gonads in *Sebastodes*. US Fish Wildl Serv Spec Sci Rep Fish 596

▶ Nichol DG, Pikitch EK (1994) Reproduction of darkblotched rockfish off the Oregon coast. Trans Am Fish Soc 123: 469–481

▶ Norton EC, MacFarlane RB (1999) Lipid class composition of the viviparous yellowtail rockfish over a reproductive cycle. J Fish Biol 54:1287–1289

▶ Parker SJ, Berkeley SA, Golden JT, Gunderson DR and others (2000) Management of Pacific rockfish. Fisheries 25:22–30

Pearcy WG (1992) Movements of acoustically-tagged yellowtail rockfish *Sebastes flavidus* on Heceta Bank, Oregon. Fish Bull 90:726–735

▶ Pereyra WT, Carvey FE Jr, Pearcy WG (1969) *Sebastodes flavidus*, a shelf rockfish feeding on mesopelagic fauna, with consideration of the ecological implications. J Fish Res Board Can 26:2211–2215

Rodgveller CJ, Lunsford CR, Fujioka JT (2012) Effects of maternal age and size on embryonic energy reserves, developmental timing, and fecundity in quillback rockfish (*Sebastes maliger*). Fish Bull 110:36–45

▶ Sogard SM, Berkeley SA, Fisher R (2008) Maternal effects in rockfishes *Sebastes* spp.: a comparison among species. Mar Ecol Prog Ser 360:227–236

▶ Spencer PD, Dorn MW (2013) Incorporation of weight-specific relative fecundity and maternal effects in larval survival into stock assessments. Fish Res 138:159–167

Wyllie Echeverria T (1987) Thirty-four species of California rockfishes: maturity and seasonality of reproduction. Fish Bull 85:229–250

▶ Yamada J, Kusakari M (1991) Staging and the time course of embryonic development in kurosoi, *Sebastes schlegeli*. Environ Biol Fishes 30:103–111

Normal and histopathological organization of the opercular bone and vertebrae in gilthead sea bream *Sparus aurata*

Juan B. Ortiz-Delgado[1,*], Ignacio Fernández[2,3], Carmen Sarasquete[1], Enric Gisbert[2]

[1]Instituto de Ciencias Marinas de Andalucía-ICMAN/CSIC, Campus Universitario Río San Pedro, Apdo. Oficial, 11510, Puerto Real, Cádiz, Spain

[2]IRTA, Centre de Sant Carles de la Ràpita (IRTA-SCR), Unitat de Cultius Experimentals, Crta. del Poble Nou s/n, 43540 Sant Carles de la Ràpita, Spain

[3]Centre of Marine Sciences (CCMAR), University of Algarve, Campus de Gambelas, 8005-139 Faro, Portugal

ABSTRACT: This study provides a comprehensive description of the tissue organization of non-deformed and deformed opercula and vertebrae from gilthead sea bream *Sparus aurata* juveniles by means of histological, histochemical and immunohistochemical approaches. Two types of opercular anomalies are described: the folding of the opercle and subopercle into the gill chamber, starting at the upper corner of the branchial cleft and extending down to its lower third; and the partial lack of the operculum (opercle, subopercle, interopercle and preopercle underdeveloped) with a regression of the loose edge extending down to its lower third. Histological observations revealed a rare type of bone remodelling process in the opercular structure, which consisted of the coalescence of contacting bone tissues (presumably from the preopercle and opercle), resulting in skeletal tissue with a trabecular aspect filled by a single-cell epithelium of cubic osteoblastic-like cells. Differences in collagen fiber thickness and its 3-dimensional arrangement between normal and deformed opercula were also found. Lordotic vertebrae were characterized by the formation of fibrous cartilage in the haemal and/or neural sides, indicating that a metaplastic shift occurred during the process of lordosis. Another major histomorphological change found in lordotic vertebrae was the complete loss of notochordal sheath integrity. Histological alterations were coupled with an imbalance of cell death and cell proliferation processes in lordotic vertebrae as well as that of bone formation/resorption, and extracellular matrix deposition activity differences which might have resulted from the remodelling process occurring in lordotic vertebrae. Altogether, these results provide an increase in our basic knowledge of bone disorders that contribute to our understanding of the mechanisms by which these skeletal anomalies appear in this fish species and which hamper its production efficiency.

KEY WORDS: Bone · Extracellular matrix · Histology · Histochemistry · Immunohistochemistry · Skeletal deformities

INTRODUCTION

During the last decades, the finfish larviculture industry has considerably improved its rearing methods under intensive conditions, but larval quality remains one of the main problems for the proper sustainable success of this productive sector (Koumoundouros 2010, Boglione et al. 2013a). The quality of farmed fish is directly related to the quality of the fry, and depends on both organoleptic and morphological characteristics that should be as similar as possible to that of wild fish, which is considered to be the

*Corresponding author: juanbosco.ortiz@icman.csic.es

quality reference by the consumer. In this sense, an anomalous external morphology (even of a few fish) could substantially decrease the consumers' overall perception of aquaculture products (Boglione et al. 2013a). Skeletal deformities have been reported in natural environments (Gavaia et al. 2009, Diggles 2013), but with lower prevalence than in intensive aquaculture. The high incidence of skeletal deformities found in farmed fish (affecting up to 30% of production) might be due to (1) the absence of predators and the constant and high availability of food, which maximizes fish survival, or (2) that industrial hatcheries still lack the proper knowledge of how to produce high-quality and healthy fry under intensive rearing conditions in order to reduce the causative factors by which skeletal deformities are induced (Boglione et al. 2013a,b). However, since in all reared batches the skeletal deformity incidences are largely different, whether the fish come from the same or from different industrial hatcheries, the second hypothesis seems more plausible. Nevertheless, skeletal deformities are one of the most significant biological and recurrent problems affecting worldwide finfish aquaculture (see reviews in Koumoundouros 2010, Boglione et al. 2013a, Cobcroft & Battaglene 2013).

Fish skeletogenesis involves the activity of 3 main cell types: chondrocytes, osteoblasts and osteoclasts (see review in Boglione et al. 2013b). The first 2 cell types secrete the extracellular matrix proteins of the cartilage and bone, respectively, while the third type produces cathepsins, matrix metalloproteinases and tartrate-resistant acid phosphatase (TRAP), providing an acidic environment in which the mineralized matrix is broken down (Ytteborg et al. 2010). In addition, skeletogenesis involves a fourth type of cell: the osteocytes, which are involved in maintaining bone matrix and acting as mechanical load receptors. In teleosts, the lack of osteocytes (acellular bone, a feature of the *Sparidae* family) implies that induced bone remodelling by mechanical load is triggered by cell types other than osteocytes (reviewed in Boglione et al. 2013a). Bone formation is brought about by a complex set of tightly regulated molecular pathways, involving extracellular matrix (ECM) constituents (e.g. alkaline phosphatase, ALP) or non-collagenous proteins such as the matrix Gla protein (MGP) or osteocalcin (OC), signalling molecules (e.g. hedgehogs and bone morphogenetic proteins) and transcription factors (Karsenty 2001, Boglione et al. 2013b). Hence, the perturbation of some of the above-mentioned factors in controlling bone development and homeostasis, as well as the incapacity of homeorhetic mechanisms to compensate for stressful environmental conditions, may disrupt skeletogenesis, ultimately leading to the appearance of skeletal deformities (Boglione et al. 2013a). With the exception of environmental contaminants and pathogens (which are generally controlled under rearing conditions), different studies suggest that unfavourable environmental biotic and abiotic conditions, inappropriate nutrition and/or genetic factors are the most probable causative agents of skeletal abnormalities in reared fish (reviewed in Boglione et al. 2013b). Previous studies have indicated that skeletal anomalies are mainly induced in early stages of development (i.e. during embryonic and larval periods), occurring long before osteological deformities are externally visible (Daoulas et al. 1991, Koumoundouros et al. 1997a,b, Gavaia et al. 2009, Boglione et al. 2013b). Nevertheless, skeletal deformities could also be induced at later stages during the nursery phase (Boglione et al. 2013a). Regardless of when skeletal deformities are diagnosed, at juvenile or later stages of development (on-growing phase), it is difficult to gather information about their onset and aetiology since this is considered a multifactorial problem (Witten et al. 2005, 2006, Boglione et al. 2013a).

Although gilthead sea bream *Sparus aurata* L. is one of the most important species in European aquaculture, cranial and spinal abnormalities are still one of the major problems that hinder the efficiency of the production cycle of this sparid species. Vertebral deformities related to the non-inflation of the swimbladder (mostly lordosis) were the most frequent abnormalities recorded at the beginning of gilthead sea bream intensive culture (1980s and early 1990s) (Chatain 1994, Andrades et al. 1996). Although the improvement and refinement of rearing techniques have considerably reduced the frequency of these types of deformities (Boglione et al. 2013a), lordosis and kyphosis are still 2 of the most relevant osteological abnormalities observed in the axial skeleton of reared gilthead sea bream, while a partial and/or total lack of operculum is the most frequent anomaly affecting its cranial region (Koumoundouros et al. 1997b, Boglione et al. 2001, 2013b, Beraldo et al. 2003, Verhaegen et al. 2007, Castro et al. 2008, Koumoundouros 2010, Prestinicola et al. 2013). Several studies have described skeletogenesis in gilthead sea bream (Faustino & Power 1998, 1999, 2001) and the typology of fish skeletal deformities (Witten et al. 2009, Boglione et al. 2013b, Prestinicola et al. 2013, among others) in order to better understand the phases of skeletogenesis and the effects of biotic and abiotic factors on the proper development of bone and the appearance of skeletal anomalies. Various procedures have been used as simple and rapid diagnostic tools for studying skeletal deformities in fish, such as X-rays, double staining or even computer tomography; allowing the identification of abnormal growth in different skeletal structures (see reviews in Witten et al. 2009, Koumoundouros 2010, Boglione et al. 2013b). Although there is a great variety of research in fish skeletal histology (reviewed in Meunier 2011), there is limited information about the histological organization of skeletal tissues in gilthead sea bream (Galeotti et al. 2000, Beraldo et al. 2003, Fernández et al. 2012). The use of histological procedures is considered a valuable tool for providing basic knowledge on bone formation, as well as providing new insights into the structural changes occurring in deformed skeletal structures (Ytteborg et al. 2012).

The aim of this study was to perform a comparative analysis of the tissue organization and distribution, as well as changes in ECM composition, between normal and deformed opercular bones and vertebrae from hatchery-reared gilthead sea bream juveniles. This description was conducted by means of histological, histochemical and immunohistochemical approaches in order to unveil the organization of these skeletal structures and the possible mechanisms by which the above-mentioned skeletal disorders develop.

MATERIALS AND METHODS

Animals

Gilthead sea bream larvae were reared under standard conditions in two 500 l cylindroconical tanks (initial density of 100 larvae l^{-1}) up to 60 d post hatch (dph) as described in Fernández et al. (2008). At that age (ca. 80 to 100 mg wet body weight), fry were reared in a 2000 l tank connected to a recirculation unit (IRTAmar™) until they weighted ca. 2 g (90 dph). During this period, fish were fed commercial diets following standard procedures according to Ortega (2008). Water conditions were as follows: 18 to 19°C; 35 ppt salinity; pH between 7.8 and 8.2; daily water renewal in the rearing tank was 20%, and the tank had gentle aeration (oxygen levels = 6.2 ± 1.1 mg l^{-1}). Photoperiod was 12:12 h light:dark, and the light intensity at water surface was 500 lux. From 90 dph, fish were kept in an outdoor open-flow 14 000 l tank until they were 10 mo old (35.5 ± 15 g body weight, 10.5 ± 1.5 cm standard length). During this period, fish were subjected to natural conditions in terms of thermo- and photoperiod regimes. The overall survival rate of the complete rearing process was estimated at ca. 15 to 20%.

After a visual examination of 1064 live specimens in sorting tasks, normal and deformed specimens with underdeveloped operculum and lordosis were sampled (7 specimens for each morphotype), sacrificed with an overdose of anaesthetic (tricaine methanesulfonate, MS-222; Sigma-Aldrich) and dissected.

Histological, histochemical and immunohistochemical procedures

Normal and deformed opercula and vertebral centra were fixed in 4% buffered paraformaldehyde (in phosphate buffer saline, PBS; pH 7.4). After fixation, samples were washed in PTW (1% PBS, 0.1% Tween-20) 3 times for 30 min each, then either preserved in methanol at −20°C or immersed for decalcification in a 10% EDTA/2% formaldehyde solution at 4°C for 7 d, followed by washes in PTW as described above. After fixation and decalcification, samples were washed, embedded in paraffin and serially sectioned at 5 to 6 µm thickness as previously described (Ortiz-Delgado et al. 2005). Undecalcified samples were subjected to the von Kossa staining method: a precipitation process in which silver ions react with phosphate in the presence of acidic medium (Clark 1981). The von Kossa staining method alone may not be sufficient to confirm mineralization, since it does not necessarily imply the presence of calcium or of hydroxyapatite, the mineral phase of bone (Bonewald et al. 2003). Although further evidence of bone mineralization may be needed using alternative procedures (Bonewald et al. 2003), in the present study, and considering morphological and molecular data from Tiago et al. (2011), we used von Kossa staining as a first approximation for whether structures were mineralized or not.

For studying ECM components of bone tissue and cartilage, adjacent sections were stained with the following techniques: haematoxylin and eosin (H/E) for histomorphological observation, orcein for elastic fibers, van Gieson trichrome and picrosirius red staining (PRS) for collagen, Mallory and haematoxylin/VOF (H/VOF) trichrome (Gutiérrez 1967) for connective tissue and Alcian Blue (AB) 8GX (Sigma) at pH 2.5/periodic acid-Schiff (PAS) for acidic and neutral mucopolysaccharides. Dye affinity or intensity of the above-mentioned staining techniques from different regions of studied tissues was ranked by visual scoring by at least 2 trained observers (10 slices per technique and 5 sections per slice) as: negative (−), very weak (+−), weak (+), moderate (++) and strong (+++). Values were scored considering the accumulation levels of the dye for each considered vertebral part, and ranged from the absence of staining (negative) to the highest stain intensity (strong), whereas intermediate scoring staining categories were established as intermediate ranges among the above-mentioned extreme scores. These protocols were conducted according to monographs by Martoja et al. (1970), Pearse (1985), Bancroft & Stevens (1990) and Sarasquete & Gutiérrez (2005). Additionally, PRS sections were observed under polarized light according to Junqueira et al. (1979) and Kiernan (2008). Briefly, after PRS and under bright field microscopy, collagen appears red on a pale yellow background. A BX41 Olympus light microscope equipped with 2 kinds of filters (a polarizer and an analyzer) was used to provide linearly polarized illumination. Digital images were obtained with a C3030 Olympus digital camera.

Bone remodelling/homeostasis, a dynamic process that relies on the correct balance between bone resorption by osteoclasts and bone deposition by osteoblasts (Boglione et al. 2013b) was described by means of staining for TRAP and ALP activities, respectively. Both enzyme activities were visualized with the acid and alkaline phosphatase leukocyte kits (Sigma-Aldrich) according to the instructions of the manufacturers.

Immunohistochemistry for ECM proteins was performed using a rabbit polyclonal antibody developed against OC and MGP from meagre (*Argyrosomus regius* Asso, 1801) as primary antibodies. Anti meagre-OC and -MGP antibodies presented a specific cross reactivity for gilthead sea bream-OC and -MGP (Simes et al. 2004). Immunohistochemical staining was performed with peroxidase-conjugated anti-rabbit IgG (Sigma) as secondary antibody, using the methodology proposed by Simes et al. (2003, 2004). Briefly, serial sections were incubated overnight with anti-OC and anti-MGP primary antibodies at appropriate dilutions (1:500 for OC and 1:250 for MGP). After several washes in PBS, sections were incubated for 1 h with goat anti-rabbit IgG peroxidase-conjugated, and stained with 3-3'-diaminobenzidine-DAB (Sigma). The endogenous peroxidase activity was blocked with 3% H_2O_2 before staining. Sections incubated with normal fish serum instead of the primary antibody were used as negative controls.

Cell proliferation and apoptosis processes were also assessed by immunohistochemical detection of proliferating cell nuclear antigen (PCNA) and cleaved caspase-3, respectively. The former assays were performed as described by Bakke-McKellep et al. (2007) for PCNA and Sanden et al. (2005) for caspase-3, using a mouse anti-PCNA (Santa Cruz Biotechno-

logy) and cleaved caspase-3 antibody (Cell Signalling), respectively. Sections incubated with normal fish serum instead of the primary antibody or without substrate incubation were used as negative controls. After staining, all sections were dehydrated by air drying and mounted with EUKITT (Labolan).

RESULTS

Macroscopically, we detected 2 types of opercular anomalies according to the classification previously described by Beraldo et al. (2003): (1) the folding of the operculum (opercle and subopercle) into the gill chamber, starting at the upper corner of the branchial cleft and extending down to its lower third (Type I; Fig. 1a), and (2) the partial lack of the operculum (lack of development of the opercle, subopercle, interopercle and preopercle) with a regression of the loose edge extending down to its lower third (Type II; Fig. 1b). The overall incidence of fish with deformed operculum was 6.3%. These opercular abnormalities unilaterally affected both sides of the head: the right (3.3%) and left (3.0%) equally; whereas the bilateral abnormality in the operculum only affected 1.1% of the reared fish.

Fish with a severe external body shape deformity (Fig. 1c) showed internal lesions characterised by an accentuated ventral curvature (lordosis) of the vertebral column (Fig. 1d). The degree of these pathological symptoms varied along the vertebral column axis and mainly affected vertebrae located between the limit of pre-haemal and haemal areas of the vertebral column (position 11 to 15 from urostyle upwards). The incidence of fish affected by lordosis was 10.1%.

Histological organization of normal and deformed operculum

The operculum is a complex structure composed of 4 distinct and articulated bony plates: the opercle, preopercle, interopercle and subopercle. At a histological level, the opercular complex of gilthead sea bream shows a regular succession of tissue layers: outer and inner skin layers, and an intermediate layer of bone (Fig. 2a). The posterior rim of the operculum is a flexible ribbed structure, which closes the gill cavity like a lid.

Transversal sections at the mid region of normal operculum stained with the PRS technique showed a preopercle mainly composed of highly trabeculated bony tissue, and a plate-like opercle with solid bone

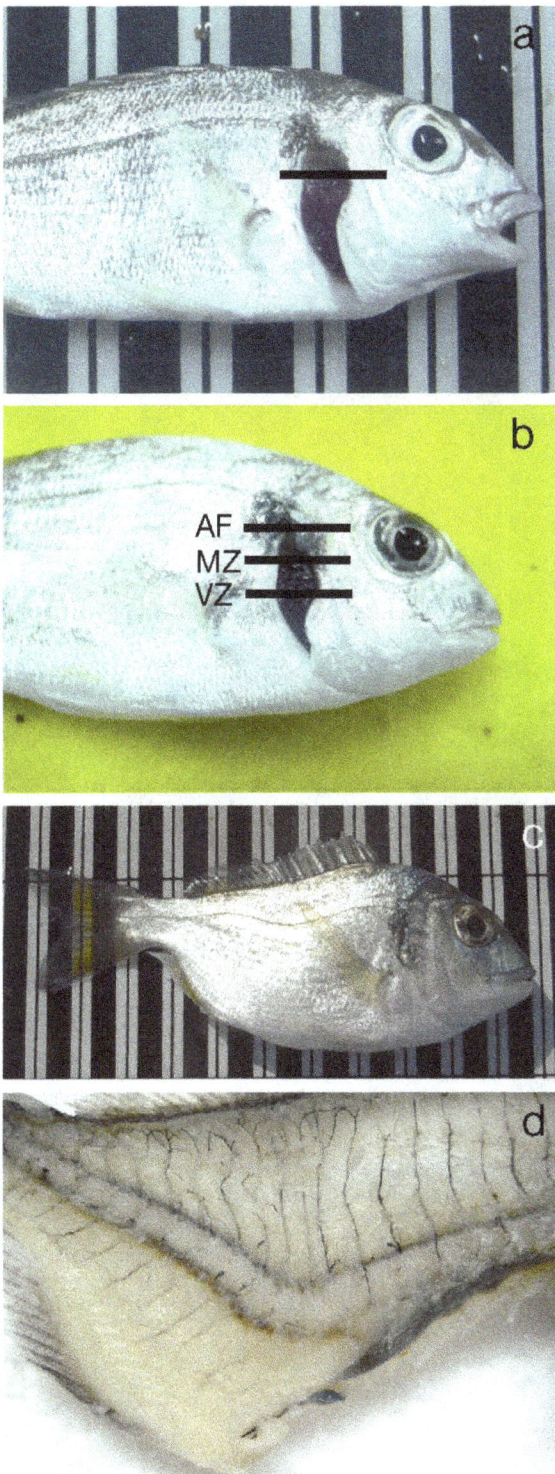

Fig. 1. Macroscopical appearance of deformed operculum and vertebra in gilthead sea bream *Sparus aurata* specimens. (a) Type I and (b) Type II opercular malformations and (c) external and (d) internal appearance of a lordotic fish. Note the accentuated ventral curvature and the upwards bending of vertebral column. Bars indicate the level of transversal dissected regions. Abbreviations: AF, articulatory facet; MZ, medial zone; VZ, ventral zone

and homogeneous mineralisation levels (Fig. 2b–e, Table 1). The anterior rim showed medially directed trabecular projections at the level of the articulatory facet that were absent in the caudal direction (Fig. 2b,c). In addition, the PRS evidenced some differences in composition of both preopercle and opercle bones. While the preopercle had a predominant green colour with scarce areas of red (orange under polarized light), the opercle showed reticular fibers with a homogeneous greenish staining (green under polarized light) (Fig. 2d,e).

The 2 different patterns of opercular deformities reported in this study showed some differences with regards to the tissue layers involved in the osteological anomaly. The opercular Type I deformity corresponded to a folded operculum, directed towards the inside of the gill cavity. Although the deformed operculum affected all tissue layers, a severely curled opercle and/or subopercle concomitant with a proliferation of epithelial tissue within the folded area were detected (Fig. 2f,g). Additionally, hypoplasia of muscular fibers and enlargement of trabecular spaces in the rostral portion of the opercle were also observed. Polarized light revealed some differences in bone composition of normal and deformed opercula (e.g. changes in colour stain); the opercle and subopercle from deformed fish presented a brighter green colour than normal ones (Fig. 2g). In contrast, the opercular Type II deformity (Fig. 2h–k) corresponded to a hypoplasic operculum in which some bone structures had disappeared depending on the dissected region (articulatory facet, medial and ventral zones; see Fig. 1b). For instance, the dissected operculum at the ventral zone showed a regression of the subopercle and a complete lack of the opercle. In this case, fibrous and epithelial cells proliferated and filled the empty space; thus, as a result of a lack of bone support, the outer and inner skin layers of the opercle folded into the gill cavity (Fig. 2h). At the medial opercular zone, the formation of an outward folding of the opercle was detected (H/VOF; Fig. 2i), as well as a thickening of the skin epithelium. Finally, the dissection of the dorsal area of the operculum revealed a modified articulatory facet in which a coalescence of contacting bony tissues was detected, as well as the formation of a chondroid bone-like structure. The trabecular structure of the opercle and subopercle was also modified; the trabecular spaces were filled with a single-cell epithelium (presumably osteoblasts) (Fig. 2j,k). All skeletal tissue involved in this deformity was ossified, as revealed by von Kossa staining (data not shown).

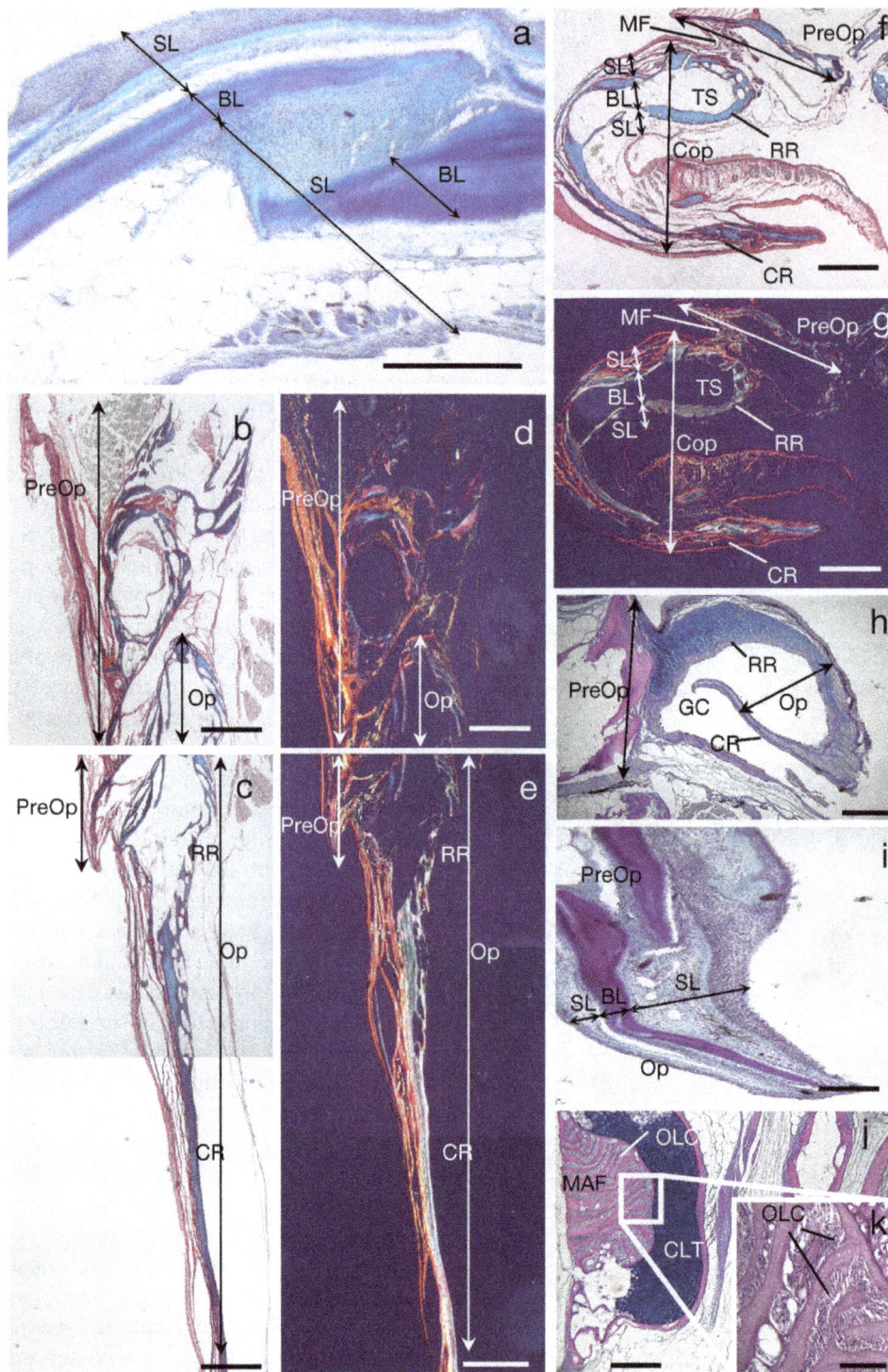

Fig. 2. Histological sections of (a–e) normal and (f–k) deformed opercula in gilthead sea bream *Sparus aurata* specimens. (a) Regular succession of skin and bone layers stained with H/VOF; (b,c) transversal sections at the mid region of normal operculum showing the trabecular bone forming the preopercle and the rostral portion of opercle, and the plate-like solid bone forming the caudal region of the opercle; (d,e) mixture of red and green colour of the preopercular bone and the homogeneous greenish staining of the opercle under polarized light; micrographs of opercula deformities Type I and Type 2 with: picrosirius red staining under (f) non-polarized and (g) polarized light; H/VOF staining at (h) ventral and (i) medial zones, AB pH 2.5/PAS staining of (j) the modified articulatory facet and (k) its magnification of deformed *S. aurata* opercula. Scale bars = 50 µm. Abbreviations: BL, bone layer; Cop, curled operculum; CLT, cartilage-like tissue; CR, caudal region; GC, gill chamber; MF, muscle fibers; MAF, modified articulatory facet; OLC, osteoblastic-like cells; Op, opercle; PreOp, preopercle; RR, rostral region; SL, skin layer; TS, trabecular spaces

Table 1. Staining properties of each part of the operculum in gilthead sea bream *Sparus aurata*. Staining intensity: (−) negative, (+−) very weak, (+) weak, (++) moderate, (+++) strong. See 'Materials and methods' for details

	H/VOF	van Gieson	PAS	AB	von Kossa	Mallory	Orcein
Preoperculum							
Outer layer	Light green	Pink	−/+−	−	−	Red	Brown
Intermediate layer	Purple	Deep red	−/+−	−	+++	Blue	Pink
Inner layer	Ligth green	Pink	+−	++	−	Red	Brown
Operculum							
Outer layer	Light green	Pink	−	−	−	Red	Brown
Intemediate layer							
Rostral portion	Purple	Deep red	+	+−	++/+++	Blue	Pink
Caudal portion	Light purple	Red	+	+−	++	Blue	Pink
Inner layer	Light green	Pink	−	−	−	Red	Brown
Ribbed structure	Green	Light red	−	−	−	Orange	Brown

Table 2. Staining properties of each part of the vertebra in gilthead sea bream *Sparus aurata*. Staining intensity: (−) negative, (+−) very weak, (+) weak, (++) moderate, (+++) strong. See 'Materials and methods' for details

	H/VOF	van Gieson	PAS	AB	von Kossa	Mallory	Orcein
Notochord	Light green	Pink	−/+−	−	−	Red	−
Surrounding notochordal epithelial layer	Green	Light red	−/+−	−	−	Red	−
Secondary chordal sheath	Light Green	Pink	−/+−	−	+/++	Blue	Brown
Primary chordal sheath	Green	Red	+−	++	−	Red	−
Centrum	Purple	Deep red	+−/+	−	+++	Deep blue	Pink
Intervertebral ligament							
Notochordal sheath	Light green	Pink	+	+	−	Pink	−
Elastic membrane	Purple	Pink	−	−	−	Blue	Brown
Collagenous ligament	Light green	Red	−	−	−	Pink/red	Pink
Osteoblasts	Purple	Red	−	−	−	Red	−
Periosteum	Light green	Pink	−	+−	−	Red	Brown
Arches	Blue	Deep red	+	+	++	Deep blue	Pink

Histological organization of normal and deformed vertebrae

The normal histological structure of transversal sections of the vertebral centra of gilthead sea bream stained with Mallory, van Gieson, H/VOF trichrome and PRS techniques showed a succession of tissue layers from the inner to the outer region. A central notochordal tissue (composed of chordoblasts and chordocytes) is surrounded of 2 fibrous layers (notochordal sheath), one composed of elastic fibers — the primary chorda sheath (Mallory trichrome-positive), and the other with mucopolysaccharides (AB-positive) and sparse collagen fibers—the secondary chorda sheath (van Gieson trichrome-positive) (Table 2, Fig. 3a–c). The notochord sheath is enclosed by 2 different bone layers: a laminar and compact layer (internal; light green under polarized light) and a cancellous layer (external; green with large orange areas under polarized light). An osteogeneous tissue layer (osteoblast layer) and the periosteum complete the skeletal structure of the vertebral body (Fig. 3a–c). Neural and hemal arches are composed of chondral bone. These arches end in a spine composed of fibrous bone that projects into the myosepta (Fig. 3d). PRS stainings of transversal sections of normal vertebra visualized under polarized light exhibited some orange surfaces embedded in a bright green colour. The fibrous layer surrounding the bony structure also contained fine and sparse green colour regions (Fig. 3e).

Transversal sections of deformed vertebrae showed an alteration of the layered disposition with fibro/cell-rich cartilage displacing cancellous and compact bone. Enlargement of trabecular spaces were also detected in lordotic vertebrae (Fig. 3f,g). Moreover, the notochordal lumen was reduced, with distorted chordocytes showing compression with a clear reduction in size (Mallory staining; Fig. 3f). Additionally, the PRS staining revealed some differences in the composition of the bone matrix between normal and deformed vertebrae. In general, and under a bright microscope, compact bone of deformed vertebra showed a red colour (immature bone) in contrast to normal vertebrae, in which compact bone stained green (mature bone) (Fig. 3d,e for normal vertebrae; Fig. 3h,i for deformed vertebrae). Moreover, under polarized light, deformed vertebra showed weak greenish staining (Fig. 3i).

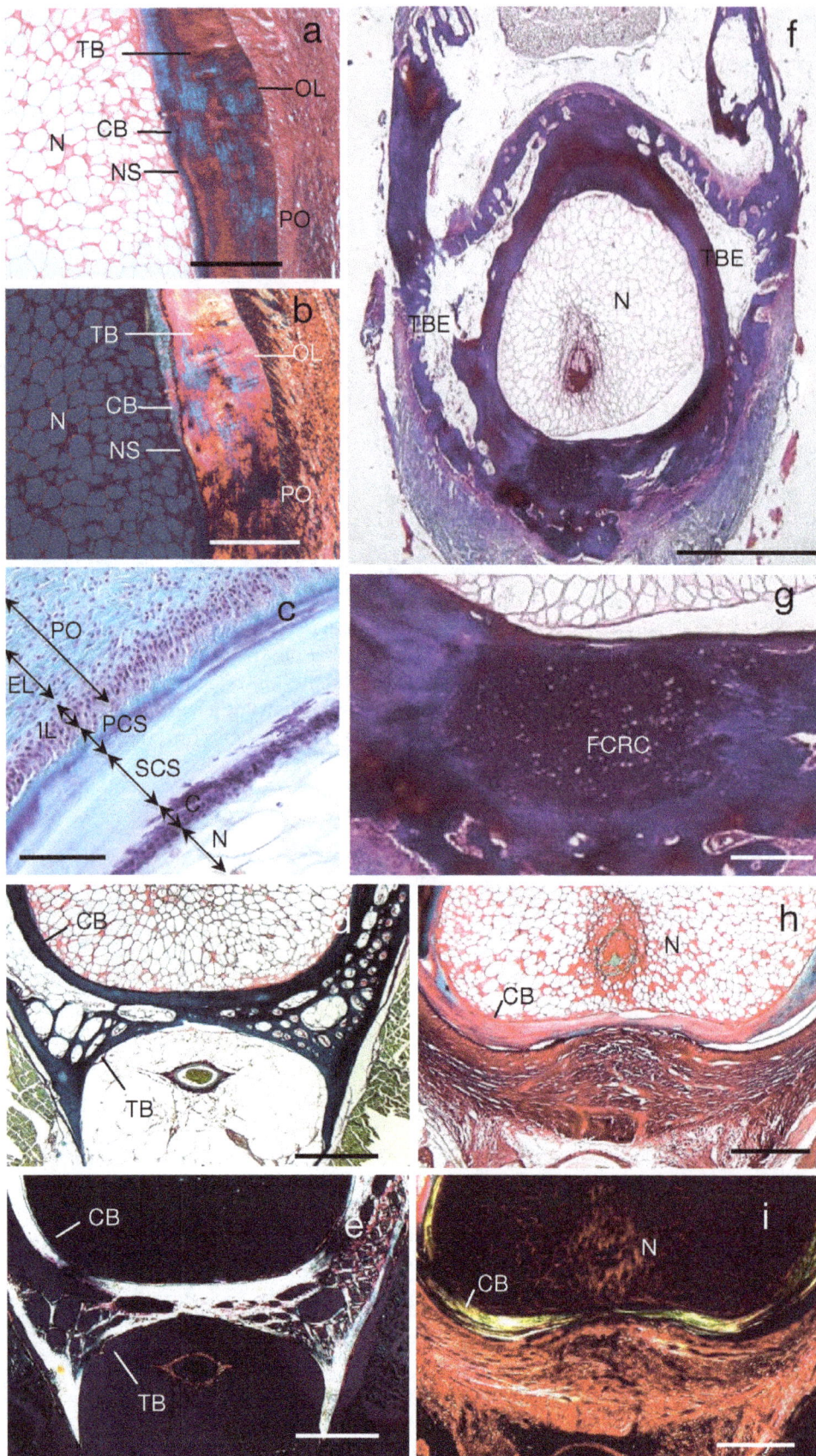

Fig. 3. Transversal sections of (a–e) normal and (f–i) deformed structure of vertebral tissue of gilthead sea bream *Sparus aurata* specimens. (a) Detailed visualization of the succession of tissue layers surrounding the notochord by picrosirius red staining under non-polarized light; (b) vertebral tissue stained with picrosirius red and visualized under polarized light showing internal (compact bone, light green) and external bony layers (trabecular bone, green with big orange areas); (c) detail of the layered distribution of the chorda sheath surrounding the notochord with H/VOF staining and (d) of the trabecular bone forming the vertebral arches stained with picrosirius red and visualized under non-polarized light. Note (e) the presence of some orange surfaces embedded in a bright green colour staining in vertebral arches with picrosirius red technique visualized under polarized light. (f,g) Transversal section of deformed vertebra with Mallory staining. Note the presence of a fibrous cartilage replacing both cancellous and compact bone. Picrosirius red staining of deformed vertebra showing a weak greenish staining visualized under (h) non-polarized and (i) polarized light. Scale bars = 50 μm. Abbreviations: C, chordoblasts; CB, compact bone; CL, collagenous ligament; EL, external layer; FCRC, fibro/cell-rich cartilage; IL, internal layer; N, notochord; NS, notochordal sheath; OL, osteogenic layer; PO, periosteum; PCS, primary chorda sheath; SCS, secondary chorda sheath; TB, trabecular bone; TBE, trabecular space enlargement

Mediosagittal sections of normal vertebral centra stained with Mallory, van Gieson, H/VOF, orcein and PRS techniques (Fig. 4a–f) showed a central part of cancellous bone with a mesh of longitudinal and transverse trabecula, funnel-shaped vertebral end-plates and the so-called intervertebral ligaments

(Fig. 4a). The vertebral centrum was composed of a bone matrix with osteoblasts within cancellous lacunae, a periosteum consisting of an external layer of connective tissue rich in fibrillar elements, and an internal layer rich in cellular elements (osteogeneus tissue composed of osteoblasts; Fig. 4b). The inter-

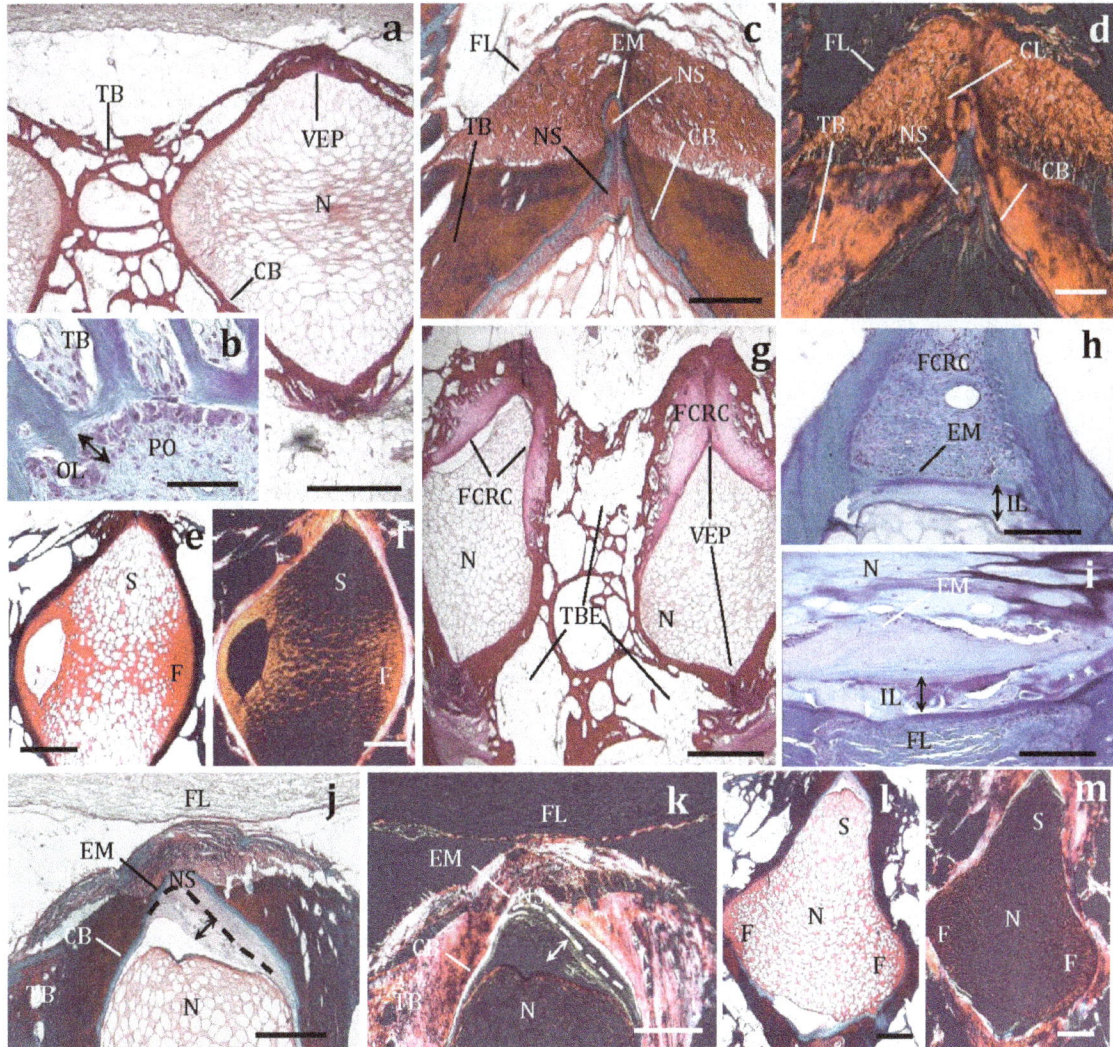

Fig. 4. Mediosagittal sections of (a–f) normal and (g–m) deformed vertebra of *Sparus aurata*. (a) Central part of cancellous bone and vertebral end plates stained with van Gieson; (b) detail of the vertebral body comprising the bone matrix with osteoblasts within cancellous lacunae and the periosteum stained with H/VOF; detailed pictures of the intervertebral region showing its different layers stained with picrosirius red and visualized under (c) non-polarized and (d) polarized light (note the heterogeneous staining of the trabecular bone in contrast with the compact one). Detailed pictures of the fibrous and spongy layers of the notochord stained with picrosirius red and visualized under (e) non-polarized and (f) polarized light (note the presence of orange stained fibers within notochordal fibrous and spongy layers). Detail of the altered vertebral core with enlarged trabecular spaces and the presence of a fibrous connective tissue in the intervertebral region with (g) van Gieson and (h,i) H/VOF stainings. Detailed micrograph of the intervertebral ligament of deformed vertebra showing a modification of the intervertebral ligament (double arrow) and of the fibrous layer stained with picrosirius red and visualized under (j) non-polarized and (k) polarized light. Note (l) the lesser proportion of the notochordal spongy layer in deformed vertebrae in comparison with normal ones with picrosirius red staining and visualized under non-polarized light and (m) the lesser orange staining with picrosirius red technique and visualized under polarized light. Scale bars = 50 µm. Abbreviations: CB, compact bone; CL, collagenous ligament; EM, elastic membrane; F, notochordal fibrous layer; FCRC, fibro/cell-rich cartilage; FL, fibrous layer; IL, intervertebral ligament; N, notochord; NS, notochordal sheath; OL, osteogenic layer; PO, periosteum; S, notochordal spongy layer; TB, trabecular bone; TBE, trabecular space enlargement; VEP, vertebral end-plates

vertebral ligament was composed of 3 acellular structural components: an inner notochordal sheath, a medial elastic membrane (orcein positive; Table 2) and an external sclerotome-derived collagenous ligament (Fig. 4c,d). Between each vertebra, the notochord tissue was composed of fibrous and spongy layers (Figs. 4e,f). Trabecular bone samples stained with PRS and visualized under polarized light had a heterogeneous composition consisting of a mixture of orange and green stainings, whereas compact bone exhibited a homogeneous composition (Fig. 4d). Notochordal fibrous and spongy layers also showed orange staining (Fig. 4f).

In comparison with normal vertebrae, the histological analysis of deformed vertebrae allowed the identification of a different tissue. Mediosagittal sections of deformed vertebrae showed an altered central part with enlarged trabecular spaces compared to normal vertebra. Moreover, affected vertebrae showed a pathological formation of fibro/cell-rich cartilage (stained pink with van Gieson's technique; Fig. 4g) replacing the notochordal and the cancellous acellular bone in the haemal or neural sides (H/VOF; Fig. 4h). In the opposite area of the same deformed vertebrae, a loss of the typical funnel-shaped bone end-plates, and a thickening of the intervertebral ligament and the external elastic membrane were also detected (H/VOF; Fig. 4i). In the intervertebral ligament of deformed vertebrae, an additional layer rich in orange staining was also visible filling the notochordal space. Moreover, abnormal vertebrae showed a lower orange staining surface on the fibrous layer of the growth zone of the bone end-plates in comparison to normal vertebrae (Fig. 4j,k), and a lesser proportion of the notochordal spongy layer was detected in deformed vertebrae in comparison with normal ones (Fig. 4l,m).

Histochemical and immunohistochemical analysis of OC and MGP proteins

Immunohistochemical analyses revealed that OC and MGP proteins showed similar distribution in mineralized bone matrix in the normal (Fig. 5a) and deformed Type I (Fig. 5b,c) and Type II (Fig. 5d–f) opercula. Additionally, MGP was found in both mineralized bone matrix and cellular elements of the modified articulatory facet from the Type II deformity (Fig. 5e).

Considering the vertebral tissue, OC and MGP protein distribution presented some differences between normal and deformed structures both in the

mineralized matrix and the notochordal tissue (OC or MGP immunostaining counterstained with AB pH 2.5 and H/VOF; Fig. 6a–p). In normal vertebrae, OC was homogeneously accumulated in the compact and cancellous bone matrix from the periosteal zone of the vertebral centra and in the arches, as well as in some osteoblasts located in the growth zone (Fig. 6a–c). In contrast, MGP preferentially accumulated in the notochordal cells, with minimal presence in the different bone layers (Fig. 6d). Deformed vertebrae showed a preferential deposition of OC in the compact bone layer compared with the cancellous one (Fig. 6e), as well as in the notochord (Fig. 6f), which coincided with von Kossa positive staining (Fig. 6g). In addition, a replacement of bone by cartilaginous tissue, showing glycosamineglycans instead of OC deposition, was detected in the lordotic compressed region (Fig. 6h). Regarding the vertebral bone endplates, some differences in OC distribution were also detected between normal and deformed vertebrae: OC was deposited in the intervertebral ligament and in the ring-like structure of non-deformed vertebrae, whereas it was undetectable in the same areas from deformed vertebrae. In addition to the thickening of the interverbebral ligament and the external elastic membrane, an accumulation of glycosamineglycans (AB positive) was also detected in these areas. Additionally, some differences in OC deposition were also detected in the fibrous layer of the growth zone and in the trabecular bone, being lower in OC content in deformed vertebra than in normal ones (Fig. 6i,j). Concerning MGP distribution in deformed vertebrae, a slight increase of staining intensity was detected in notochordal cells of deformed vertebrae (Fig. 6k,l).

Finally, histological observations revealed that osseous tissue was replaced by cartilaginous tissue rich in glycosaminoglycans in the intervertebral space of lordotic vertebrae (Fig. 6m,n). In a more advanced stage, the recently formed cartilaginous tissue accumulated OC, and a new calcified matrix was formed (incipient fusion stage) (Fig. 6o,p).

Bone homeostasis markers

We analysed ALP and TRAP activities in both normal and deformed bone structures in order to identify cells responsible for bone deposition and resorption, respectively. However, our results did not show clear differences between non-deformed and deformed skeletal structures with regards to ALP activity (data not shown). TRAP activity in non-deformed vertebrae was distributed in the bone matrix of the trabec-

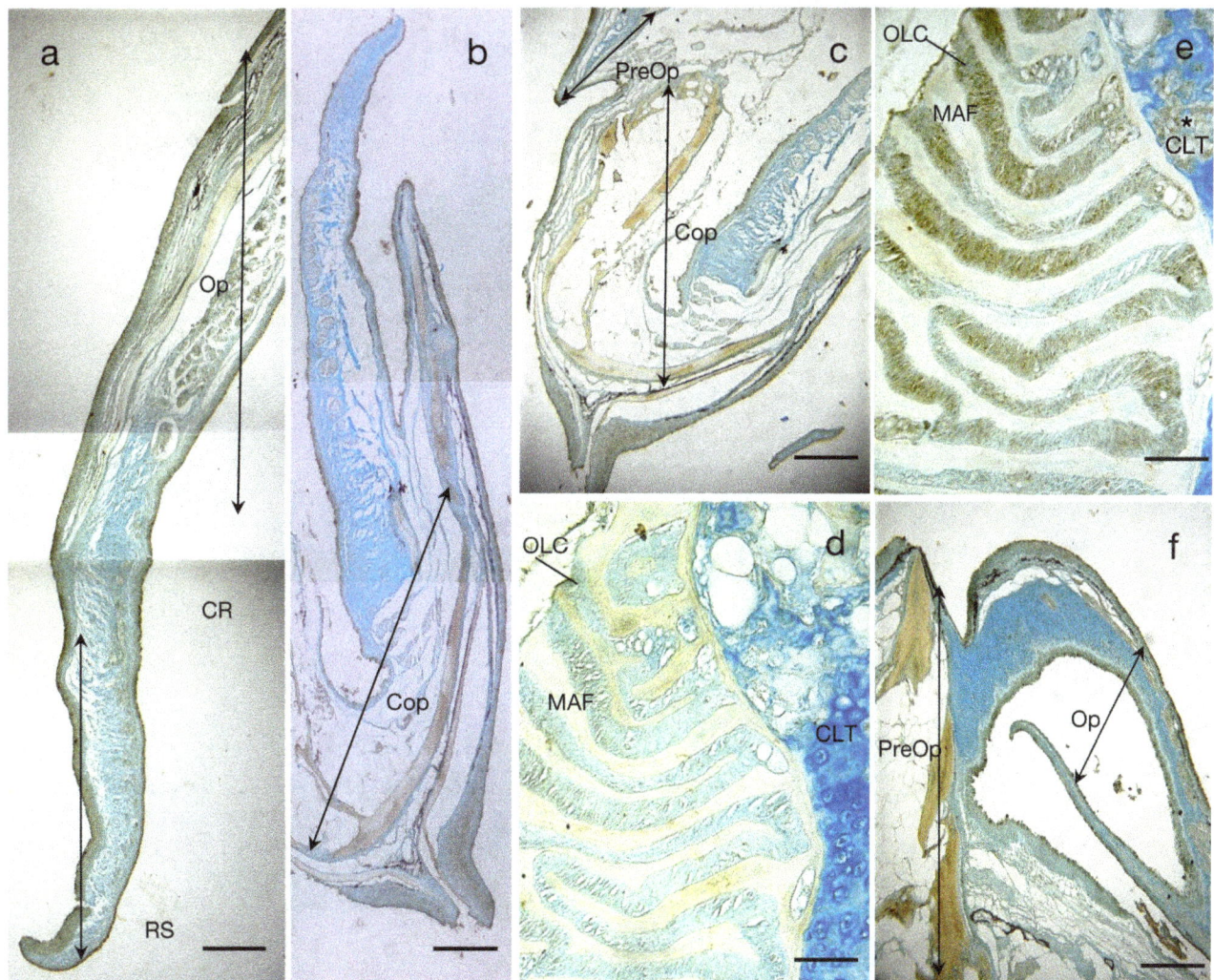

Fig. 5. Osteocalcin (OC) and matrix Gla protein (MGP) immunoreactivity counterstained with AB pH 2.5 in normal and deformed *Sparus aurata* opercular structures. (a) OC immunostaining in non-deformed operculum; (b) OC and (c) MGP immunostaining in bone opercular structures from Type I deformity. Note that no differences in OC or MGP protein distribution within the bone matrix could be detected. (d) OC and (e) MGP immunostaining in modified articulatory facet in an opercular Type II deformity. Note the presence of MGP protein both in cellular elements (osteoblastic-like cells from the articulatory facet and cells within cartilage-like tissue, *) and in calcified matrix from the modified articulatory facet. (f) OC immunostaining in an opercular Type II deformity at level of ventral zone. Scale bars = 50 µm. Abbreviations: CLT, cartilage-like tissue; Cop, curled operculum; CR, caudal region; MAF, modified articulatory facet; OLC, osteoblastic–like cells; Op, operculum; PreOp, preoperculum; RS, ribbed structure

ular region (Fig. 7a,b), as well as the osteoclasts within the trabecula lacunae and bone-reabsorbing fronts of normal vertebrae (Fig. 7c,d), but not in deformed ones (Fig. 7e). Moreover, an intense reaction for TRAP activity was exclusively detected in the chondrocytic areas of deformed vertebrae (Fig. 7f).

To determine whether the skeletal anomalies found in gilthead sea bream could be linked to an imbalance in cell cycling rather than to unbalanced cell activity, PCNA immunostaining was performed to measure cell proliferation, and a caspase-3 assay was conducted to detect processes related to cellular apoptosis. Immunohistochemistry of PCNA showed differential distribution between normal and deformed vertebral bodies and operculum, with a marked increase of immunopositive cells in the growth zone of bone end-plates (Fig. 7g,h) in the notochordal cells (Fig. 7i) and the compression zone of vertebral ligaments (Fig. 7j) from deformed vertebrae. However, caspase-3 as an immunomarker of cell apoptosis did not show such differences (data not shown).

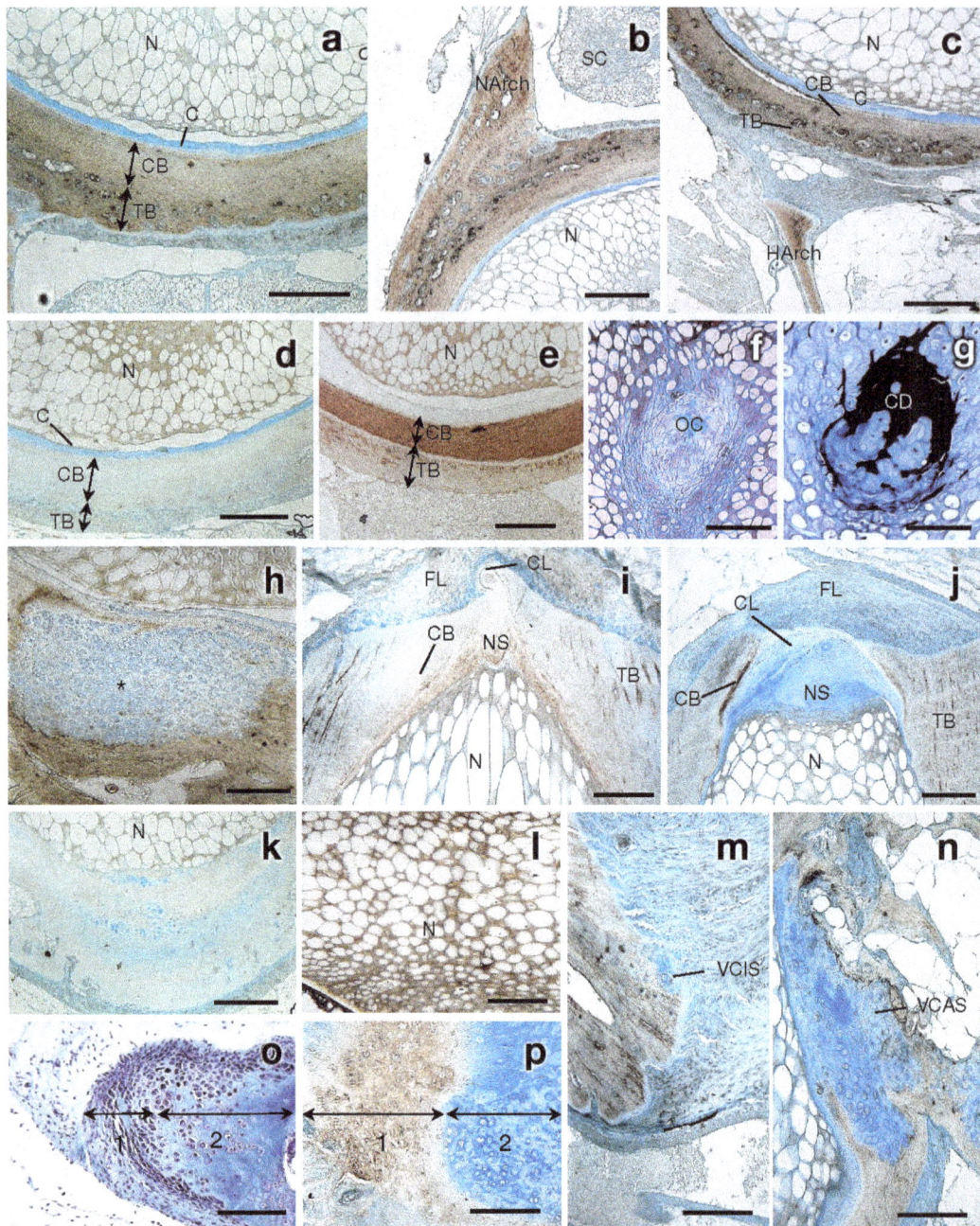

Fig. 6. Osteocalcin (OC) and matrix Gla protein (MGP) distribution counterstained with AB pH 2.5 in control and deformed vertebra of *Sparus aurata*. Homogeneous OC distribution both in compact and trabecular bone from the periosteal zones of (a) the vertebral centra and from (b) neural and (c) haemal arches. (d) Preferential accumulation of MGP in notochord with minimal presence in the different bone layers. Preferential accumulation of OC in (e) compact bone and (f) in notochordal cells from deformed fish coinciding with (g) a calcium deposition (von Kossa staining) in the same area. (h) Fibrous-like structure in deformed vertebra showing glycosaminoglycans (*) instead of OC deposition. (i) OC accumulation in the zone of vertebral end-plates from un-deformed fish and (j) OC deposition in the fibrous layer of the growth zone and notochordal sheath in deformed fish. (k) MGP distribution in deformed vertebra with a scarce presence of MGP in bone matrix and (l) a slight increase in staining intensity in altered notochordal cells. Detail of a compression area in the vertebral end-plates from a deformed vertebra, showing (m) initial and (n) more advanced stages in which osseous tissue is gradually replaced by cartilaginous. OC accumulation in (1) the periphery of (2) the newly formed cartilaginous tissue from the compression area (incipient vertebral fusion) stained with (o) H/VOF or (p) not. Scale bars = 50 µm. Abbreviations: C, chordoblasts; CB, compact bone; CD, calcium deposition; CL, collagenous ligament; FL, fibrous layer; HArch, haemal arch, N, notochord; NArch, neural arch; NS, notochordal sheath; OC, osteocalcin; SC, spinal cord; TB, trabecular bone, VCIS, vertebral compression initial stage; VCAS vertebral compression in an advanced stage

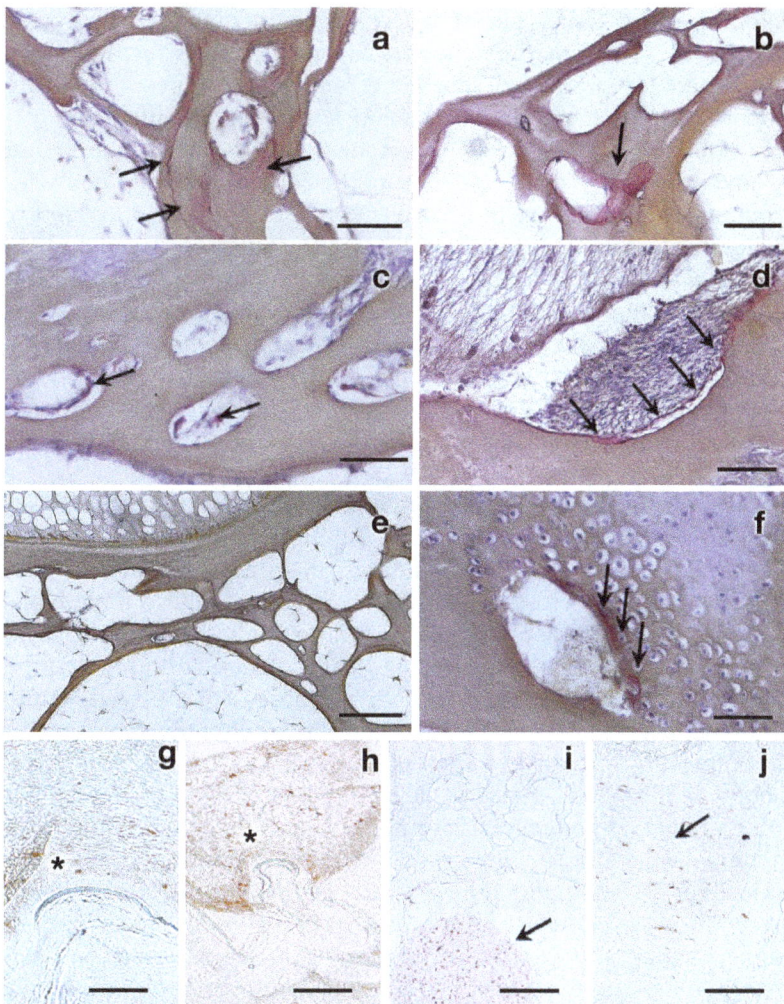

Fig. 7. (a–f) Tartrate-resistant acid phosphatase (TRAP) activity and (g–j) proliferating cell nuclear antigen (PCNA) distribution in (a–d) normal and (e–f) deformed vertebra. TRAP enzymatic activity both in (a,b) bone matrix (arrows) and osteoclasts (c) within trabecula and (d) in reabsorbing front. Note (e) absence of TRAP activity bone matrix in deformed vertebra and (f) the presence of enzymatic activity in the fibrous-like structures of deformed vertebra. Differences in PCNA distribution (∗) in the growth zones of the vertebral end-plates from (g) normal and (h) deformed vertebra. In deformed vertebra, cell proliferation is also detected in (i) the notochordal cells and (j) in compression zone of vertebral ligament (arrows). Scale bars = 50 μm

DISCUSSION

In this study, the incidence of opercular anomalies in gilthead sea bream *Sparus aurata* was moderate (up to 6.3 %) in comparison with previously reported frequencies in farmed population of this species (up to 80 %; Paperna 1978, Barahona-Fernándes 1982, Chatain 1994, Andrades et al. 1996). However, such differences may be attributed to different rearing conditions and improvements in the rearing procedures over the last 2 decades (Koumoundouros 2010,

Prestinicola et al. 2013). Opercular anomalies detected in gilthead sea bream juveniles were randomly distributed on both sides of the head, which was in agreement with previous data reported by Beraldo et al. (2003). The external appearance of opercular deformities found in this study were in accordance with previous studies which suggested that opercular anomalies were due to the inside folding of the opercle and subopercle bones or to their shortening/atrophy (Koumoundouros et al. 1997a, Galeotti et al. 2000, Beraldo et al. 2003, Morel et al. 2010). Histological results from our study revealed that the reduced opercular surface in the Type I deformity was mainly due to the folding of the edge of the opercle from the superior corner towards the gill chamber. The conservation of the different opercular bones that compose the operculum suggested an effect of rearing conditions during sensitive developmental periods, coinciding with the beginning of skeletogenesis of the opercular complex, which is formed by intramembranous ossification from a condensed core of mesenchymal cells (Koumoundouros et al. 1997b, Galeotti et al. 2000, Beraldo et al. 2003). Considering that the supportive tissue of the operculum was not yet formed during early development, mechanical damage (such as that caused by excessive water movement in the growing tanks) could cause the above-mentioned opercular malformations (Galeotti et al. 2000). Regarding opercular deformity Type II, our results revealed that the coalescence between opercular bone areas might have caused semi-rigidity and tissue fusion of the opercular structures, giving the appearance of underdeveloped or incomplete tissue. Depending on the dissected region (articulatory facet, medial or ventral zone), underdevelopment of the opercle, subopercle and/or both were clearly visible. Sectioning the ventral zone of the operculum showed a regression of the subopercle and a complete loss of the opercle, although the covering tissue layers of this skeletal structure were maintained and folded into the gill cavity. Furthermore, opercular sectioning at the medial level showed an outward folding of the opercle and a thickening of the epithelial cell layers. According to Galeotti et al. (2000), the above-mentioned opercular disorders might be mechanically-induced by forced opercular movements occurring during ventilation or food ingestion processes, although the putative involvement of nutritional factors and/or environmental pollutants should not be neglected (Boglione et al. 2013a).

Specimens displaying the opercular Type II deformity also showed a novel type of opercular anomaly that consisted of a coalescence of contacting bone tissues, presumably from the preopercle and opercle. The above-mentioned anomaly had a trabecular aspect filled by a single cell epithelium of cubic osteoblastic-like cells. The origin of this anomaly might be a disorganization of the blastema (aggregate of mesenchymal cells), the precursor of the opercular elements formed by intramembranous ossification at early developmental stages, brought about as a result of inadequate rearing conditions such as unbalanced diet, inadequate aeration or unfavourable temperature during the embryonic and larval stage (Loizides et al. 2013). Regardless of the cause of these types of deformities, there is evidence that opercular abnormalities compromise the welfare (Noble et al. 2012) and product quality (Cobcroft & Battaglene 2013) of the fish stock, as well as the biological performance of the fish, reducing their resistance to oxygen stress and predisposing the gills to pathological infections (Koumoundouros et al. 1997b). Moreover, a clear connection has been established between opercular or vertebral abnormalities and high mortality rates (Andrades et al. 1996, Loizides et al. 2013).

Opercular bones stained with picrosirius red were visualized under polarized light, which enhances the normal birefringence of collagen fibers. The birefringence colour appears to be a measure of the thickness of the collagen fibers, the density of their packing and their spatial arrangement. As fiber thickness increases, their colour changes from green (thinner collagen fibers) to bright orange/red (larger collagen fibers; Junqueira et al. 1982, Hiss et al. 1988, Dayan et al. 1989). Thus, we were able to determine potential differences in ECM composition. In non-deformed opercula, this technique showed 2 populations of collagen fibers: one group of orange, strongly birefringent fibres (thick collagen fibers) and another group of green, weakly birefringent fibers (thin fibers). A comparison between normal and deformed opercula revealed differences in collagen fiber thickness and in their 3-dimensional arrangement, consisting of reduced and scattered areas of orange dye (fewer big collagen fibers) in deformed opercular bones (opercle and subopercle). This fact has been suggested as a bone remodelling process linked to changes in bone composition and degradation of collagen fibers (Fernández et al. 2012). Considering that the mechanical properties of the opercular bone depend on the composition and structure of its matrix (Totland et al. 2011), the thickness of collagen fibers in deformed opercula might indicate a weakness and vulnerability to mechanical disturbances of this skeletal structure, since collagen content confers certain tissue elasticity (Prades et al. 2010), and therefore changes in its proportion could reduce tissue flexibility and plasticity (Canavese & Colitti 1996, Ytteborg et al. 2012).

Vertebral column malformations may occur frequently in the wild, but are rarely detected (Gavaia et al. 2009) whereas they are quite commonly detected under intensive aquaculture conditions (Kranenbarg et al. 2005, Koumoundouros 2010, Prestinicola et al. 2013, among others). In this study, the frequency of lordotic vertebrae (10.1%) was within the normal range found in other studies and hatcheries (Andrades et al. 1996, Koumoundouros et al. 2002, Prestinicola et al. 2013). The aetiology of this skeletal disorder might vary depending on the rearing and environmental conditions (see reviews in Koumoundouros 2010, Boglione et al. 2013a). Independent of the causative factor, lordosis may occur at different developmental stages. For instance, some authors have reported that vertebral deformities may appear during the notochord segmentation and vertebral centrum differentiation processes (Fernández et al. 2008, Haga et al. 2009). Others have suggested that they are a consequence of dysfunctions in collagen metabolism at notochordal and perinotochordal collagen sheets during early development (Santamaría et al. 1994). Moreover, vertebral deformities may also occur later in ontogeny (i.e. during the ongrowing period), at which point they are generally induced by mechanical overloads (Kranenbarg et al. 2005) or by the curvature of the vertebral axis (Gorman et al. 2010), or a combination of both (Gavaia et al. 2002, Cardeira et al. 2012). Irrespective of the cause of vertebral anomalies, they can compromise the biological performance of fish. For example, in European sea bass, pre-haemal kyphosis was shown to induce lethargic behaviour and subsequent heavy mortality during vertebral axis osteogenesis, as a result of the compression of the neural tube by the deformed vertebrae (Koumoundouros et al. 2002). Furthermore, Basaran et al. (2007) showed that lordosis significantly decreased the endurance and critical swimming speed in European sea bass juveniles. In the present study, gilthead sea bream lordotic vertebrae showed the formation of fibrous cartilage in the haemal and/or neural sides of the vertebral centrum. This chondroid tissue might meet the demand for an accelerated growth rate and/or the demand for a shear-resistant support (Huysseune 2000, Cardeira et al. 2012). Moreover, in the central notochord of affected vertebrae, regions of densely packed chordocytes lacking vacuoles were also observed, whereas calcium deposition within de-vacuolated chordocytes was detected as reported in salmonids (Ytteborg et al. 2012). Thus, the change in chordocyte morphology from vacuolated to hyperdense, and the increase in calcium deposition in the affected region of gilthead sea bream lordotic vertebrae might indicate that a metaplastic shift was involved, as it has been previously described in spinal fusions in fish (Witten et al. 2005, Ytteborg et al. 2012, Boglione et al. 2013b). Another major histomorphological change in lordotic gilthead sea bream vertebrae was the disorganization of the intervertebral region, which lead to a complete loss of notochordal sheath integrity. These results were in concordance with those reported by other authors for Atlantic salmon *Salmon salar* reared at high temperatures (Ytteborg et al. 2010), gilthead sea bream fed hypervitaminosis A (Fernández et al. 2012) and guppy *Poecilia reticulata* displaying a curveback syndrome (Gorman et al. 2010). Major histomorphological changes of lordotic vertebrae in gilthead sea bream specimens from this study were linked to the loss of the integrity of notochordal cells, which was associated with the presence of modified chordocytes in which calcium deposition was noticeable. Similarly, Loizides et al. (2013) pointed out the ectopic presence of altered chordocytes as well as increased notochordal sheath production as the causes of the vertebral compression and fusion (VCF) syndrome in gilthead sea bream.

PCNA immunohistochemistry indicated that osteoblasts in the growth zone of the bone end-plates showed a marked increase in cell proliferation in deformed vertebral centra, which was not counteracted by an increase in cell death as revealed by caspase-3 immunostaining. In addition, a marked increase in cell proliferation (PCNA staining), but not of cell apoptosis was also detected in notochordal cells and in the compression zone of vertebral ligament in lordotic vertebrae. Several studies in higher vertebrates have suggested that changes in the balance between cell death and cell proliferation are involved in bone and cartilage defects, which could ultimately lead to skeletal deformities (Cockroft

& New 1978, Miura et al. 2004). In this sense, the above-mentioned imbalanced cell cycling detected in lordotic vertebrae might also explain the presence of dense packaged chordocytes, occupying most of the intervertebral space without vacuolation, as was described by Ytteborg et al. (2010) in Atlantic salmon vertebral fusions. Regarding the bone homeostasis measured by ALP and TRAP immunostaining, although this procedure may not be considered to be a quantitative assessment of bone formation and resorption, the present data might indicate that lordotic vertebrae were subjected to a more localized bone resorption process than non-deformed ones, since TRAP activity was only found in the chondrocytic areas of deformed vertebrae but was homogeneously distributed in the trabecular spaces of normal vertebrae. Local and intense TRAP activity in deformed vertebrae is in agreement with the reduced thickness of collagen fibers and the previously suggested bone remodelling process occurring in this deformed skeletal structure. These results differ from those previously described in spinal fusions by Ytteborg et al. (2012) and Boglione et al. (2013b).

In the vertebral growth zones of the teleosts, ECM-producing cells (osteoblasts and bone lining cells) express a combination of proteins having distinct functions in mineralisation and/or deposition of the osteoid (Krossøy et al. 2009, Boglione et al. 2013b). Among these proteins, OC and MGP have been reported to be involved in the mineralization of the ECM. In this sense, OC has been proposed as a regulator of bone maturation (Krossøy et al. 2009), whereas MGP acts as an inhibitor of calcification (Luo et al. 1997). In this study, a preferential accumulation of OC in compact bone and notochordal cells of lordotic vertebrae was detected, which might be attributed to the remodelling process occurring in lordotic vertebrae (Boglione et al. 2013b) as well as the above-mentioned cartilage formation by metaplasic transformation of bone, forming cells that would be latter mineralized and remodelled into bone (Witten et al. 2005). Concerning MGP, a preferential deposition was detected in notochordal cells of non-deformed vertebrae compared with deformed ones. Higher levels of OC and MGP deposition in notochordal tissue of deformed vertebrae were similar to those reported by Fernández et al. (2012) in gilthead sea bream juveniles fed hypervitaminosis A. These results are in agreement with previous studies in which an increased co-transcription of both chondrogenic and osteogenic markers were found in the notochord of Atlantic salmon displaying spinal fusions (Ytteborg et al. 2010, 2012). Furthermore, bone tissue with lower levels of OC might demineralize more easily than that with higher OC content (Krossøy et al. 2009), suggesting that lordotic vertebrae may be more fragile than non-deformed ones. However, this hypothesis requires further corroboration by additional mineral content analyses.

CONCLUSIONS

This study provided a comprehensive description of the main morphological and histological features of normal and deformed opercula and vertebrae from gilthead sea bream in order to increase basic knowledge of bone disorders in this species. Our data revealed that important histological, histochemical and immunohistochemical differences were found between non-deformed and deformed opercula and vertebrae. A rare type of tissue remodelling process was described in fish displaying the Type II opercular deformity, which consisted of the coalescence of contacting bone tissues, presumably from the preopercle and opercle, resulting in skeletal tissue with a trabecular aspect filled by a single cell epithelium of cubic osteoblastic-like cells. Additionally, there were differences in collagen fiber thickness and 3-dimensional arrangement between normal and deformed opercula, as well as hypoplasia of muscle fibers, which might affect the flexibility of deformed opercula. Lordotic gilthead sea bream vertebrae showed the formation of fibrous cartilage in their haemal and/or neural sides, indicating that a metaplastic shift occurred during the process of lordosis. Another major histomorphological change in lordotic vertebrae was the complete loss of notochordal sheath integrity. These alterations were coupled with an imbalance between cell death and cell proliferation processes in lordotic vertebrae, as well as in bone formation/resorption and ECM deposition activity, which might have resulted from the remodelling process occurring in lordotic vertebrae. Altogether, these results provide an increase in the basic knowledge of bone disorders that will add to our understanding of the mechanisms by which these skeletal disorders appear in this fish species and which hamper its production efficiency.

Acknowledgements. The authors express their gratitude to I. Viaña for providing technical assistance and to Dr. D. Simes for providing the OC and MGP antibodies. This work was funded by Ministry of Science and Innovation (MICIIN) of the Spanish government (projects AGL2008-03897-C01/C04 and AGL2010-15951). J.B.O-D. was supported by the Programa Ramón y Cajal (MICINN, Spain). I.F. was supported by a predoctoral Spanish MICINN fellowship (reference, BES- 2006-12650) and a postdoctoral fellowship (SFRH/BDP/82049/2011) from Fundação para a Ciência e Tecnologia (FCT), Portugal.

LITERATURE CITED

▶ Andrades JA, Becerra J, Fernández-Llebrez P (1996) Skeletal deformities in larval, juvenile and adult stages of cultured gilthead sea bream (*Sparus aurata* L.). Aquaculture 141:1–11

▶ Bakke-McKellep AM, Froystad MK, Lilleeng E, Dapra F, Refstie S, Krogdahl A, Landsverk T (2007) Response to soy: T-cell-like reactivity in the intestine of Atlantic salmon, *Salmo salar* L. J Fish Dis 30:13–25

Bancroft JD, Stevens A (1990) Theory and practice of histological techniques. Churchill Livingstone, London

▶ Barahona-Fernándes MH (1982) Body deformation in hatchery reared European sea bass *Dicentrachus labrax* (L). Types, prevalence and effect on fish survival. J Fish Biol 21:239–249

▶ Basaran F, Ozbilgin H, Ozbilgin YD (2007) Effect of lordosis on the swimming performance of juvenile sea bass (*Dicentrarchus labrax* L.). Aquacult Res 38:870–876

▶ Beraldo P, Pinosa M, Tibaldi E, Canavese B (2003) Abnormalities of the operculum in gilthead sea bream (*Sparus aurata*): morphological description. Aquaculture 220:89–99

▶ Boglione C, Gagliardi F, Scardi M, Cataudella S (2001) Skeletal descriptors and quality assessment in larvae and post-larvae of wild-caught and hatchery-reared gilthead sea bream (*Sparus aurata* L. 1758). Aquaculture 192:1–22

▶ Boglione C, Gisbert E, Gavaia P, Witten PE, Moren M, Fontagné S, Koumoundouros G (2013a) A review on skeletal anomalies in reared European larvae and juveniles. Part 2: Main typologies, occurrences and causative factors. Rev Aquac 5:S121–S167

▶ Boglione C, Gavaia P, Koumoundouros G, Gisbert E, Moren M, Fontagné S, Witten PE (2013b) A review on skeletal anomalies in reared European fishes. Part 1: Normal and anomalous skeletogenic processes. Rev Aquac 5:S99–S120

▶ Bonewald LF, Harris SE, Rosser J, Dallas MR and others (2003) Von Kossa staining alone is not sufficient to confirm that mineralization *in vitro* represents bone formation. Calcif Tissue Int 72:537–547

Canavese B, Colitti M (1996) Observations under LM and SEM of opercle malformations in sea bream (*Sparus aurata*). Teratology 53:27A (Abstract)

▶ Cardeira J, Bensimon-Brito A, Pousão-Ferreira P, Cancela ML, Gavaia PJ (2012) Lordotic-kyphotic vertebrae develop ectopic cartilage-like tissue in Senegalese sole (*Solea senegalensis*). J Appl Ichthyology 28:460–463

▶ Castro J, Pino-Querido A, Hermida M, Chavarrias D and others (2008) Heritability of skeleton abnormalities (lordosis, lack of operculum) in gilthead seabream (*Sparus aurata*) supported by microsatellite family data. Aquaculture 279:18–22

▶ Chatain B (1994) Abnormal swimbladder development and lordosis in sea bass (*Dicentrachus labrax*) and sea bream (*Sparus aurata*). Aquaculture 119:371–379

Clark G (1981) Staining procedures. Williams & Wilkins, Baltimore, MD

▶ Cobcroft JM, Battaglene SC (2013) Skeletal malformations

in Australian marine finfish hatcheries. Aquaculture 396-399:51–58

► Cockroft DL, New DAT (1978) Abnormalities induced in cultured rat embryos by hypertermia. Teratology 17: 277–283

► Daoulas C, Economou AN, Bantavas I (1991) Osteological abnormalities in laboratory reared sea-bass (Dicentrarchus labrax) fingerlings. Aquaculture 97:169–180

► Dayan D, Hiss Y, Hirshberg A, Bubis JJ, Wolman M (1989) Are the polarization colors of Picrosirius red-stained collagen determined only by the diameter of the fibers? Histochemistry 93:27–29

► Diggles BK (2013) Saddleback deformities in yellowfin bream, Acanthopagrus australis (Günther), from South East Queensland. J Fish Dis 36:521–527

► Faustino M, Power DM (1998) Development of osteological structures in the sea bream: vertebral column and caudal fin complex. J Fish Biol 52:11–22

► Faustino M, Power DM (1999) Development of the pectoral, pelvic, dorsal and anal fins in cultured sea bream. J Fish Biol 54:1094–1110

► Faustino M, Power DM (2001) Osteologic development of the viscerocranial skeleton in sea bream: alternative ossification strategies in teleost fish. J Fish Biol 58:537–572

► Fernández I, Hontoria F, Ortiz-Delgado JB, Kotzamanis Y, Estévez A, Zambonino-Infante J, Gisbert E (2008) Larval performance and skeletal deformities in farmed gilthead sea bream (Sparus aurata) fed with graded levels of Vitamin A enriched rotifers (Brachionus plicatilis). Aquaculture 283:102–115

► Fernández I, Ortiz-Delgado JB, Sarasquete C, Gisbert E (2012) Vitamin A effects on vertebral bone tissue homeostasis in gilthead sea bream (Sparus aurata) juveniles. J Appl Ichthyology 28:419–426

► Galeotti M, Beraldo P, de Dominis S, D'Angelo L and others (2000) A preliminary histological and ultrastructural study of opercular anomalies in gilthead sea bream larvae (Sparus aurata). Fish Physiol Biochem 22:151–157

► Gavaia PJ, Dinis MT, Cancela ML (2002) Osteological development and abnormalities of the vertebral column and caudal skeleton in larval and juvenile stages of hatchery-reared Senegal sole (Solea senegalensis). Aquaculture 211:305–323

► Gavaia PJ, Domingues S, Engrola S, Drake P, Sarasquete C, Dinis MT, Cancela ML (2009) Comparing skeletal development of wild and hatchery-reared Senegalese sole (Solea senegalensis, Kaup 1985): evaluation in larval and postlarval stages. Aquacult Res 40:1585–1593

► Gorman KF, Handrigan GR, Jin G, Wallis R, Breden F (2010) Structural and micro-anatomical changes in vertebrae associated with idiopathic-type spinal curvature in the curveback guppy model. Scoliosis 5:10

Gutiérrez M (1967) Coloración histológical para ovarios de peces, crustáceos y moluscos. Inv Pesq 31:265–271

► Haga Y, Dominique VJ, Du SJ (2009) Analyzing notochord segmentation and intervertebral disc formation using the twhh:gfp transgenic zebrafish model. Transgenic Res 18: 669–683

► Hiss J, Hirshberg A, Fundoiano-Dayan D, Bubis JJ, Wolman H (1988) Aging of wound healing in an experimental model in mice. Am J Forensic Med Pathol 9:310–312

Huysseune A (2000) Skeletal system. In: Ostrander GK (ed) The laboratory fish. Academic Press, San Diego, CA, p 307–317

► Junqueira LCU, Bignolas G, Brentani RR (1979) Picrosirius staining plus polarization microscopy, a specific method for collagen detection in tissue sections. Histochem J 11: 447–455

► Junqueira LCU, Montes GS, Sanchez EM (1982) The influence of tissue sections thickness on the study of collagen by the picrosirius-polarization method. Histochemistry 74:153–156

► Karsenty G (2001) Transcriptional control of osteoblast differentiation. Endocrinology 142:2731–2733

Kiernan JA (2008) Methods for connective tissue. In: Kiernan JA (ed) Histological and histochemical methods: theory and practice, 4th edn. Scion, Bloxham, p 190–213

Koumoundouros G (2010) Morpho-anatomical abnormalities in Mediterranean marine aquaculture. In: Koumoundouros G (ed) Recent advances in aquaculture research. Transworld Research Network, Kerala, p 125–148

► Koumoundouros G, Gagliardi F, Divanach P, Boglione D, Cataudella S, Kentouri M (1997a) Normal and abnormal osteological development of caudal fin in Sparus aurata L. fry. Aquaculture 149:215–226

► Koumoundouros G, Oran G, Divanach P, Stefanakis S, Kentouri M (1997b) The opercular complex deformity in intensive gilthead sea bream (Sparus aurata L.) larviculture. Moment of apparition and description. Aquaculture 156:165–177

► Koumoundouros G, Maingot E, Divanach P, Kentouri M (2002) Kyphosis in reared sea bass (Dicentrarchus labrax L.): ontogeny and effects on mortality. Aquaculture 209: 49–58

► Kranenbarg S, Waarsing JH, Muller M, Weinans H, van Leeuwen JL (2005) Lordotic vertebrae in sea bass (Dicentrarchus labrax L.) are adapted to increased loads. J Biomech 38:1239–1246

► Krossøy C, Ornsrud R, Wargelius A (2009) Differential gene expression of bgp and mgp in trabecular and compact bone of Atlantic salmon (Salmo salar L.) vertebrae. J Anat 215:663–672

► Loizides M, Georgiou AN, Somarakis S, Witten PE, Koumoundouros G (2013) A new type of lordosis and vertebral body compression in Gilthead seabream (Sparus aurata Linnaeus, 1758): aetiology, anatomy and consequences for survival. J Fish Dis, doi:10.1111/jfd.12189

► Luo G, Ducy P, Mckee MD, Pinenro GJ, Loyer E, Behringer RR, Karsenty G (1997) Spontaneous calcification of arteries and cartilage in mice lacking matrix GLA protein. Nature 386:78–81

Martoja R, Martoja-Pierson M (1970) Técnicas de histología animal. Toray-Masson, Barcelona

► Meunier FJ (2011) The osteichtyes, from the paleozoic to the extant time, through histology and palaeohistology of bony tissues. C R Pal 10:347–355

► Miura M, Chen XD, Allen MR, Bi YM, Gronthos S, Seo BM (2004) A crucial role of caspase-3 in osteogenic differentiation of bone marrow stromal stem cells. J Clin Invest 114:1704–1713

► Morel C, Adriaens D, Boone M, De Wolf T, Van Hoorebeke L, Sorgeloos P (2010) Visualizing mineralization in deformed opercular bones of larval gilthead sea bream (Sparus aurata). J Appl Ichthyology 26:278–279

► Noble C, Cañon Jones H, Damsgård B, Flood M and others (2012) Injuries and deformities in fish: their potential impacts upon aquacultural production and welfare. Fish Physiol Biochem 38:61–83

Ortega A (2008) Cultivo de Dorada (Sparus aurata). In: Espinosa de los Monteros J (ed) Cuadernos de Acuicultura. Fundación Observatorio Español de Acuicultura, Madrid

▶ Ortiz-Delgado JB, Simes DC, Gavaia P, Sarasquete C, Cancela ML (2005) Osteocalcin and matrix GLA protein in developing teleost teeth: identification of sites of mRNA and protein accumulation at single cell resolution. Histochem Cell Biol 124:123–130

▶ Paperna I (1978) Swimbladder and skeletal deformation in hatchery bred *Sparus aurata*. J Fish Biol 12:109–114

Pearse AGE (1985) Histochemistry: theoretical and applied, 4th edn. Vol 2: analytical technology. Churchill Livinstone, Edinburgh

▶ Prades JM, Dumollard JM, Duband S, Timoshenko A and others (2010) Lamina propria of the human vocal fold: histomorphometric study of collagen fibers. Surg Radiol Anat 32:377–382

▶ Prestinicola L, Boglione C, Makridis P, Spanò A and others (2013) Environmental conditioning of skeletal anomalies typology and frequency in gilthead seabream (*Sparus aurata* L., 1758) juveniles. PLoS ONE 8:e55736

▶ Sanden M, Berntssen MHG, Krogdahl A, Hemre GI, Bakke-McKellep AM (2005) An examination of the intestinal tract of Atlantic salmon, *Salmo salar* L., parr fed different varieties of soy and maize. J Fish Dis 28:317–330

▶ Sanatamaría JA, Andrades JA, Herráez P, Fernández-Llebrez P, Becerra J (1994) Perinotochordal connective sheet of gilthead sea bream larvae (*Sparus aurata*, L.) affected by axial malformations: a histochemical and immunocytochemical study. Anat Rec 240:248–254

Sarasquete C, Gutiérrez M (2005) New tetrachromic VOF stain (Type III-G.S) for normal and pathological fish tissues. Eur J Histochem 49:105–114

▶ Simes DC, Williamson M, Ortiz-Delgado JB, Viegas SCB, Price PA, Cancela ML (2003) Purification of matrix Gla protein from a marine teleost fish, *Argyrosomus regius*: calcified cartilage and not bone as the primary site of MGP accumulation in fish. J Bone Miner Res 18:244–259

▶ Simes DC, Williamson M, Schaff BJ, Gavaia PJ, Ingleton PM, Price PA, Cancela ML (2004) Charaterization of osteocalcin (BGP) and matrix Gla protein (MGP) fish specific antibodies: validation for immunodetection studies in lower vertebrates. Calcif Tissue Int 74:170–180

▶ Tiago DM, Laizé V, Bargelloni L, Ferraresso S, Romualdi C, Cancela ML (2011) Global analysis of gene expression in mineralizing fish vertebra-derived cell lines: new insights into anti-mineralogenic effect of vanadate. BMC Genomics 12:310

▶ Totland GK, Fjelldal PG, Kryvi H, Løkka G and others (2011) Sustained swimming increases the mineral content and osteocyte density of salmon vertebral bone. J Anat 219: 490–501

▶ Verhaegen Y, Adriaens D, De Wolf T, Dhert P, Sorgeloos P (2007) Deformities in larval gilthead sea bream (*Sparus aurata*): a qualitative and quantitative analysis using geometric morphometrics. Aquaculture 268:156–168

▶ Witten PE, Gil-Martens L, Hall BK, Huysseune A, Obach A (2005) Compressed vertebrae in Atlantic salmon *Salmo salar*: evidence for metaplastic chondrogenesis as a skeletogenic response late in ontogeny. Dis Aquat Org 64:237–246

▶ Witten PE, Obach A, Huyseune A, Baeverfjord G (2006) Vertebrae fusion in Atlantic salmon (*Salmo salar*): development aggravation and pathways of containment. Aquaculture 258:164–172

▶ Witten PE, Gil-Martens L, Huysseune A, Takle H, Hjelde K (2009) Towards a classification and an understanding of developmental relationships of vertebral body malformations in Atlantic salmon (*Salmo salar* L.). Aquaculture 295:6–14

▶ Ytteborg E, Baeverfjord G, Torgersen J, Hjelde K, Takle H (2010) Molecular pathology of vertebral deformities in hyperthermic Atlantic salmon (*Salmo salar*). BMC Physiol 10:12

▶ Ytteborg E, Baeverfjord G, Takle H (2012) Four stages characterizing vertebral fusions in Atlantic salmon. J Appl Ichthyology 28:453–459

3

Habitat use of the gnomefishes *Scombrops boops* and *S. gilberti* in the northwestern Pacific Ocean in relation to reproductive strategy

N. Takai*, Y. Kozuka, T. Tanabe, Y. Sagara, M. Ichihashi, S. Nakai, M. Suzuki, N. Mano, S. Itoi, K. Asahina, T. Kojima, H. Sugita

Department of Marine Science and Resources, College of Bioresource Sciences, Nihon University, Fujisawa 252-0880, Japan

ABSTRACT: The Japanese gnomefishes *Scombrops boops* and *S. gilberti* closely resemble each other but their geographical distributions in the northwestern Pacific Ocean are markedly different. In order to understand the determinants of reproductive isolation, we examined habitat use in relation to the reproductive strategy employed by these species in the Izu-Islands region, where adults of both species coexist. We examined the species compositions of the gnomefishes based on mitochondrial sequence differences, and the maturation process of gonads using the gonado-somatic index (GSI) and by histological observation of the gonads. Genetic analysis showed that large individuals (>400 mm SL) of both species were present in the region, whereas all small gnomefish (<400 mm SL) were identified only as *S. boops*. Thus, the Izu-Islands region is likely utilized as a nursery by *S. boops* but not by *S. gilberti*. The GSI of adult *S. gilberti* showed significantly higher values in March compared to the consistently low values detected for *S. boops*. In addition, the GSI-elevated *S. gilberti* had mature ovaries and testes. These results suggest that the Izu-Islands region is utilized as a spawning ground by *S. gilberti* but not by *S. boops*. It appears that *S. boops* spawns in the upper reaches of the Kuroshio Current and that the offspring are transported to the Izu-Islands region by the current. The differences between the species-specific spawning grounds, in conjunction with the transport provided by the currents, likely determine the interspecific differences in geographical distribution.

KEY WORDS: Reproductive isolation · Mesopelagic fish · Spawning · Nursery · Kuroshio · Izu · mtDNA · Geographical distribution

INTRODUCTION

The gnomefishes *Scombrops boops* and *S. gilberti* are commercially important species inhabiting the coastal waters of the Japanese Archipelago. Since the beginning of the 20th century, there has been speculation that these may actually be the same species because of the extreme similarity of their appearances (Tanaka 1931, Matsubara 1971, Yasuda et al. 1971). Recently, however, Itoi et al. (2008) found clear interspecific differences in the sequences of the control region and the 16S rRNA gene of mitochondrial DNA (mtDNA) for *S. boops* and *S. gilberti*, suggesting that these species are genetically different.

These fishes show similar life history patterns; both species inhabit the shallow waters in the sublittoral zone during the juvenile and young stages, and later move to the rocky bottom in the dysphotic zone at a depth of 200 to 700 m (Hayashi 2002). They are both carnivores, preying on fish, squid and shrimp (Mochizuki 1997). However, in contrast to the similarity in morphological characteristics and life history patterns, the geographical distributions of the 2 gnomefishes are markedly different. *S. boops* is broadly distributed, from the waters off southern Hokkaido (Japan) to Taiwan, whereas *S. gilberti* is found in a relatively narrow region of the Pacific from the waters off southern Hokkaido to the coastal

*Corresponding author: takai@brs.nihon-u.ac.jp

waters of the Izu Islands (see Fig. 1, Hayashi 2002). The interspecific difference in geographical distribution is more marked at the early life stages. The fishing grounds for young *S. gilberti* are mainly restricted to the coastal waters off the Sanriku coast, the northeastern district of the Japanese mainland. Young *S. gilberti* are rarely collected in the coastal waters of the Izu Islands, in spite of the abundant catch of adults there (Itoi et al. 2010). By contrast, young *S. boops* are broadly distributed along the Japanese Archipelago on both sides — the Pacific Ocean and the Sea of Japan (Itoi et al. 2010). At present, little is known about the reasons for such a marked interspecific difference in geographical distribution.

Recently, a few studies have presented genetic evidence to explain the disparate distributions of the 2 gnomefishes. Noguchi et al. (2012) sequenced the mitochondrial gene, cytochrome *b*, in *S. boops* collected from 7 sampling stations around the Japanese Archipelago and found that a single haplotype (a combination of alleles closely located on a chromosome or mitochondria) in each station. This suggested that *S. boops* consisted of a single population in this region, probably with a major spawning ground to the south of Kyushu Island (see Fig. 1). A northeastward transport of eggs and planktonic larvae born in the southern region by the Kuroshio Current and the Tsushima Warm Current could distribute *S. boops* throughout the Japanese Archipelago (see Fig. 1). On the other hand, Itoi et al. (2011) sequenced the cytochrome *b* gene for *S. gilberti* collected from the coastal waters of the Pacific side of eastern Japan and showed that genetic differences were not significant between adults from the Izu-Islands region and young from the Sanriku region. Itoi et al. (2011) presented the hypothesis that *S. gilberti* utilizes the Izu-Islands region for spawning and the Sanriku region as a nursery. The Kuroshio Current and its warm-core rings can transport eggs and planktonic larvae to the Sanriku Coast, although this transport system is somewhat unstable (Muto 1985). The results of these studies indicate that the geographical distributions of these gnomefishes could possibly be related to interspecific differences in reproductive patterns, including the location of the species-specific spawning grounds and the transport of eggs and planktonic larvae by ocean currents.

In the present study, we focused on the habitat use of these 2 gnomefish species in the Izu-Islands region. As mentioned above, the geographical distributions of the adults of the 2 species overlap in this region only (Itoi et al. 2011). If the inference by the genetic studies is correct, it can be assumed that the Izu-Islands region is the spawning ground for *S. gilberti* but not for *S. boops*. We examined the species compositions of the gnomefishes using a genetic discrimination method and the development process of the gonads in adults in the Izu-Islands region in order to understand their habitat use in relation to the reproductive strategy employed by these species in the region.

MATERIALS AND METHODS

Life history of the gnomefishes

Scombrops boops and *S. gilberti* are both large-sized species, with adults reaching a maximum of 1 m or more in total length (Mochizuki 1997). Both species are considered to mature at a body size of about 380 mm total length when they are 3 yr old (Mochizuki 1977, 1997). In captivity, the adults spawn isolated pelagic eggs with a diameter of about 1.2 mm and the larvae hatch about 3 d after the spawning in a rearing environment with a controlled water temperature of about 15°C (Yamada 1995). Juveniles settle on the rocky bottom in the shallow coastal waters after the planktonic stage, and remain near shore during the juvenile and young life stages. Their habitat shifts to deeper waters as they grow, with adults settling in the dysphotic zone at a depth of 200 to 700 m (Hayashi 2002).

According to Okiyama (1988) and Iwai (2005), their life history stages can be defined as follows. (1) Larva: fewer fin rays than the adults. (2) Juvenile: same number of fin rays as the adults, but with a markedly different external appearance from that of adults. (3) Young: external appearance similar to the adult form, but the proportion of some body parts differs from those of the adults. (4) Immature: external appearance almost the same as that of adults, but the gonads are not developed. (5) Adult: large body size and well-developed characters, with high reproductive ability.

Fish sampling

We performed 3 types of investigations (Investigation I, II, III) in the Izu Island region (Fig. 1) to examine (I) the temporal and spatial changes in species composition of the different sizes of gnomefish, (II) the temporal change in species composition of young gnomefish at a fixed station, and (III) the maturation

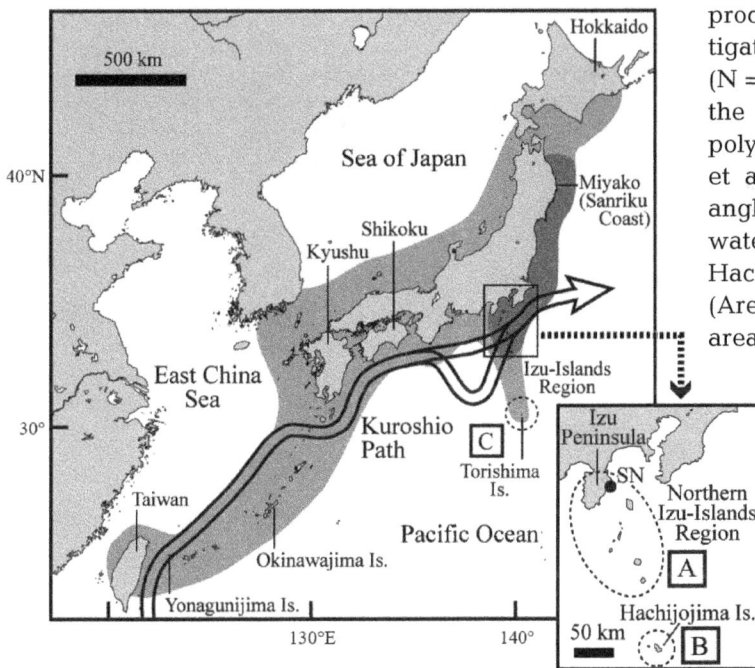

Fig. 1. Study area showing the 3 regions (dotted circles) where gnomefish were collected: (A) in the coastal waters of the northern Izu Islands, (B) Hachijojima Island, and (C) Torishima Island. The young for Investigation II were sampled at a set-net station (SN, solid circle). The distribution of gnomefishes in the northwestern Pacific Ocean are shaded gray according to Itoi et al. (2010). *Scombrops boops* is found in the overall gray zones, whereas *S. gilberti* is found in the dark gray zone only

process of the gonads of adult gnomefish. For Investigation I, we collected different-sized gnomefish (N = 367) from May 2007 to July 2008 and identified the species using PCR-restriction fragment length polymorphism (PCR-RFLP) of mtDNA, based on Itoi et al. (2008) (Table 1). The fish were caught by angling, trawling and set-net fisheries in the coastal waters of the northern Izu-Islands region (Area A), Hachijojima Island (Area B), and Torishima Island (Area C) (Fig. 1). We divided the study area into 3 areas on the basis of the positional relation to the Kuroshio Current, since recent genetic studies have suggested that the transport of eggs and planktonic larvae by the Kuroshio Current might be a determining factor for the geographical distributions of the 2 gnomefishes (Itoi et al. 2011, Noguchi et al. 2012). Areas A and B are located in the main region and the southern margin of the Kuroshio Current, respectively, while Area C is about 300 km south of the Kuroshio region. The fish caught by angling and trawling were mostly collected from a depth zone of 150 to 600 m. A set-net is a type of fixed fishing gear which compels fish to swim into a bag net located in the innermost chamber. Portions of the results from Investigation I (May 2007 to January 2008) have

Table 1. The total number (N) of gnomefish collected from the Izu-Islands region from May 2007 to July 2008, and the number (n) and range of standard lengths (SL, mm) of *Scombrops boops* and *S. gilberti* identified by PCR-restriction fragment length polymorphism (PCR-RFLP) of mtDNA

Month	N	S. boops Female		S. boops Male		S. gilberti Female		S. gilberti Male	
		n	SL	n	SL	n	SL	n	SL
2007									
May	13	8	198.2–357.8	5	207.4–338.4	0	–	0	–
Jun	48	16	215.3–489.0	20	219.9–506.4	5	422.0–466.7	7	447.5–468.9
Jul	10	5	280.9–351.4	5	251.9–485.3	0	–	0	–
Aug	39	25	214.9–290.0	10	217.7–268.3	1	487.7	3	485.7–510.0
Sep	23	7	285.4–327.0	9	242.0–304.8	3	456.6–482.7	4	452.2–480.0
Oct	40	20	236.5–360.8	20	234.8–365.2	0	–	0	–
Nov	25	11	186.4–396.6	13	176.6–372.2	1	571.5	0	–
Dec	17	10	265.2–424.6	7	252.1–561.0	0	–	0	–
2008									
Jan	17	7	305.3–526.0	10	282.0–528.6	0	–	0	–
Feb	18	8	281.3–446.5	9	277.0–334.6	1	487.3	0	–
Mar	18	10	297.6–483.3	8	276.2–490.7	0	–	0	–
Apr	21	11	248.6–494.5	9	241.8–352.5	1	433.2	0	–
May	41	23	200.1–406.8	16	197.5–344.0	0	–	2	466.2–501.3
Jun	15	4	211.7–333.2	5	209.8–353.3	2	493.5–530.8	4	426.3–492.5
Jul	22	11	215.2–443.4	10	203.2–455.0	1	426.1	0	–
Total	367	176	186.4–526.0	156	176.6–561.0	15	422.0–571.5	20	426.3–510.0

already been reported by Itoi et al. (2010). Here, we present new results from Investigation I.

For Investigation II, we sampled young gnomefish every month from a fishing boat at a set-net station in the coastal waters of the eastern Izu Peninsula, from May 2009 to June 2010 (Stn SN in Fig. 1, Table 2). The set-net had a bag net mesh size of 15 to 30 mm, and was laid at a depth of 5 to 50 m. A total of 289 young were collected throughout the sampling period. For Investigation III, a total of 70 gnomefish caught by angling were collected from the coastal waters of the northern Izu-Islands region and Hachijojima Island from November 2009 to May 2010 (Fig. 1, Table 3). We collected 10 adults every month except in April and May 2010. Only 7 fish were obtained in April, and therefore 3 additional fish were collected in May so as to total 20 fish during April and May. In Investigations II and III, we also used PCR-RFLP on the mtDNA to identify the species.

Mochizuki (1977) inferred that the spawning season of gnomefish in the coastal waters of eastern Japan is December to April for *S. boops* and March to May for *S. gilberti*, based on seasonal variations in the gonadosomatic index (GSI). Therefore, we collected the adults between November and May to examine the maturation process for Investigation III.

All specimens in Investigations I to III were measured to the nearest 0.1 mm standard length (SL).

Table 2. Total number (N) and range of standard lengths (SL, mm) of young gnomefish collected monthly by the set-net sampling in the coastal waters of the eastern Izu Peninsula from May 2009 to June 2010, and the number of *Scombrops boops* and *S. gilberti* identified by PCR-restriction fragment length polymorphism (PCR-RFLP) of mtDNA

Month	N	SL	*S. boops*	*S. gilberti*
30 May 09	42	47.8–119.8	42	0
26 Jun 09	38	70.0–134.6	38	0
29 Jul 09	16	100.3–145.6	16	0
19 Aug 09	50	79.2–139.7	50	0
18 Sep 09	3	115.2–122.6	3	0
15 Oct 09	28	132.2–164.2	28	0
26 Nov 09	16	127.4–174.9	16	0
20 Dec 09	0	–	–	–
14 Jan 10	11	163.2–175.6	11	0
17 Feb 10	0	–	–	–
15 Mar 10	0	–	–	–
22 Apr 10	0	–	–	–
27 May 10	50	57.6–106.1	50	0
20 Jun 10	35	78.6–121.0	35	0
Total	289	47.8–175.6	289	0

Table 3. Total number (N) and range of standard lengths (SL, mm) of adult gnomefish collected from the Izu-Islands region from November 2009 to March 2010, and the number of *Scrombrobs boops* and *S. gilberti* identified by PCR-restriction fragment length polymorphism (PCR-RFLP) of mtDNA

Month	N	SL	*S. boops*	*S. gilberti*
2009				
Nov	10	435.1–567.0	5	5
Dec	10	493.4–576.1	5	5
2010				
Jan	10	444.1–615.3	5	5
Feb	10	426.3–526.6	7	3
Mar	10	464.9–572.2	5	5
Apr	7	440.4–675.7	7	0
May	13	481.0–625.1	4	9
Total	70	426.3–675.7	38	32

DNA extraction and PCR amplification

A small portion of skeletal muscle was excised from every individual. Total genomic DNA was extracted from the muscle using the method of Sezaki et al. (1999). Partial fragments of 16S rRNA gene on mtDNA were amplified by PCR using the primers 16SAR-L (5'-CGC CTG TTT ATC AAA AAC AT-3') and ftRLeu_R (5'-CTG TTB RAA GGG CTT AGG BCT TTT GC-3'), following Palumbi et al. (1991) and Itoi et al. (2007) respectively. PCR amplification was done in a 20 µl reaction mixture containing genomic DNA as template, 4.0 µl of 5× *GoTaq* DNA polymerase buffer, 0.8 µl of 10 µM primers, 2.0 µl of 2 mM dNTP and 1 unit of *GoTaq* DNA polymerase (Promega). The thermal cycling profile for the PCR consisted of initial denaturation at 94°C for 3 min followed by 40 cycles of denaturation at 94°C for 15 s, annealing at 50°C for 20 s and extension at 72°C for 45 s.

PCR-RFLP

The PCR conditions employed for RFLP analysis were the same as described above. PCR products for the 16S rRNA gene of the gnomefish were digested with 10 units of restriction enzymes *Eco*NI (New England Biolabs; www.neb.com) and *Mva*I (Toyobo; www.toyobo.co.jp) in a 10 µl mixture containing 1.0 µl of 10× buffer supplied with the kit and 5.0 µl of PCR products. The digested products were electrophoresed on a 3% agarose gel and visualized with ethidium bromide.

Sequencing of PCR products

When an individual could not be identified to species by the PCR-RFLP analysis, the PCR products were sequenced for both strands with a 3130*xl* Genetic Analyzer (Applied Biosystems) using a BigDye Terminator v3.1 Cycle Sequencing Kit (Applied Biosystems). Alignment of partial sequences of the 16S rRNA gene for the gnomefishes obtained in this study was carried out using CLUSTAL W (Thompson et al. 1994).

Analysis of maturation process

The 70 adults collected in Investigation III were analyzed to study the maturation of the gonads. Gonads were extracted from the specimens by dissection, weighed to the nearest 0.1 g (wet weight), and fixed in 10% formalin. According to Mochizuki (1977), the GSI was calculated for both sexes by means of the following formula: $GSI = (GW / SL^3) \times 10^7$, where GW = gonad weight (g) and SL = standard length (mm). Generally, the GSI value increases with the development of the gonad. Mochizuki (1977) reported maximum GSI values of 68.3 for females and 35.5 for males (both values for *S. gilberti*).

Tissue preparation for the gonads of 70 adults in Investigation III was done using 2 methods: the paraffin sectioning method and the frozen section method. We used the paraffin sectioning method for 56 individuals with low GSI values (<20), and the frozen section method for 14 individuals with high GSI values (≥20), since it is too difficult to retain the shape of developed germ cells (≥20 in GSI) by the paraffin sectioning method. In the paraffin sectioning method, the fixed tissues were embedded in paraffin (mp 56 to 58°C), serially sectioned to 5 μm thickness, and stained with hematoxylin and eosin for microscopic observation. In the frozen section method, the fixed tissues were frozen at −80°C in hexane and dry ice, and serially sectioned to 4 μm thickness. The slices were placed on adhesive films (Kawamoto 2003) and stained with hematoxylin and eosin. We observed several parts of each gonad using light microscopes to identify the level of maturation.

Statistical analysis

For Investigations I and III, the distributions of SL were compared among groups using the Kruskal-Wallis test, followed by Dunn's multiple comparisons test for the post-hoc analysis in Investigation I. The GSI values in Investigation III were also compared among groups using the Kruskal-Wallis test, followed by Dunn's multiple comparisons test. All statistical analyses were performed using Prism 4.0c (GraphPad Software).

RESULTS

Species compositions

The PCR-RFLP analysis of mtDNA showed that both *Scombrops boops* and *S. gilberti* were included in the gnomefish collected from May 2007 to July 2008, except for 18 individuals from the waters of Torishima Island (Table 1, Fig. 2). The collected fish consisted of 256 *S. boops* and 27 *S. gilberti* from the northern Izu-Islands waters (Area A), 58 *S. boops* and 8 *S. gilberti* from the Hachijojima-Island waters (Area B) and 18 *S. boops* only from the Torishima-Island waters (Area C). Torishima Island is located about 480 km away from the Izu Peninsula, and catching gnomefish is rare in this area. Therefore, we obtained 18 specimens in March 2008 only.

Body size distribution was markedly different between the 2 species (Table 1, Fig. 2). The body size of *S. boops* showed a wide variation, ranging from 176.6 to 561.0 mm SL, whereas all *S. gilberti* individuals were larger than 422.0 mm SL, with a maximum of 571.5 mm SL. The SL was significantly different among 4 groups (i.e. the females and the males of each species) (Kruskal-Wallis test, $H = 81.82$, p < 0.0001). According to Dunn's multiple comparisons test, significant differences were found for every combination of *S. boops* and *S. gilberti* irrespective of sex (p < 0.001 for every combination), whereas intraspecific sexual differences were not significant for either species (p > 0.05).

Small individuals of *S. gilberti* (<400 mm SL) were not collected by set-net sampling (Table 2); all 289 young fish collected during the period from May 2009 to June 2010, ranging in size from 47.8 to 175.6 mm SL, were identified as *S. boops* by the PCR-RFLP analysis of mtDNA. Young fish were collected on most sampling days from May to November, but not from December to April (except for 11 individuals in January).

The adults collected from November 2009 to May 2010 consisted of 6 individuals of *S. boops* and 29 *S. gilberti* from the northern Izu-Islands waters (Area A), as well as 32 individuals of *S. boops* and 3 *S. gilberti* from the Hachijojima Island waters (Area B),

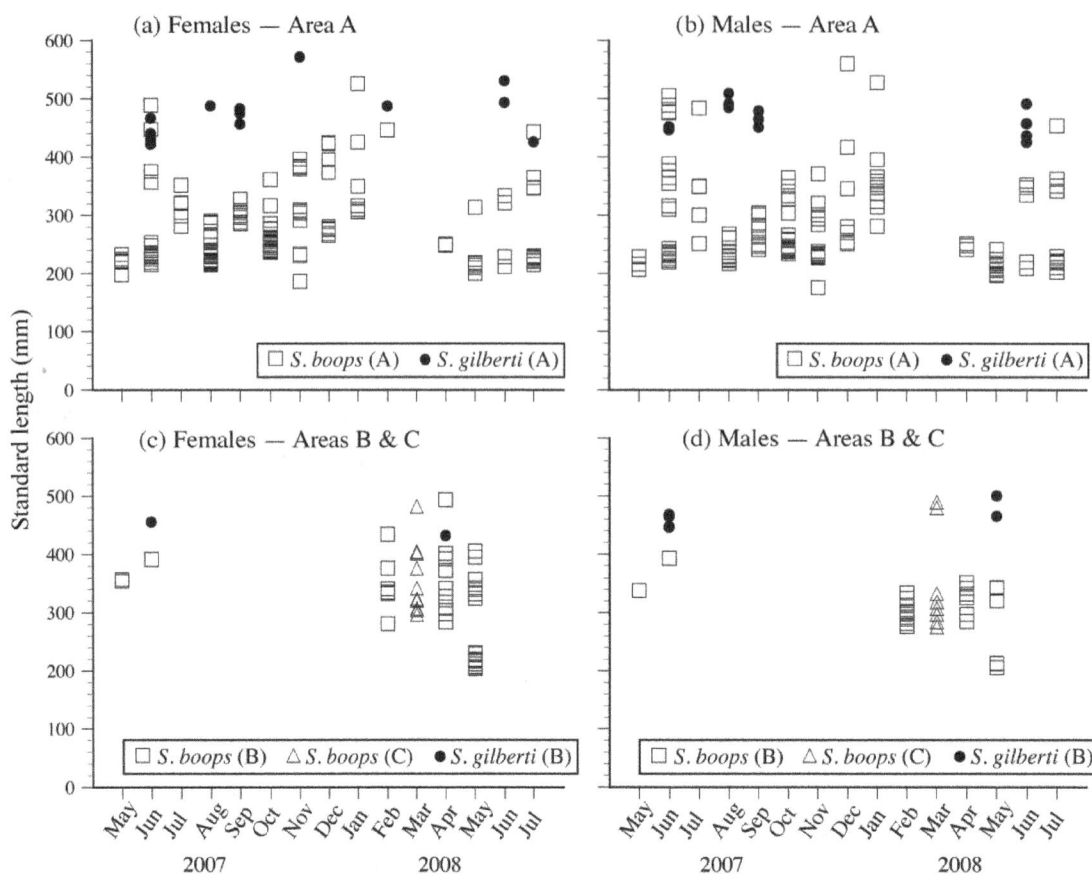

Fig. 2. Distributions of standard length (mm) of gnomefishes *Scrombrobs boops* and *S. gilberti* collected from May 2007 to July 2008 (Investigation I). (a) Females from the northern Izu-Islands waters (Area A); (b) males from Area A; (c) females from the Hachijojima-Island waters (Area B) and the Torishima-Island waters (Area C); (d) males in Areas B and C. Results from Areas B and C are graphed together to minimize the number of panels (due to smaller sample sizes), since data from these areas do not overlap

as shown by the PCR-RFLP analysis of mtDNA (Table 3). The SL of the adults ranged from 434.3 to 601.1 mm for females and 426.3 to 675.7 mm for males of *S. boops*, and from 464.9 to 612.1 mm for females and 435.1 to 567.0 mm for males of *S. gilberti*. There was no significant difference in the SL among the females and males of each species (Kruskal-Wallis test, $H = 5.455$, $p > 0.05$).

GSI

The GSI values of the different-sized gnomefish collected in Investigation I (May 2007 to July 2008) ranged from 0.0 to 3.0 for *S. boops* and 0.4 to 24.0 for *S. gilberti*. Relatively high GSI values of >20 were found for 2 female *S. gilberti* caught in April 2008 (24.0) and February 2008 (20.5); all other gnomefish showed low GSI values of ≤5.1.

The GSI values of the adult gnomefish collected in Investigation III (November 2009 to May 2010)

were markedly different between the species in both the northern Izu-Islands waters and the Hachijojima Island waters (Fig. 3). The GSI of *S. boops* ranged from 0.2 to 3.0 except for one male (12.4) from the Hachijojima Island waters. By contrast, many individuals of *S. gilberti* showed high GSI values of ≥20 (9 females and 5 males). The difference in GSI was significant among the females and males of each species (Kruskal-Wallis test, $H = 48.23$, $p < 0.0001$). According to Dunn's multiple comparison test, significant differences ($p < 0.001$) were found for every combination of *S. boops* and *S. gilberti*, irrespective of sex, except between *S. gilberti* males and *S. boops* females ($p < 0.01$). No intraspecific sexual differences were found for either species ($p > 0.05$).

The GSI values of *S. gilberti* were elevated from November to March in both females and males (Fig. 3). The highest values were found in March (females, 47.6; males, 36.6), which subsequently decreased, with a range of 3.7 to 33.3 in May.

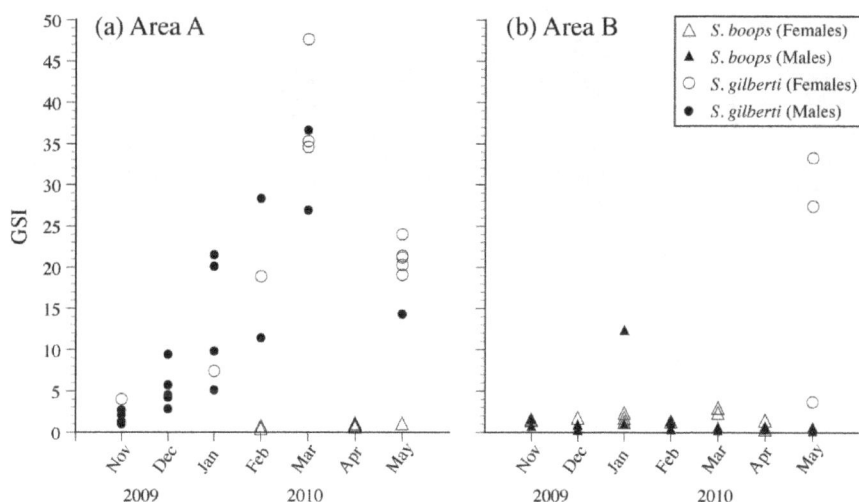

Fig. 3. Monthly changes in gonadosomatic index (GSI) of adult gnomefishes *Scrombrobs boops* and *S. gilberti* from (a) the northern Izu-Islands waters (Area A), and (b) the Hachijojima-Island waters (Area B)

Oogenesis

Oocytes were found to vary in size and form in the ovaries of the adult *S. gilberti* collected from November 2009 to May 2010 (Fig. 4). In the present study, the oocytes were classified into 7 stages (O1–O7), mainly referring to Takano (1989): chromatin-nucleolus stage (O1), perinucleolus stage (O2), yolk vesicle stage (O3), oil droplet stage (O4), vitellogenic stage (O5), migratory nucleus stage (O6) and maturation stage (O7).

At the beginning of oogenesis (Stage O1), the size of the nucleoli in the nucleus increases slightly (Fig. 4a); by Stage O2, the nucleoli are located at the periphery of the nucleus (Fig. 4a). In the next phase, yolk vesicles appear in the cytoplasm (O3, Fig. 4a) and subsequently, oil droplets surround the nucleus (O4, Fig. 4b). In Stage O5, yolk globules appear in the cytoplasm and the oil droplets are scattered throughout the cytoplasm (Fig. 4c). Maturation of the oocyte leads to the migration of the nucleus toward the animal pole at Stage O6 (Fig. 4d), and finally to the disappearance of the nucleus, and the conglomeration of the oil droplets at Stage O7 (Fig. 4e).

We identified the maturation levels for the respective females on the basis of the oocytes at the most advanced stage. The maturation levels of female *S. boops* were identified to be at the early stages (from O2 to O4 in almost all individuals), except for one female that was categorized as O6 (Fig. 5). This exceptional female showed a markedly low GSI value of 3.0, and its ovary included various oocytes at different stages, from primary oocytes to apparently regressed oocytes posterior to maturation.

By contrast, the ovaries of *S. gilberti* included mature oocytes at stages O6 and O7 in 8 of 14 females (Fig. 5). These mature females were collected in March and May (no *S. gilberti* were found in the adult gnomefish collected in April). The GSI of the female *S. gilberti* appeared to peak in March as mentioned above, whereas mature oocytes were more frequently found in May (Fig. 5).

Spermatogenesis

In the present study, spermatogenesis was classified into 6 stages (T1 to T6), mainly referring to Takahashi (1989) and Asahina et al. (1980): growth stage (T1), growth–maturation stage (T2), functional maturation stage (T3), maturation–post-spawn stage (T4), post-spawn stage (T5), and testicular quiescent stage (T6) (Fig. 6).

During Stage T1, seminal lobules are filled with spermatocytes by the growth of spermatogonia (Fig. 6a). In Stage T2, the amount of spermatids and sperm increases (Fig. 6b), and the seminal lobules are filled with sperm at Stage T3 (Fig. 6c). The amount of sperm decreases during Stage T4, and empty spaces become conspicuous in the seminal lobules (Fig. 6d). Sperm still remains at Stage T5, but the empty space increases and the walls bordering the cysts thicken (Fig. 6e). Eventually, the sperm almost completely disappears at Stage T6 (Fig. 6f).

In contrast to the ovaries, the testes of *S. boops* were identified to be mature (Stages T2 to T4) in as many as 8 males (Fig. 7). In particular, the male with a relatively high GSI value of 12.4 was identified to be in functional maturation Stage T3, generally regarded as the most mature stage (Takahashi 1989). In addition, immature testes at Stages T5 and T6, characterized as regressed testes, were also found in 7 *S. boops* individuals.

On the other hand, all the testes of *S. gilberti* were identified as T1 and T2 only (Fig. 7). Five males with relatively high GSI values of 11.4 to 36.6, collected between February and May, were all identified as T2.

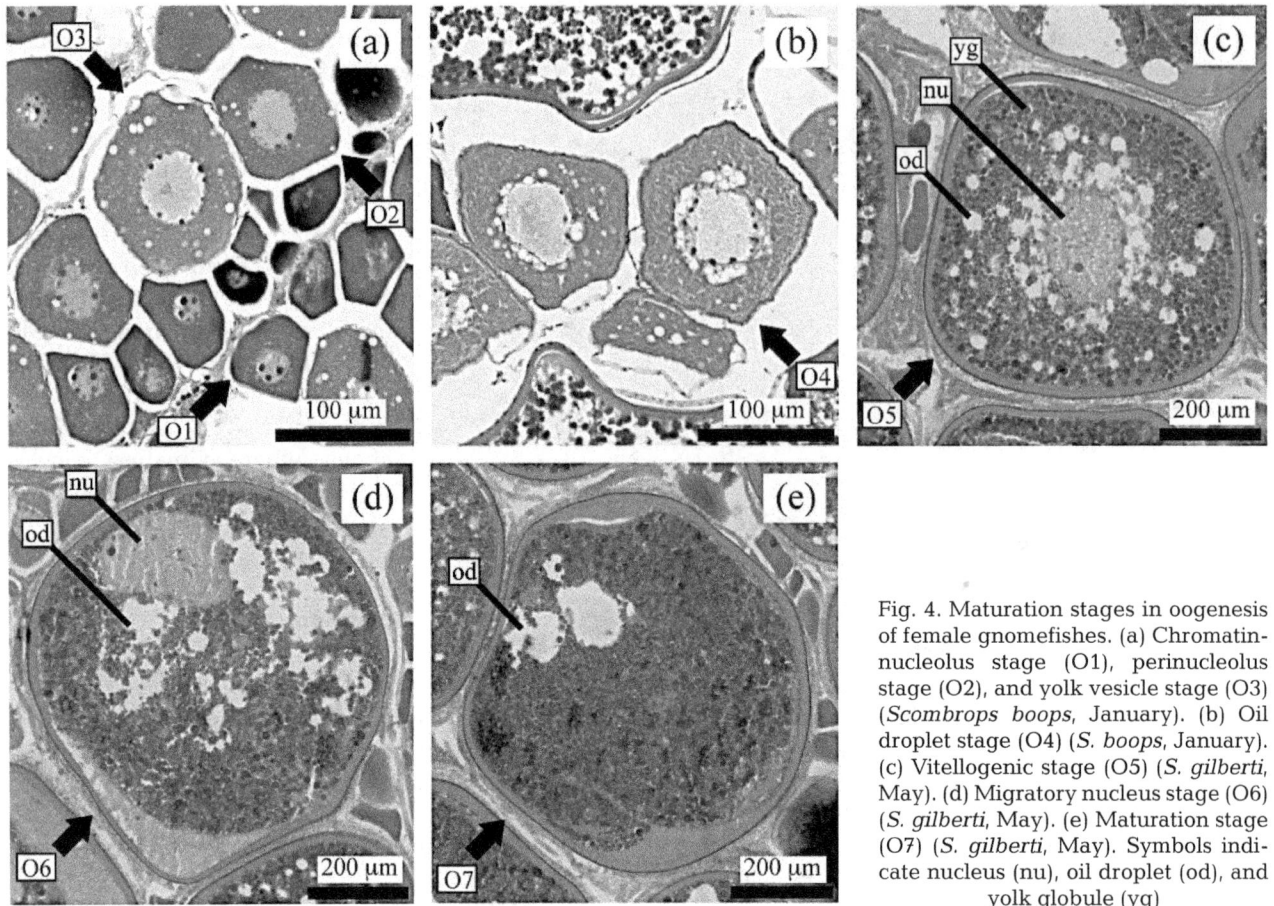

Fig. 4. Maturation stages in oogenesis of female gnomefishes. (a) Chromatin-nucleolus stage (O1), perinucleolus stage (O2), and yolk vesicle stage (O3) (*Scombrops boops*, January). (b) Oil droplet stage (O4) (*S. boops*, January). (c) Vitellogenic stage (O5) (*S. gilberti*, May). (d) Migratory nucleus stage (O6) (*S. gilberti*, May). (e) Maturation stage (O7) (*S. gilberti*, May). Symbols indicate nucleus (nu), oil droplet (od), and yolk globule (yg)

DISCUSSION

Habitat use in the Izu-Islands region

The gnomefish collected from May 2007 to July 2008 consisted of 332 *S. boops*, and only 35 *S. gilberti* (Investigation I, Table 1). It is noteworthy that all 35 *S. gilberti* were large (>400 mm SL), whereas *S. boops* showed a wide range in SL (Fig. 2). Furthermore, during the set-net sampling between May 2009 and June 2010, 289 young gnomefish were collected, all of which were *S. boops* (i.e. no young *S. gilberti* were collected; Investigation II, Table 2). Previous genetic studies of gnomefish in the Izu-Islands region have reported that almost all the small gnome-fish (<400 mm SL) were identified as *S. boops* (Itoi et al. 2008, 2010), but little is known about seasonal variations in species composition of the young in the region.

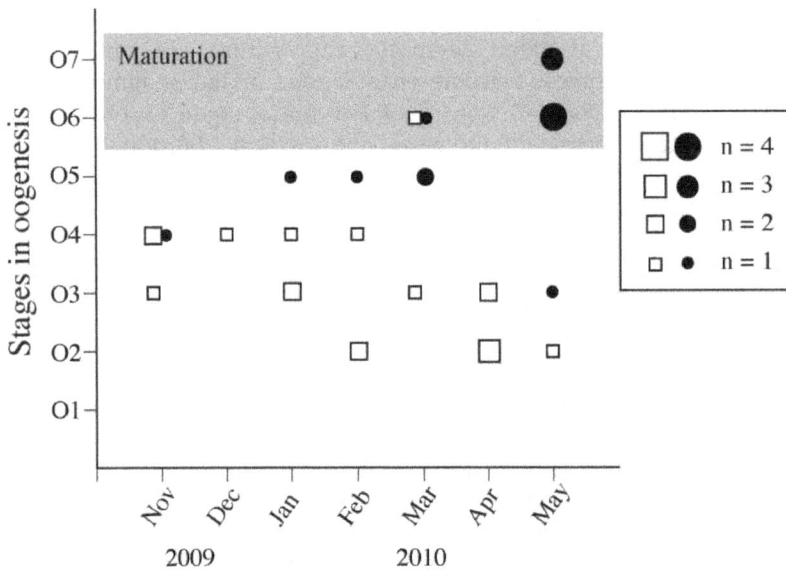

Fig. 5. Monthly changes in the stages of oogenesis for *Scombrops boops* (□) and *S. gilberti* (●) from Nov 2009 to May 2010. The size of the open squares and solid circles reflects the number of individuals at each stage

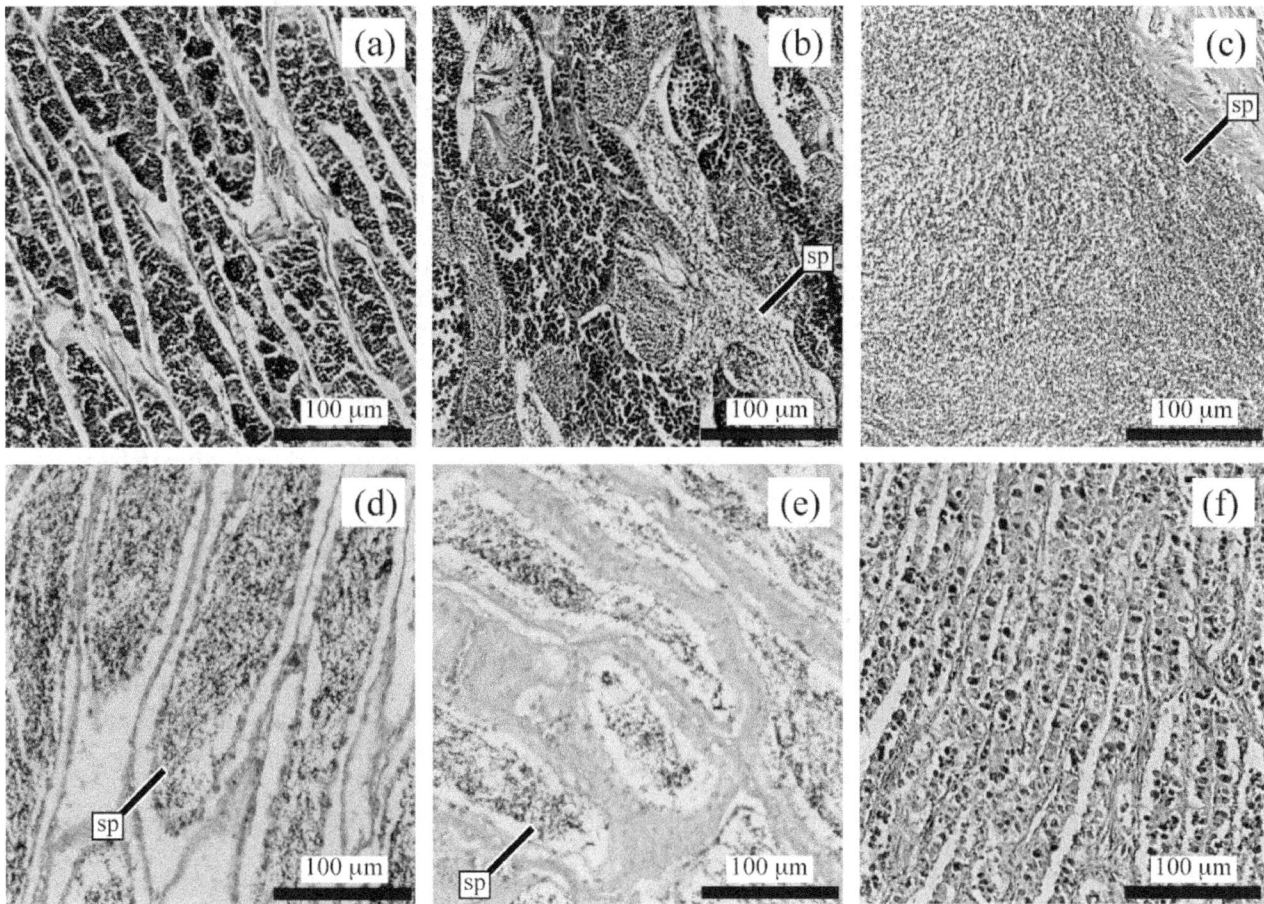

Fig. 6. Maturation stages in spermatogenesis of male gnomefishes. Spermatogenesis was classified into 6 stages: (a) growth stage (T1, *Scombrops gilberti*, December), (b) growth–maturation stage (T2, *S. boops*, December), (c) functional maturation stage (T3, *S. boops*, January), (d) maturation–post-spawn stage (T4, *S. boops*, March), (e) post-spawn stage (T5, *S. boops*, March), and (f) testicular quiescent stage (T6, *S. boops*, May). Sperm is indicated by the symbol (sp)

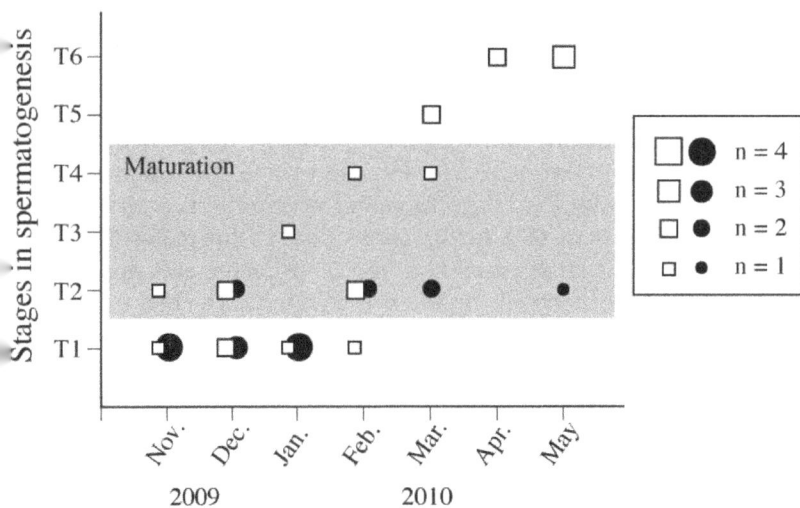

Fig. 7. Monthly changes in the stage of spermatogenesis for *Scombrops boops* (□) and *S. gilberti* (●) from Nov 2009 to May 2010. The size of the open squares and solid circles reflects the number of individuals at each stage

The year-round results in the present study demonstrate that only *S. boops* is present in the Izu-Islands region in young and immature stages. In other words, only *S. boops* utilizes this region as a nursery.

The GSI values of the gnomefish collected from November 2009 to May 2010 were markedly different between *S. boops* and *S. gilberti*, for both females and males (Investigation III, Fig. 3). All 5 individuals of *S. gilberti* collected in March showed high GSI values of 26.9 to 47.6, whereas all *S. boops* individuals showed low GSI values (ranging from 0.4 to 3.0 in females and 0.2 to 12.4 in males). The results of the histological observations of the ovaries were consistent with the seasonal variation in GSI. The ovaries of *S. gilberti* were identified as mature (Stages O6 and O7) in 8 females collected between March and May (Fig. 5). It appears that the mature ovaries of *S. gilberti* frequently occurred in May, posterior to the peak in the GSI. In contrast, the ovaries of *S. boops* were identified as immature (Stages O2 to O4), except for one female that had some mature oocytes at Stage O6, and an abundance of immature oocytes at Stages O2 and O3. This exceptional ovary was extracted from a female with a markedly low GSI value of 3.0, and it was inferred that this female had already completed spawning in this season. Yamada (1995) experimentally reared 79 young *S. boops* (150 to 170 mm in total length) caught in the northern Izu-Islands region in September 1989, and reported that 9 individuals survived until March 1994 in a concrete pond with a capacity of 50 m^3, growing to 380 to 400 mm in total length. These individuals reportedly spawned in the pond 5 times between March 2 and 25. The result in this experimental study showed that *S. boops* can spawn at a size of >400 mm SL in early spring. Accordingly, the low GSI values and immature ovaries in large *S. boops* (>400 mm SL) collected in the present study from November to May demonstrate that *S. boops* does not spawn in the Izu-Islands region. Only *S. gilberti* likely utilizes this region as a spawning ground in the spring, probably from March to May.

In contrast, the results of the histological observations of the testes were inconsistent with the seasonal variation in GSI (Fig. 7). Eight *S. boops* males had mature testes at Stages T2 to T4, in spite of having low GSI values ranging from 0.2 to 12.4, compared with the higher GSI values of >20 in *S. gilberti*. This result suggests that the testes of *S. boops* in the Izu-Islands region fully develop in the maturation process but do not increase in size. In any case, these male *S. boops* do not mate in the Izu-Islands region, since mature females of *S. boops* are likely absent from the region as mentioned above.

In *S. gilberti*, mature testes at Stage T2 occurred for half a year from December 2009 to May 2010 (Fig. 7). The T2 testes from fish collected in December showed low GSI values of 4.5 to 9.4, whereas the T2 testes during February to May showed high GSI values of 11.4 to 36.6. It appears that the testes of *S. gilberti* increased in size during the T2 stage. We are unable to make a generalization about the maturation process of male *S. gilberti* in April and May, since only one male *S. gilberti* was found among the adult gnomefish collected during this study period. In future studies, it would be necessary to collect and analyze more males of *S. gilberti* in that season.

The nursery ground for *S. gilberti*

In the present study, no small *S. gilberti* (<400 mm SL) were collected in the Izu-Islands region, results that are similar to previous studies (Tables 1 & 2; Itoi et al. 2008, 2010). This indicates that most eggs and planktonic larvae born in the region are transported to other, distant locations. They are likely transported to the Sanriku region in northeastern Japan by the Kuroshio Current and its warm-core rings, and subsequently settle on the sublittoral bottom as inferred by Itoi et al. (2011). On 8 June 2012, a juvenile gnomefish with 32.4 mm SL was seined in a seagrass bed in Miyako Bay in the Sanriku region, and subsequently identified as *S. gilberti* by mtDNA sequencing (S. Itoi unpubl. data). Itoi et al. (2010) reported that 106 young *S. gilberti* (106 to 181 mm SL), genetically identified, were collected off the Miyako Bay from 18 September to 21 November 2008. Thus, this paper suggests that (1) adult *S. gilberti* spawn in spring (March to May) in the Izu-Islands region, and (2) young *S. gilberti* grow in the nursery of the Sanriku region between summer and autumn after being transported there by the Kuroshio Current.

Spawning ground of *S. boops*

In this study, mature *S. boops* adults with high GSI values were not found in the Izu-Islands region (Fig. 3). Mochizuki (1977) suggested that *S. boops* spawn from December to April in the coastal waters of eastern Japan, and from October to February in the coastal waters of western Japan, based on seasonal variations in the GSI of the gnomefish collected from a broad area on the Pacific side. However, the maximum GSI values of *S. boops* reported by Mochizuki (1977) were 24.3 for females and 18.2 for males, much lower than the GSI values of *S. gilberti*, with a maximum of 68.3 for females and 35.5 for males in the same study, and 47.6 and 36.6 for females and males, respectively, in the present study. The ovaries of *S. boops* collected by Mochizuki (1977) were likely immature.

In early December 2012, we collected an adult *S. boops* with a high GSI value of 30.2 from the coastal waters of the Yonagunijima Island in the southernmost shelf break of the East China Sea. We also collected 2 females with developed ovaries at Stage O6 in that month, although the GSI values from that individual were relatively low (11.1 to 16.8) (N. Takai unpubl. data). The Nansei Regional

Fisheries Research Laboratory (1969) reported that planktonic larvae and juvenile *S. boops* were collected from late February to early March from the surface layer in the shelf break zone of the East China Sea. Thus, we expect that *S. boops* spawns in the region during the winter and are transported to the Izu-Islands region over a great distance. Consequently, the geographical differences in species-specific spawning grounds and in the current transport system likely determine the interspecific differences in the geographical distributions of *S. boops* and *S. gilberti*.

It is generally supposed that the shelf break of the East China Sea is the major spawning ground for many pelagic fishes and invertebrates, and that the warm currents of the Kuroshio Current and the Tsushima Warm Current transport eggs, larvae, and juveniles to the general area of the Japanese Archipelago on both sides: the Pacific Ocean and the Sea of Japan. Some examples are Pacific bluefin tuna *Thunnus orientalis* (Kitagawa et al. 2010), Japanese horse mackerel *Trachurus japonicus*, Yellowtail *Seriola quinqueradiata* (Ochiai & Tanaka 1986) and Japanese common squid *Todarodes pacificus* (Okutani 1995). The reproductive strategy of *S. boops* appears to be exactly the same as that of these pelagic species, although the adult gnomefish is a mesopelagic fish inhabiting the layer 200 to 700 m deep. The shelf break of the East China Sea might be important for the reproductive strategy of mesopelagic fishes as well as that of many pelagic fishes.

Acknowledgements. We thank Akazawa-Gyogyo for their cooperation in set-net sampling of the fish. This study was financially supported by Research Grants for 2007 and 2008 from College of Bioresource Sciences, Nihon University.

LITERATURE CITED

► Asahina K, Iwashita I, Hanyu I, Hibiya T (1980) Annual reproductive cycle of a bitterling, *Rhodeus ocellatus ocellatus*. Bull Jpn Soc Sci Fish 46:299–305

Hayashi M (2002) Scombropidae. In: Nakabo T (ed) Fishes of Japan with pictorial keys to the species, English edn. Tokai University Press, Tokyo, p 786

► Itoi S, Saito T, Washio S, Shimojo M, Takai N, Yoshihara K, Sugita H (2007) Speciation of two sympatric coastal fish species, *Girella punctata* and *Girella leonina* (Perciformes, Kyphosidae). Org Divers Evol 7:12–19

► Itoi S, Takai N, Naya S, Dairiki K and others (2008) Species identification method for *Scombrops boops* and *Scombrops gilberti* based on polymerase chain reaction-restriction fragment length polymorphism analysis of mitochondrial DNA. Fish Sci 74:503–510

► Itoi S, Odaka J, Yuasa K, Akeno S and others (2010) Distribution and species composition of juvenile and adult scombropids (Teleostei, Scombropidae) in Japanese coastal waters. J Fish Biol 76:369–378

► Itoi S, Odaka J, Noguchi S, Noda T and others (2011) Genetic homogeneity between adult and juvenile populations of *Scombrops gilberti* (Percoid, Scombropidae) in the Pacific Ocean off the Japanese Islands. Fish Sci 77:975–981

Iwai T (2005) Introduction to ichthyology. Koseisha-Koseikaku, Tokyo

► Kawamoto T (2003) Use of a new adhesive film for the preparation of multi-purpose fresh-frozen sections from hard tissues, whole-animals, insects and plants. Arch Histol Cytol 66:123–143

► Kitagawa T, Kato Y, Miller MJ, Sasai Y, Sasaki H, Kimura S (2010) The restricted spawning area and season of Pacific bluefin tuna facilitate use of nursery areas: a modeling approach to larval and juvenile dispersal processes. J Exp Mar Biol Ecol 393:23–31

Matsubara K (1971) Pomatomidae (Scombropidae). Fish morphology and hierarchy, Part 1. Ishizaki-Shoten, Tokyo

Mochizuki K (1977) Systematics, distribution, growth and reproduction of scombropid fishes. PhD dissertation, Tokyo University

Mochizuki K (1997) Scombropidae. In: Okamura O, Amaoka K (eds) Sea fishes of Japan. Yama-Kei Publishers, Tokyo, p 310

Muto S (1985) The coastal waters of the Sanriku region. II. Physical environment. In: Oceanographical Society of Japan (ed) Coastal Oceanography of Japanese Islands. Tokai University Press, Tokyo, p 220–231

► Noguchi S, Itoi S, Takai N, Noda T, Myojin T, Yoshihara K, Sugita H (2012) Population genetic structure of *Scombrops boops* (Percoid, Scombropidae) around the Japanese archipelago inferred from the cytochrome *b* gene sequence in mitochondrial DNA. Mitochondrial DNA 23:223–229

Ochiai A, Tanaka M (1986) Ichthyology, Vol 2. Koseisha-Koseikaku, Tokyo

Okiyama M (1988) An atlas of the early stage fishes in Japan. Tokai University Press, Tokyo

Okutani T (1995) Cuttlefish and squids of the world in color. Okumura Printing, Tokyo

Palumbi S, Martin A, Romano S, McMillan WO, Stice L, Grabowski G (1991) The simple fool's guide to PCR, version 2. University of Hawaii, Honolulu, HI

Sezaki K, Begum RA, Wongrat P, Srivastava MP and others (1999) Molecular phylogeny of Asian freshwater and marine stingrays based on the DNA nucleotide and deduced amino acid sequences of the cytochrome *b* gene. Fish Sci 65:563–570

Takahashi H (1989) Structure and gametogenesis of the testis. In: Takajima F, Hanyu I (eds) Suizoku-Hanshoku-gaku (Reproductive biology of aquatic animals). Midori-Shobo, Tokyo, p 35–64

Takano K (1989) Structure and gametogenesis of the ovary. In: Takajima F, Hanyu I (eds) Suizoku-Hanshokugaku (Reproductive biology of aquatic animals). Midori-Shobo, Tokyo, p 3–34

Tanaka S (1931) On the distribution of fishes in Japanese waters. J Fac Sci Tokyo Univ 4 3:1–90

Nansei Regional Fisheries Research Laboratory (1969) The summary of Shunyo-Maru Cruise for the marine biological research in the area between the southern Shikoku and eastern Taiwan. Nansei Regional Fisheries Research Laboratory, Kochi

► Thompson JD, Higgins DG, Gibson TJ (1994) CLUSTAL W: improving the sensitivity of progressive multiple sequence alignment through sequence weighting, position-specific gap penalties and weight matrix choice. Nucleic Acids Res 22:4673–4680

Yamada T (1995) Spawning and hatching of Japanese scombropid *Scombrops boops*. Saibai Giken 23:145–146 (in Japanese)

Yasuda F, Mochizuki K, Kawajiri M, Nose Y (1971) On the meristic and morphometric differences between *Scombrops boops* and *S. gilberti*. Jpn J Ichthyol 18:118–124

4

Depth interactions and reproductive ecology of sympatric Sillaginidae: *Sillago robusta* and *S. flindersi*

Charles A. Gray[1,2,3,*], **Lachlan M. Barnes**[1,4], **Dylan E. van der Meulen**[1,5],
Benjamin W. Kendall[1,6], **Faith A. Ochwada-Doyle**[1,7], **William D. Robbins**[1,8]

[1]NSW Primary Industries, Cronulla Fisheries Research Centre, Cronulla, NSW 2230, Australia

Present addresses: [2]WildFish Research, Grays Point, Sydney, NSW 2232, Australia
[3]University of New South Wales, Randwick, NSW 2052, Australia
[4]Cardno, St Leonards, NSW 2065, Australia
[5]Batemans Bay Fisheries Centre, NSW 2536, Australia
[6]Seglaregatan, Gothenburg 41457, Sweden
[7]Sydney Institute of Marine Science, Mosman, NSW 2088, Australia
[8]Wildlife Marine, Sorrento, Perth, WA 6020, Australia

ABSTRACT: This study examined whether differences existed in the depth distributions and reproductive strategies of the co-occurring *Sillago robusta* and *S. flindersi* in coastal waters off eastern Australia. Marked spatial and temporal dissimilarities in demography and reproduction were observed between the 2 species, with *S. robusta* being more abundant in the shallow (15–30 m) strata and *S. flindersi* in the mid (31–60 m) strata, with neither species being consistently abundant in the deep (61–90 m) strata. The size composition of *S. robusta* was similar across depths, but smaller and immature *S. flindersi* predominantly occurred in the shallow strata, with larger and mature individuals occurring deeper. These data indicate partitioning of habitat resources, which may aid species coexistence. Both species potentially spawned year-round, which is probably an adaptation to the region's dynamic coastal environment. However, a greater proportion of *S. robusta* was in spawning condition between September and March, whereas *S. flindersi* displayed no such temporal pattern. Maturity ogives differed significantly between sexes and locations for both species. Both species displayed similar ovarian development, with females having multiple concurrent oocyte stages, indicating potential multiple spawning events as evidenced in other Sillaginidae. For both species, estimated batch fecundity increased with fish length, but *S. robusta* had a greater fecundity at any given length than *S. flindersi*. In contrast, *S. flindersi* potentially produced larger-sized eggs and invested greater energy into gonad development than *S. robusta*, indicating the 2 species have evolved slightly different reproductive strategies. Despite this, both species are subjected to substantial trawl fisheries, which may have already impacted their reproductive ecologies.

KEY WORDS: Life history · Habitat partitioning · Population structure · Reproduction · Maturity · Spawning · Fishery exploitation

INTRODUCTION

Knowledge of the life history characteristics of co-occurring species is fundamental to understanding biotic processes that influence structure and maintenance of assemblages of organisms (Schoener 1974,

Ross 1986, Roff 1992). Closely related teleost species that coexist often display different life history characteristics, such as reproductive and recruitment strategies, diets and small-scale distributions and abundances (Ross 1977, Roff 1991, Hyndes et al. 1997, Genner et al. 1999, Colloca et al. 2010). Such strate-

*Corresponding author: charles.gray@wildfishresearch.com.au

gies maximise partitioning of resources, minimise the potential for interspecific competition, and may contribute to the coexistence of different assemblages and thus affect assemblage structure (Schoener 1974, Werner et al. 1977, Ross 1986).

Members of the teleost family Sillaginidae (whiting) inhabit subtropical and temperate coastal shelf and estuarine waters of the Indian and Western Pacific Oceans (McKay 1992, www.fishbase.org). The family contains 31 species, of which 13 occur in Australian waters, with 6 of these distributed along the east coast (McKay 1992). Sillaginids display a high degree of similarity in coloration and external morphology, having slender elongate fusiform-shaped bodies and long conical snouts. Most species attain maximum total lengths less than 40 cm and longevities under 12 yr (McKay 1992, Kendall & Gray 2009). Sillaginids are benthic carnivores that feed on small invertebrates such as polychaetes and crustaceans (Gunn & Milward 1985, Hyndes et al. 1997, Hajisamae et al. 2006) and can be a prominent component of soft-sediment benthic ichthyofaunas (Gray & Otway 1994, Chen et al. 2009, Gray et al. 2011, Nakane et al. 2013). Several species are important in commercial, recreational and artisanal fisheries (McKay 1992, Kailola et al. 1993, Gray & Kennelly 2003).

Life history and ecological aspects of several sillaginid species have been investigated; notably the coastal shelf species *Sillago sihama* in Indian waters (Radhakrishnan 1957, Reddy & Neelakantan 1992, Hajisamae et al. 2006, Shamsan & Ansari 2010), *S. aeolus* in Japanese waters (Rahman & Tachihara 2005a, b), *Sillaginodes punctata* off southern and south-western Australia (Fowler & Short 1996, Hyndes et al. 1998, Fowler et al. 1999, 2000), and *Sillago analis*, *S. burrus*, *S. vittata*, *S. robusta*, *S. schomburgkii* and *S. bassensis* off western Australia (Coulson et al. 2005, Hyndes & Potter 1996, 1997, Hyndes et al. 1996a,b, 1997). Similarly, the demographic characteristics of the estuarine-nearshore distributed *S. ciliata*, *S. maculata* and *S. analis* have been examined in eastern Australia (Cleland 1947, Burchmore et al. 1988, Kendall & Gray 2009, Stocks et al. 2011). There have been few investigations of the biological characteristics of the eastern Australian coastal shelf species *S. robusta* and *S. flindersi*, even though they are the most abundant sillaginids inhabiting these waters and are subject to substantial (approximately 2000 t per annum) commercial fisheries (Rowling et al. 2010).

The distributions of *S. robusta* and *S. flindersi* overlap off eastern Australia; *S. robusta* consists of 2 disparate populations extending between approximately 24° S and 34° S on the west and east coasts,

whereas *S. flindersi* is endemic to the east and southern mainland coasts as well as around northeastern Tasmania, occurring between approximately 25° S and 44° S (Fig. 1). Both species are an important by-product in coastal penaeid trawl fisheries (Kennelly et al. 1998, Macbeth et al. 2012), whilst *S. robusta* is targeted in a limited entry quota-based trawl fishery in its northern distribution (Butcher & Hagedoorn 2003, Zeller et al. 2012). In contrast, *S. flindersi* is taken across several multi-sector trawl and Danish-seine fisheries managed by different jurisdictions throughout its distribution (Kemp et al. 2012). Depending on the jurisdiction, different input and output controls are used as management tools for each species, including limited entry, fishing gear and vessel restrictions, spatial and temporal closures, legal length limits and total allowable catches. Concerns over discarding in some fisheries have resulted in the development of fishing gears that are more selective at retaining market-sized sillaginids (Broadhurst et al. 2005, Graham et al. 2009). Similar attention to resolving the biological parameters of these species has not taken place. Little is known about important aspects of the reproductive biology of either species, including lengths and ages at maturity, and modes, times and locations of spawning. Without such information, appropriate fishery and species management plans cannot be evaluated.

Previous studies show that sillaginids generally attain sexual maturity at young ages (1 to 3 yr), spawn multiple times over protracted spawning seasons,

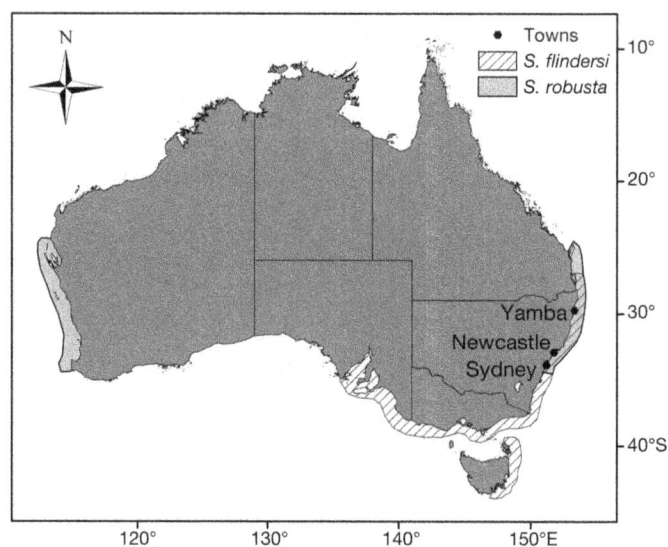

Fig. 1. Distribution of *Sillago robusta* and *S. flindersi* throughout Australia and the Yamba and Newcastle sampling locations off eastern Australia

and display depth-associated ontogenetic shifts in distribution that may assist partitioning of resources (Burchmore et al. 1988, Hyndes et al. 1996a,b, 1997, Kendall & Gray 2009). Here we investigate whether this general paradigm is applicable to *S. robusta* and *S. flindersi* where their distributions overlap in coastal waters off eastern Australia. We specifically test whether the relative abundances, length compositions and reproductive characteristics of populations of these species differ temporally, and between depths and locations.

MATERIALS AND METHODS

Study area and sampling procedures

Sampling was conducted in inner-continental shelf waters (<100 m depth) across transects adjacent to Yamba (29° 26' S, 153° 20' E) and Newcastle (32° 55' S, 151° 45' E) off eastern Australia (Fig. 1). This region is characterised by a dynamic oceanography dominated by the southward flowing East Australian Current and associated eddies (Huyer et al. 1988, Roughan & Middleton 2004, Suthers et al. 2011). The water column is usually thermally stratified in summer but not always in winter, with surface water temperatures typically oscillating between a mean maximum of 24 to 26° C in late summer (February–March) to a mean maximum of 13 to 15° C in late winter/spring (August–September).

Sampling at both locations was stratified across 3 depth ranges; 15–30 m, 31–60 m and 61–90 m; hereafter referred to as the shallow, mid and deep strata respectively. The middle of each depth range corresponded to a distance of approximately 3, 13 and 26 km offshore at Yamba and 1, 7 and 10 km at Newcastle. All sampling was done over soft substrata where commercial trawling for penaeid prawns and whiting regularly occurs. Sampling took place at night within 1 week of the full moon, every 4 weeks at both locations. Sampling extended for 2 full years between November 2005 and November 2007 at Yamba and between October 2006 and November 2007 at Newcastle. Samples were collected using a chartered ocean prawn trawl vessel from each port rigged with standard regulated 'triple' gear, in which each of the 3 nets had a headline length of 10.8 m, stretched mesh of 42 mm hung on the diamond throughout the body and cod-end, the latter which had a circumference of 100 meshes. The general selectivity of *S. flindersi* in this penaeid fishing gear configuration is reported in Broadhurst et al. (2005).

Two replicate tows, each of 60 min bottom duration at an average speed of 2.3 knots (a linear distance of approximately 4.2 km) were completed in each depth strata at both locations at each time of sampling. The depth strata first sampled each month was randomly chosen, after which either the deeper or shallower depths were progressively sampled due to logistic considerations. It took approximately 8 h to complete all 6 tows at each location. For each replicate tow (sample), the catch from all 3 nets was combined and sorted, with all sillaginids identified and kept separate for biological sampling. The number and total weight of each sillaginid species captured in each sample was determined on-board the vessel.

Processing of samples

Whole catches or random sub-samples (100 to 200 individuals) of both species from each replicate tow were counted and measured (fork length [FL] nearest 1 mm) for relative abundance and length composition. The total weight of the catch and subsample were weighed (nearest 5 g). A further subsample of 30 individuals of each species from each replicate tow at each depth was retained on ice for processing in the laboratory. These fish were measured for FL, weighed (wet weight, nearest 0.1 g) and had their gonads removed and weighed (blotted dry weight, nearest 0.1 g) to calculate the gonadosomatic index (GSI) for each individual: GSI = (gonad weight / whole fish weight) × 100. Each gonad was staged macroscopically following a development criteria based on oocyte size, colour and visibility adapted from Scott & Pankhurst (1992): for males: I = immature, II = spermatogenic, III = partially spermiated, IV = fully spermiated, V = spent; for females: I = immature, II = immature/regressed, III = vitellogenic, IV = hydrated, V = ovulated, VI = spent. The gonads from a subset of females of both species were kept (preserved in 70 % alcohol) to determine oocyte development and estimate potential batch fecundity (described below in 'Batch fecundity').

Distributions and population structure

General linear models (GLMs), assuming a Gaussian distribution with a log-transformed response variable, were used to test for differences in the relative abundance of each species across depths, seasons nested in years and years at Yamba, and across depths and seasons at Newcastle. Seasons were defined as summer (December–February), autumn (March–May), winter (June–August), spring (September–November) (allowing 2 full years between December 2005 and November 2007). Akaike information criteria (AIC) values were used to determine the most parsimonious model and probability tests (F-tests at α = 0.05) were used to determine the influence of each term in each model by comparing change in deviance when each term was included or excluded from the model (Nelder & Wedderburn 1972, Quinn & Keough 2002). Tukey's post-hoc tests (α = 0.05) were used to compare the different levels of each significant factor. Differences between depths in the length compositions of each species (pooled across seasons) at each location were tested using Kolmogorov-Smirnov (K-S) tests.

Oocyte development

Histological examination of a selection of preserved (70 % alcohol) Stage II, III and IV ovaries of both species was used to determine the development pattern of oocytes and to verify the macroscopic staging of females. Small sections were dissected from the middle of each ovary, treated in an automated tissue processor, with the resulting tissues embedded in paraffin wax and sectioned at 5 μm thickness on a rotary microtome. Sections were deparaffinised, differentiated in acidified alcohol and stained in alcoholic eosin. Histological staging was based on the most advanced cohort of oocytes in each ovary section (West 1990).

Individual oocyte development was examined by determining the size distributions of oocytes in 10 random individuals of each species with Stage III ovaries. The entire ovary was blotted dry and weighed (0.0001 g), after which 3 replicate sub-samples were taken from the mid-section of each ovary, blotted dry, weighed (0.0001 g) and placed in a sealed 70 ml sample jar containing 70 % alcohol solution. Each sub-sample was placed in a sonic bath (Unisonics FXP4) for a period no longer than 20 min to dislodge individual oocytes from surrounding connective tissue (Barnes et al. 2013). Oocytes from each sub-sample were transferred into a petri-dish, separated from each other, scanned and imaged at 1200 dpi resolution. Image analysis software (Image J, Version 1.38) was used to determine the number and size of oocytes in each sub-sample. Size-frequency plots of oocyte diameters were produced for each gonad.

Length at maturity

The estimated FL at which 50 % (L_{50}) of males and females attained reproductive maturity was determined by fitting a logistic regression model using the binomial GLM function in R to the proportions of immature (Stages 1 and II) and mature (Stages III and above) fish in each 1 mm length class. The data used in these analyses was obtained during periods of high GSI. Differences between sexes and locations (and years for Yamba) in the estimated L_{50} values of each species were tested using the 2-sampled Z technique with α = 0.05 (Gunderson 1977).

Reproductive period

Temporal changes in mean male and female GSI values and proportions of fish with each macroscopic gonad stage (of individuals larger than the estimated mean length at maturity) were used to estimate the timing of spawning. Elevated GSI values and high proportions of fish with gonads staged III to VI were interpreted as probable spawning. The GLM procedures outlined above ('Distributions and population structure') were used to examine the influence of depths, seasons and, where relevant, year and seasons nested in year, on the proportions of mature individuals of each species present at each location. These GLMs assumed a binomial distribution, treating maturity as a binary response variable (1 = mature and 0 = immature), and used chi-squared (α = 0.05) probability tests within the analysis of deviance tables.

Batch fecundity

The largest size class of oocytes (vitellogenic, >0.30 mm for S. robusta and >0.35 mm for S. flindersi) in mature, pre-spawning (Stage III) fishes were considered suitable for estimating potential batch fecundity (BF) (Hunter et al. 1985). The ovaries of up to 25 individuals from each species collected mid-spawning season at each location and in both years were examined. For each individual, the number of oocytes present was calculated using the same methodologies describe above ('Oocyte development') for investigating oocyte size–frequency distributions. Potential BF was estimated by scaling the number of oocytes present within the weighed ovarian subsample to the total preserved weight of the ovary. Log-linear models were used to describe relationships between estimated BF and FL and ANCOVA were used to test whether BF for each species differed according to location and FL.

RESULTS

Relative abundance

The GLMs showed that depth and season influenced the relative abundance of *Sillago robusta* and *S. flindersi* in different ways. Notably, *S. robusta* were most abundant in the shallow strata across most seasons at Yamba and Newcastle (Tukey's HSD tests: $p < 0.05$, Fig. 2), whereas, *S. flindersi* were consistently most abundant in the mid-depth strata across all seasons at Yamba (for which data were combined across years) and at Newcastle (Tukey's HSD tests:

$p < 0.05$, Fig. 2). Neither species was consistently caught in large numbers in the deep strata, with *S. robusta* only found deep in spring, and *S. flindersi* only in spring and winter at Yamba.

Seasonal changes in the relative abundance of each species at each location were inconsistent and dependent on depth and year (where relevant). For example, abundances of *S. robusta* at Yamba (data combined across depths) did not show any significant differences between seasons in 2006, whereas they were significantly greater in summer and autumn in 2007 (Tukey's HSD tests: $p < 0.05$, Fig. 2). At Newcastle, abundances of this species only varied between seasons in the shallow strata, where they were significantly greater in summer compared to winter (Tukey's HSD test: $p < 0.05$, Fig. 2). In contrast, abundances of *S. flindersi* at Yamba (data combined across years and depths) were greatest in winter and spring (Tukey's HSD test: $p < 0.05$, Fig. 2), whereas at Newcastle they were significantly lowest in autumn (Tukey's HSD test: $p < 0.05$, Fig. 2).

Population structure

The length compositions of samples of each species differed significantly according to depth and location (multiple K-S tests, $p < 0.05$ in all cases). Despite this, some general patterns were evident; notably a similar length range of *S. robusta* was present across all depth strata within each year at Yamba, and in the shallow and deep strata at Newcastle (Fig. 3). A cohort of small-sized (<10 cm FL) *S. robusta* was present in the shallow and mid-depth at Yamba in 2006 but not in 2007. For *S. flindersi*, a cohort of smaller-sized (< 13 cm FL) individuals predominated the shallow strata at both locations, whereas the mid and deep strata primarily contained individuals > 13 cm FL (Fig. 3). This later length cohort was also prominent in the shallow strata at Yamba in 2007.

Oocyte development

Both species displayed a similar pattern of ovarian development. Stage II ovaries contained unyolked oocytes of a variety of sizes, whereas Stage III ovaries contained a mixture of unyolked oocytes, partially yolked oocytes and oocytes that were in an advanced yolk stage of development (Fig. 4). The diameter sizes of this latter cohort of oocytes ranged from 0.30 to 0.45 mm for *S. robusta* and 0.35 to 0.50 mm for *S. flindersi* (Fig. 5). This suggested that *S. flindersi*

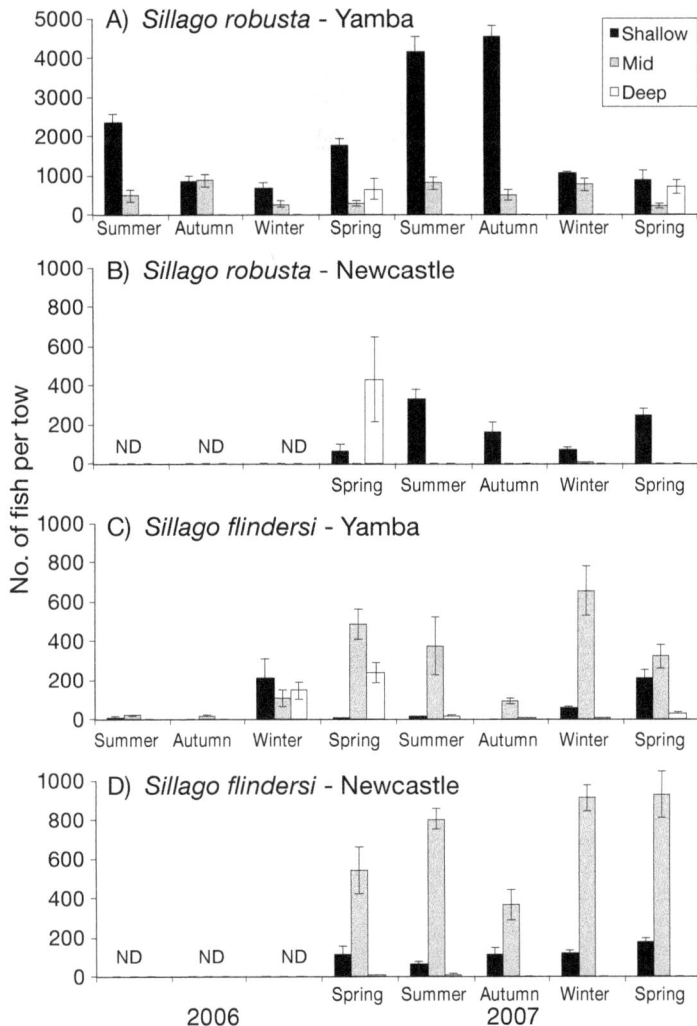

Fig. 2. Mean (±SE) number of *Sillago robusta* and *S. flindersi* caught per tow in the shallow, mid and deep depth strata in each season at Yamba and Newcastle. ND: no data

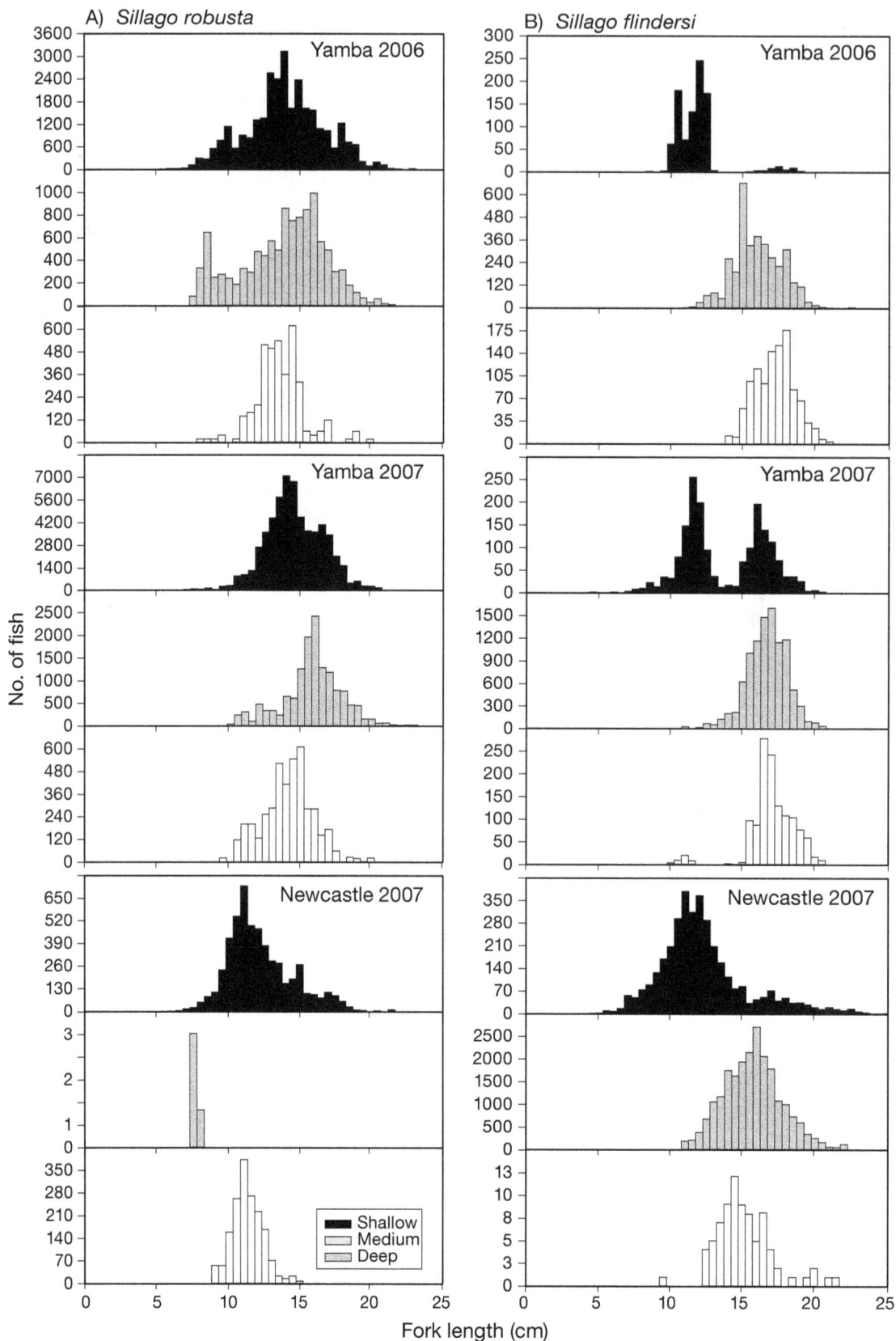

Fig. 3. Population length composition of *Sillago robusta* and *S. flindersi* in the shallow, mid and deep depth strata at Yamba and Newcastle. Data pooled across all sampling times for each year. Note different scales of *y*-axes

A) *Sillago robusta*

Stage II Stage III Stage IV

B) *Sillago flindersi*

Stage II Stage III Stage IV

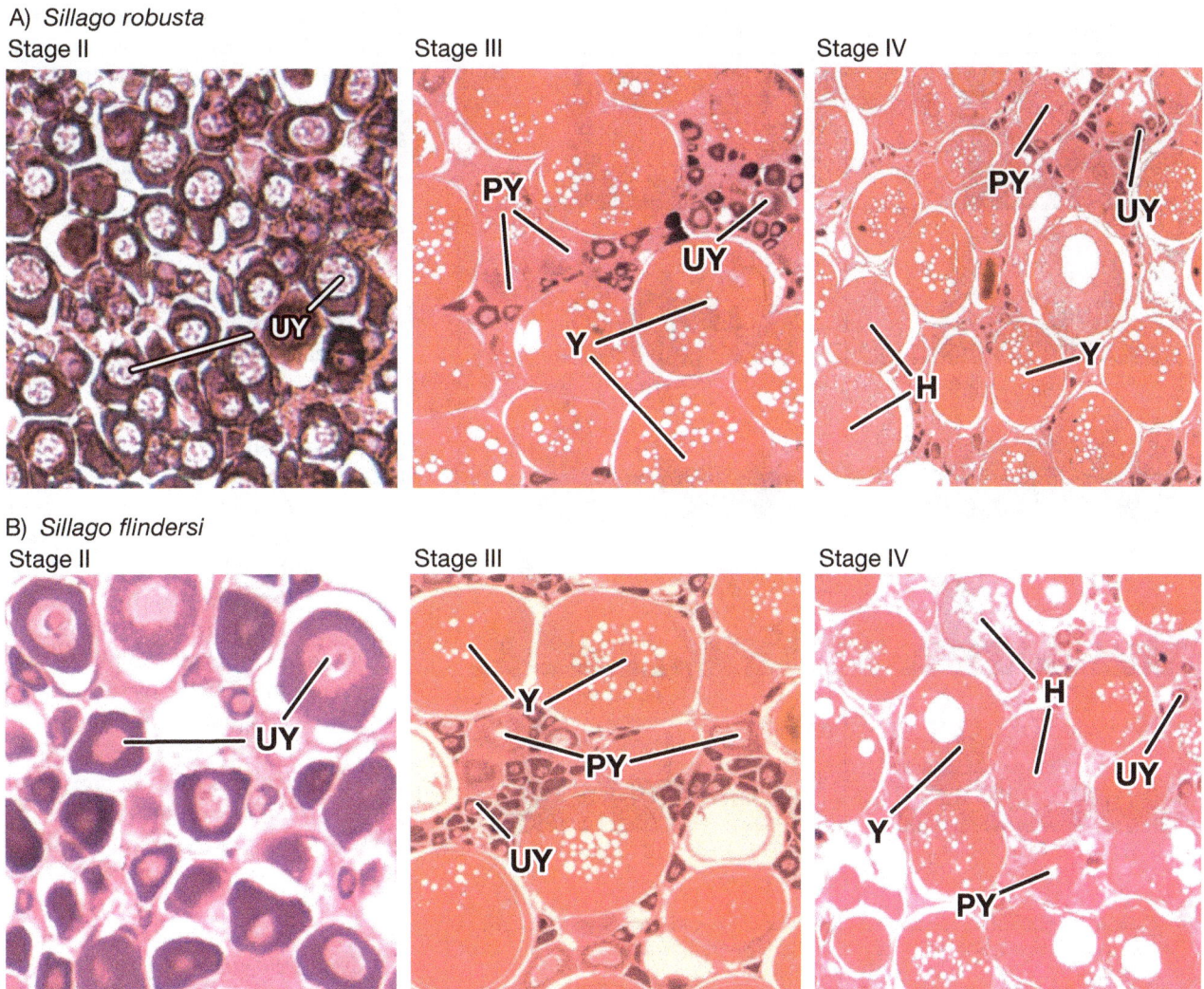

Fig. 4. Photos showing microscopic characteristics of ovary stages II, III and IV for *Sillago robusta* and *S. flindersi*. Stage II: immature ovary containing developing unyolked (UY) ooctyes, Stage III: mature ovary containing unyolked oocytes, partially yolked (PY) oocytes and advanced yolk (Y) stage oocytes, Stage IV: mature ovary containing ooctyes in all other stages of development as well as hydrated (H) oocytes

might have produced larger eggs than *S. robusta*. Stage IV ovaries contained hydrated oocytes as well as oocytes in each of the previous stages of development (Fig. 4).

Length at maturity

There was no consistent effect of sex on length at maturity for either species. Notably, the L_{50} was significantly ($p < 0.05$) smaller for males than females of both species at both locations in 2007, but this was not the case in 2006 when the L_{50} was significantly ($p < 0.05$) greater for male than for female *S. robusta* and there was no significant ($p > 0.05$) difference between sexes for *S. flindersi* (Fig. 6). Evidence of

spatial interactions in length at maturity was also apparent, with the L_{50} of *S. robusta* being significantly ($p < 0.05$) smaller at Newcastle for both sexes (Fig. 6). Significant spatial difference of length at maturity was also observed for female *S. flindersi*, which was greatest at Newcastle ($p < 0.05$), but not for males ($p > 0.05$). Males displayed temporal variations, with the L_{50} being significantly ($p < 0.05$) larger for both species at Yamba in 2006.

The estimated L_{50} values for male and female *S. robusta* ranged from 12.84 to 15.35 cm FL and from 14.08 to 14.83 cm FL, respectively (Fig. 6). Similarly, the estimated L_{50} values for male and female *S. flindersi* ranged from 13.27 to 13.96 cm FL and from 13.88 to 14.87 cm FL, respectively. The observed smallest mature male and female *S. robusta* was 11.3 and

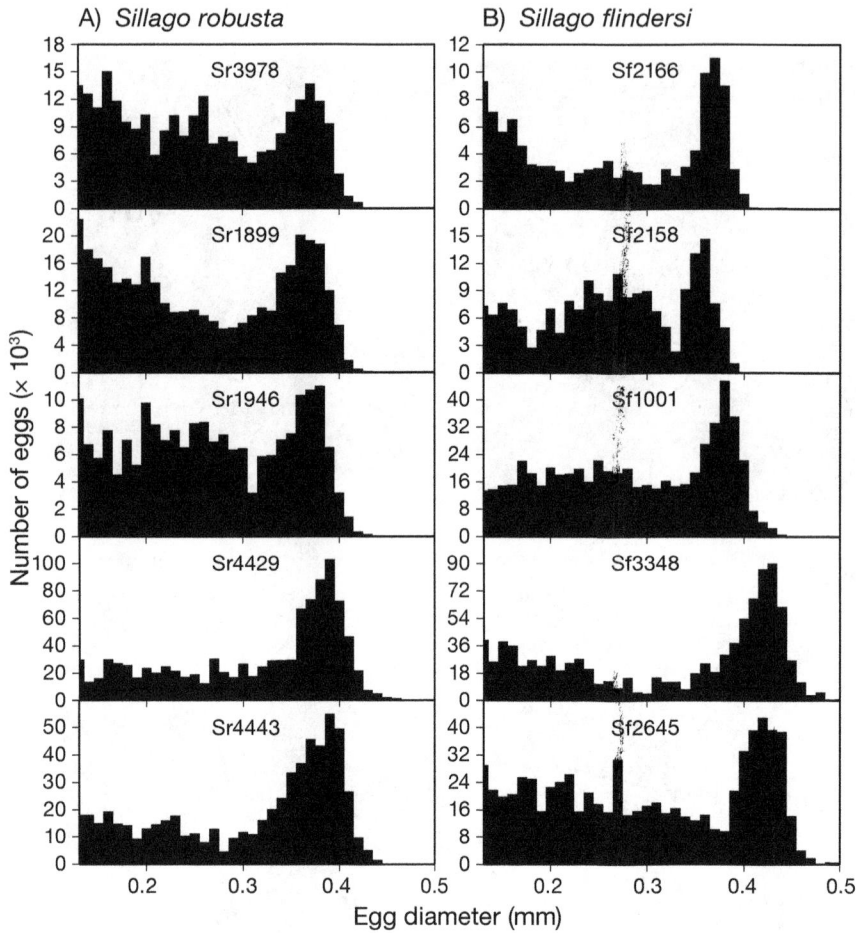

Fig. 5. Size distributions of oocytes in 5 individual Stage IV ovaries of *Sillago robusta* and *S. flindersi*. Numeric code refers to individual sample. Note different scales of *y*-axes

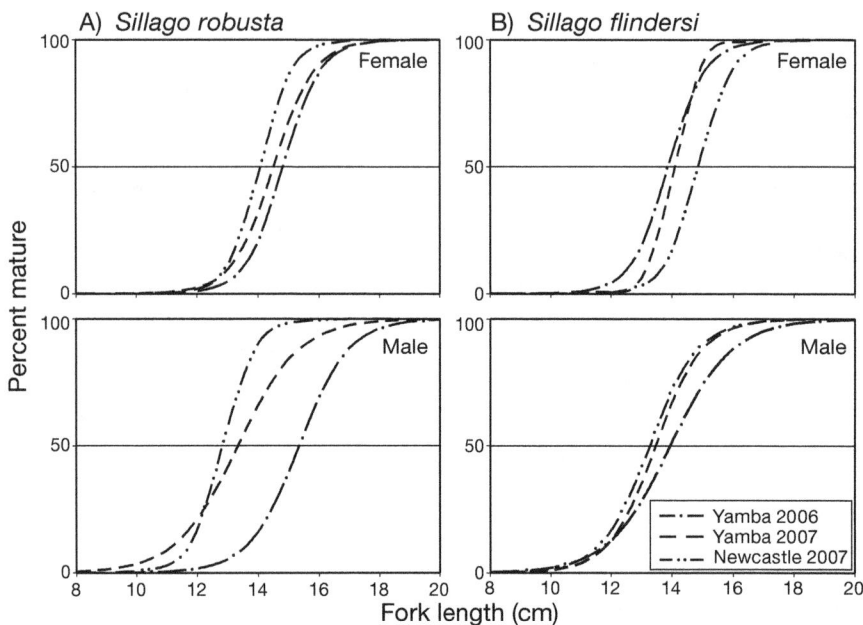

Fig. 6. Estimated maturity ogives of female and male *Sillago robusta* and *S. flindersi* at Yamba and Newcastle. Data combined across depth strata

13.0 cm FL at Yamba and 11.4 and 12.5 cm FL at Newcastle, respectively. Likewise, the observed smallest mature male and female *S. flindersi* was 11.1 and 11.6 cm FL at Yamba and 11.0 and 13.0 cm FL at Newcastle, respectively.

Spawning

The macroscopic staging of gonads and changes in mean GSI values indicated both species potentially spawned year-round at both locations. Female and male *S. robusta* and *S. flindersi* with mature gonads (Stage III and higher) were present each month at Yamba (except June 2007 for *S. flindersi*) and most months at Newcastle (Fig. 7). Further, for both species ovaries containing hydrated eggs (Stage IV) were collected across most months at Yamba, although few were observed at Newcastle. There was also no consistent pattern for either species as to the months when particular gonad stages (mature/immature) were most or least prevalent. Nevertheless, for *S. robusta* a greater proportion of Stage I individuals were present in 2006 than 2007 at Yamba (Fig. 7)

The mean GSI values of female and male *S. robusta* at Yamba displayed similar trends through time and were generally lowest in late autumn and winter (April to July in 2006 and May to July in 2007). Mean GSI values tended to be highest between September and March in both years, suggesting that potentially a greater proportion of individuals spawn throughout the austral spring and summer (Fig. 8). There was no evidence of any such trend for this species at Newcastle where mean GSI values were relatively high between April and October. The maximum mean (±SE) monthly GSI for female and male *S. robusta* was 2.86 ± 0.12 and 2.55 ± 0.23 at Yamba and 3.27 ± 0.82 and 1.85 ± 0.19 at New-

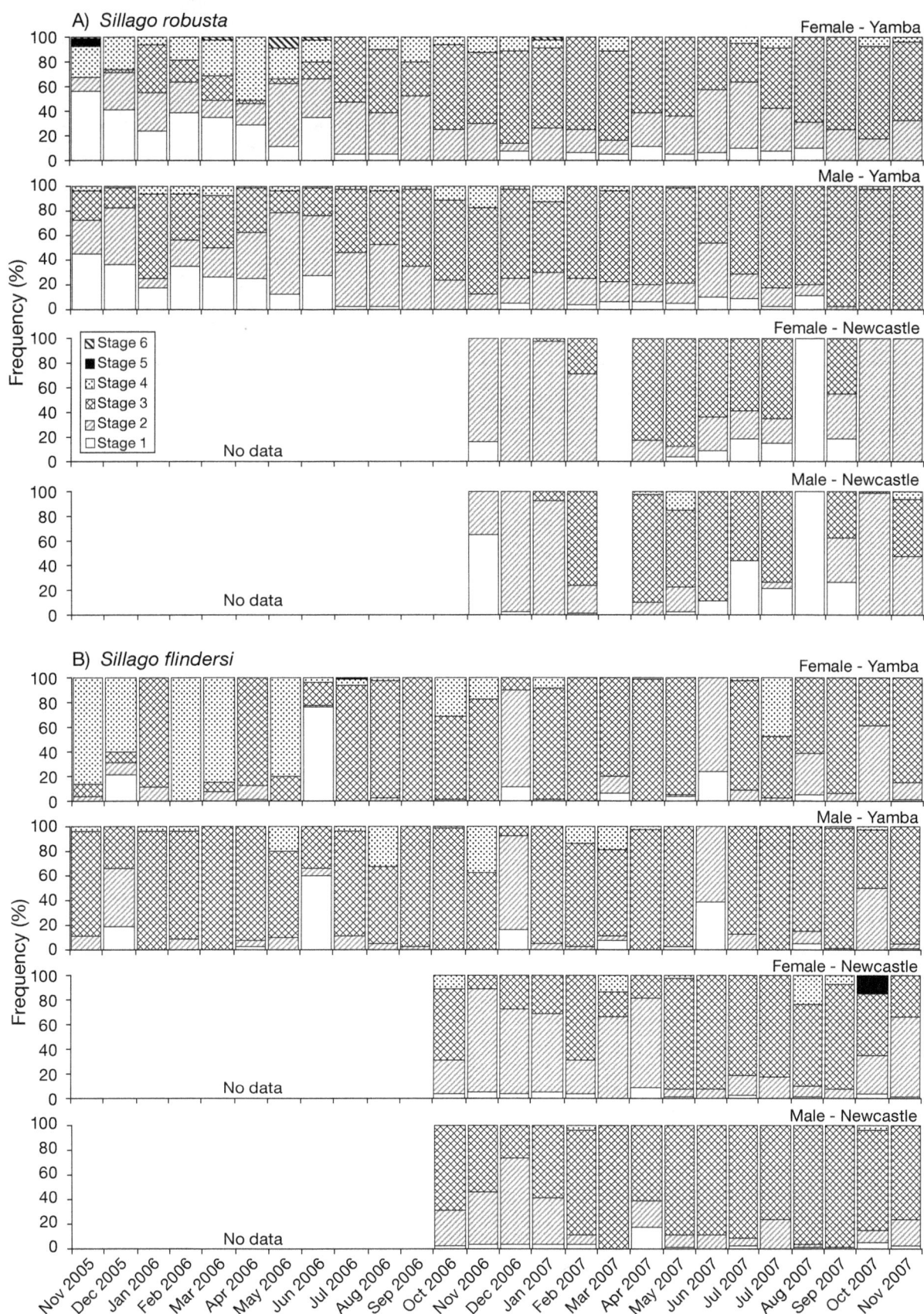

Fig. 7. Frequency of different macroscopically staged ovaries and testis of *Sillago robusta* and *S. flindersi* at Yamba and Newcastle. Data combined across depth strata

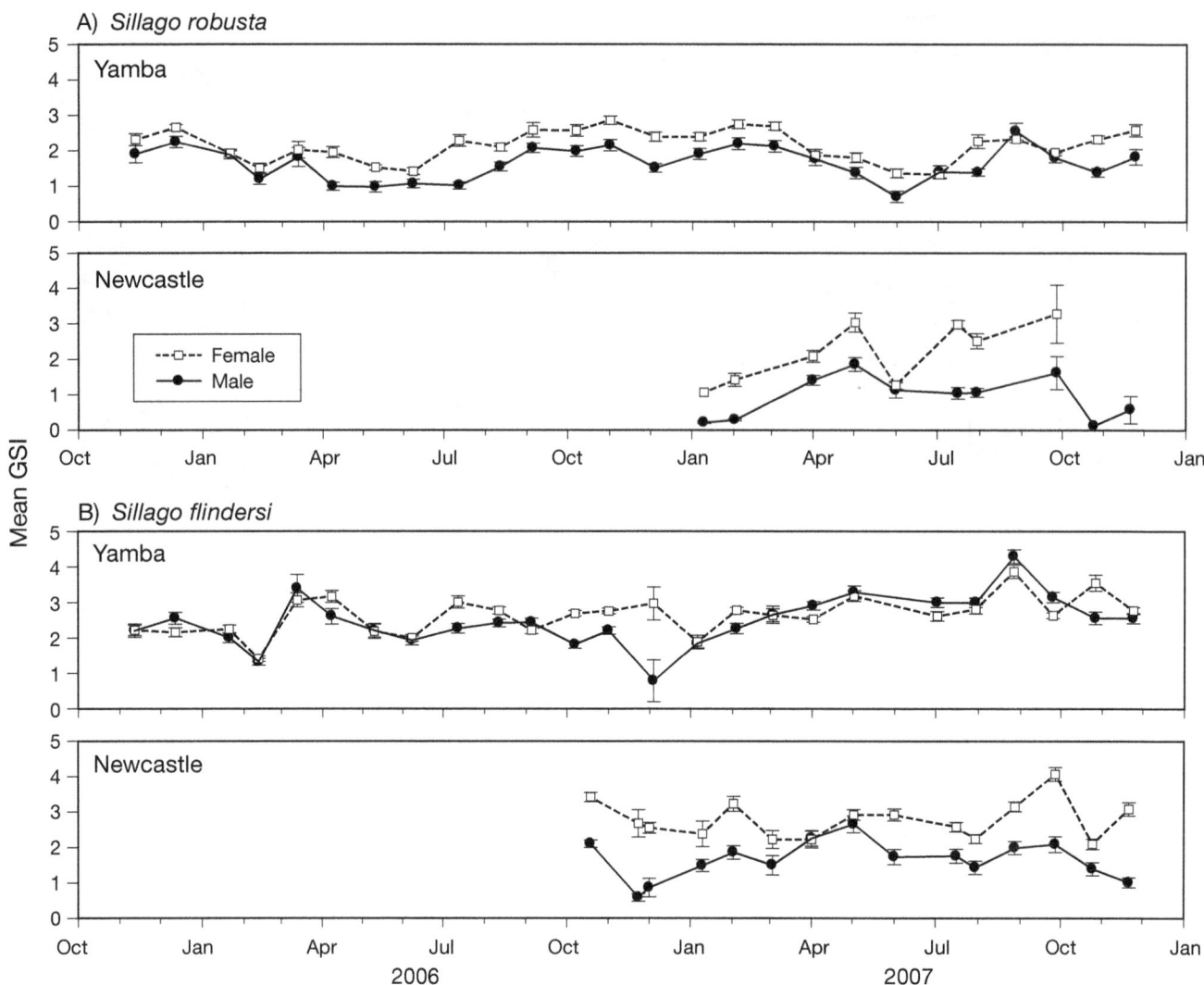

Fig. 8. Monthly variation in mean (±SE) GSI of female and male *Sillago robusta* and *S. flindersi* at Yamba and Newcastle. Data combined across depth strata

castle, respectively. The mean GSI values of male and female *S. flindersi* remained relatively stable, displaying no seasonality throughout sampling at either location (Fig. 8). The maximum mean (±SE) monthly GSI for female (male) *S. flindersi* was 3.87 ± 0.18 and 4.29 ± 0.19 at Yamba and 4.07 ± 0.19 and 2.65 ± 0.24 at Newcastle, respectively.

For both species, the effect of year could not be tested due to significant depth × season (nested in year) and depth × year interactions. The maturity data collected from Yamba were therefore analysed in separate GLMs for the 2 years. The GLMs identified that a greater proportion of mature *S. robusta* occurred in the mid compared to the shallow strata in summer, winter and spring of 2006 and in summer, autumn and spring of 2007 at Yamba (Tukey's HSD test: p < 0.05, Fig. 9; note that, for both years, this species was only caught in the deep strata dur-

ing spring). In contrast, the significant depth-related effect in the GLM reflected a predominance of mature fish in the shallow strata compared to the deep strata at Newcastle (GLM: p (>Chi) < 0.05, Fig. 9; but note that this species was primarily only caught in the shallow strata and very few were caught in the mid strata irrespective of season). A significantly greater proportion of mature *S. flindersi* occurred in the mid and deep strata compared to the shallow strata across most seasons in both years at Yamba (Tukey's HSD test: p < 0.05, Fig. 9). At Newcastle, the greatest proportion of mature *S. flindersi* occurred in the mid strata in winter and autumn (Tukey's HSD test: p < 0.05, Fig. 9; note that no *S. flindersi* were captured in the deep strata during these seasons) but in the deep strata in spring 2006 and summer 2007 (Tukey's HSD test: p < 0.05, Fig. 9).

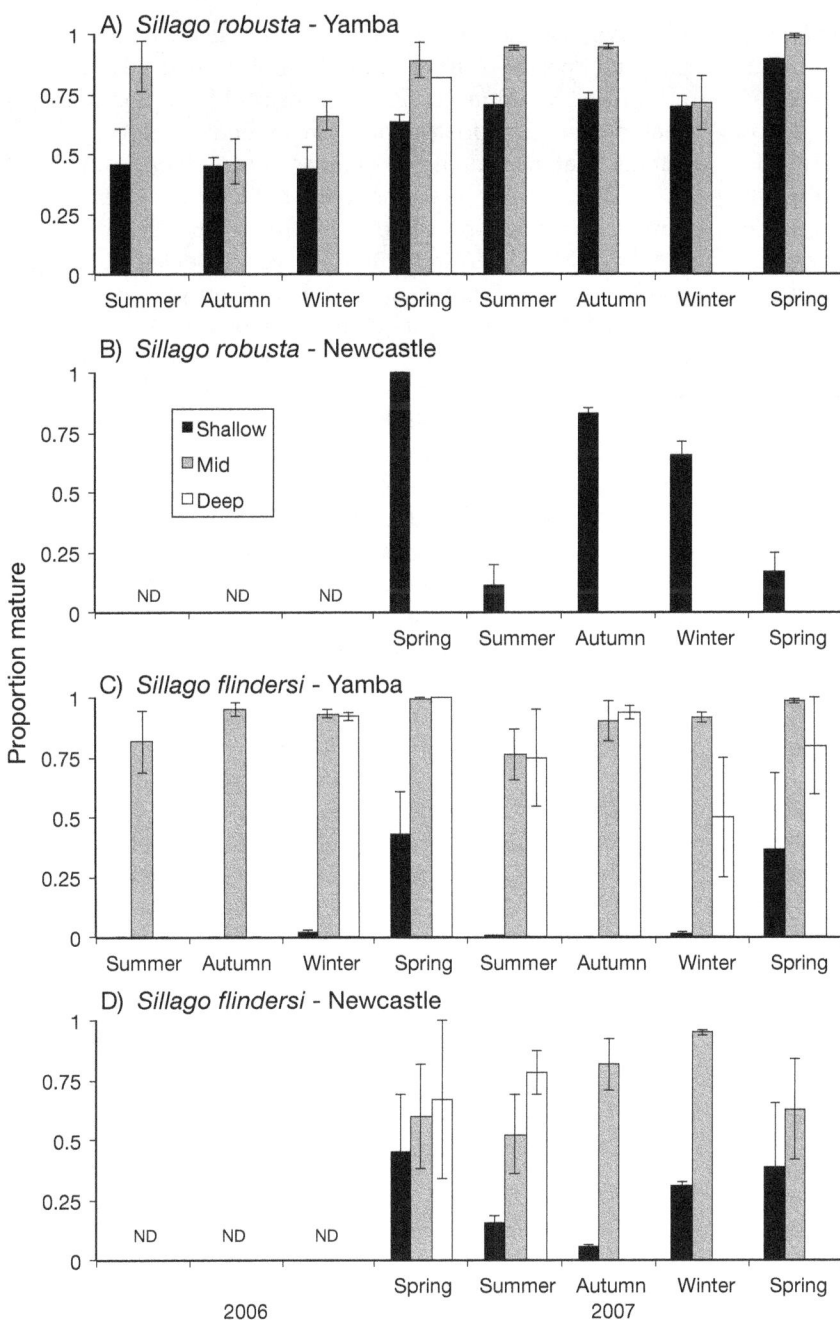

Fig. 9. Mean (±SE) proportion of mature *Sillago robusta* and *S. flindersi* in the shallow, mid and deep depth strata in each season at Yamba and Newcastle. ND: no data

Batch fecundity

There was no significant difference between locations in the relationship between estimated BF and FL for either species (ANCOVA, p > 0.05 in all cases), so data were combined across locations for each species. Estimated BF of both species significantly increased with FL (p < 0.001 in both cases); the log-linear rela-

tionship $[\log(BF) = \log(a) + b \times \log(FL)]$ was: $\log(a) = -3.0787$, b = 4.9280, $r^2 = 0.6921$, n = 25 for *S. robusta* and $\log(a) = -5.2534$, b = 5.5159, $r^2 = 0.7524$, n = 24 for *S. flindersi*. Estimated BF ranged from 7048 to 284755 eggs for *S. robusta* measuring 13.0 to 21.6 cm FL, and from 6773 to 256100 for *S. flindersi* measuring 13.1 to 23.4 cm FL. These data indicated that at a given FL, fecundity was greater in *S. robusta* than in *S. flindersi*.

DISCUSSION

Population structuring and habitat partitioning

Spatial and temporal structuring of eastern populations of *Sillago robusta* and *S. flindersi* was evident along a number of gradients suggesting possible partitioning of resources similar to other sympatric Sillaginidae (Hyndes et al. 1996a,b, 1997) and teleost families (Ross 1977, Genner et al. 1999, Barnes et al. in press). In general, *S. robusta* was more prevalent in the shallow strata whereas *S. flindersi* mostly occurred in the mid strata, while neither species consistently utilised the deep strata. Depth stratification of sillaginid and teleost populations and assemblages is widespread (Werner et al. 1977, Hyndes et al. 1999, Labropoulou et al. 2008, Gray et al. 2011).

The population length structure of *S. robusta* was similar across all depth strata, which is concordant with western populations of the species (Hyndes & Potter 1996). In contrast, smaller and immature *S. flindersi* were more prominent in the shallow strata compared to the mid and deep strata, which were dominated by larger (mature) individuals. These data support the 'smaller-shallower' phenomena (Middleton & Musick 1986, Stefanescu et al. 1992) and suggest that *S. flindersi* uses the shallow strata as a nursery area and then moves to deeper waters with growth and maturity. This is synonymous with the hypothesised life history of other Sillaginidae, including coastal *S. bassensis* and *S. vittata* (Hyndes & Potter 1996, Hyndes et al. 1996b) and estuarine *S. ciliata* and *S. maculata* (Weng 1986,

Burchmore et al. 1988). Depth-related ontogenetic shifts in distribution are common among teleosts (Macpherson & Duarte 1991, Methratta & Link 2007, Labropoulou et al. 2008), with shallow waters hypothesised to provide more food and greater protection of juveniles from predators (Ryer et al. 2010). Such stratification is also a potential mechanism to reduce intra- and interspecific competition and aid broad-scale partitioning of resources among co-occurring species (Hyndes & Potter 1996, 1997). In the present study, the shallow-water preference of small *S. flindersi* placed them directly in the preferred habitat of *S. robusta*. It is conceivable that a dietary mechanism ensured the smaller *S. flindersi* individuals were not competitively disadvantaged (Hyndes et al. 1997, Barnes et al. 2011). Alternatively, necessary resources in the shallow strata may not have been limited, reducing the need for species stratification.

Spawning

The data presented here indicate that *S. robusta* and *S. flindersi* are income-spawners (McBride et al. 2013), yet each species may have evolved slightly different reproductive strategies to deal with the dynamic coastal environment they inhabit. Despite neither species investing greatly in reproduction (low GSI values of females and males), the larger-growing *S. flindersi* generally displayed higher GSI ratios, suggesting it invests more energy into gonad production than *S. robusta*. Comparable differences in reproductive investment between other co-occurring Sillaginidae have been observed (Hyndes & Potter 1996). Further, our data suggest that *S. flindersi* produced fewer (at any given length) but potentially larger eggs than *S. robusta*. We could not ascertain here, however, the effects of such trade-offs between the potential quality and quantity of larvae produced (Duarte & Alcaraz 1989) on the reproductive success and population replenishment of either species.

Despite these apparent differences, the ovaries of mature *S. robusta* and *S. flindersi* contained oocytes of multiple sizes and developmental stages, indicating their potential to spawn several times within a given 'spawning' period (De Vlaming 1983, Hunter & Macewicz 1985, West 1990). This concurs with other sillaginids (Lee & Hirano 1985, Hyndes & Potter 1996, Fowler et al. 1999, Kendall & Gray 2009), adding further support to the hypothesis that the Sillaginidae are multiple-batch spawners. This spawning strategy is widespread among teleosts (Sarre & Potter 1999, Walsh et al. 2011, Gray et al. 2012), allowing individuals

to maximize the number of eggs produced over a particular period (Burt et al. 1988, McBride at al. 2013).

Mature individuals of both species could potentially spawn over extended periods. Fowler et al. (1999) estimated that individual *Sillaginodes punctata* spawned at least 20 times throughout their 3 mo spawning period. We could not determine the frequency and number of times an individual of either species may have released eggs throughout any given period or throughout its life. Consequently, the total number of eggs that each individual produces each year (total fecundity) could not be estimated. Nevertheless, estimated batch fecundity in both species was positively related to fish length, indicating that reproductive output is potentially greater in larger individuals, as in most teleosts (Parker 1992). Having a greater abundance of larger (and presumably older) individuals could theoretically increase the collective reproductive potential of teleost populations, and enhance larval survival (Berkeley et al. 2004). Fishing gears that allow a greater proportion of fish to reach larger sizes could enhance the reproductive potential and sustainability of these sillaginid populations.

The potential year-round spawning of *S. robusta* and *S. flindersi*, as evidenced by changes in GSI values and the macro- and microscopic staging of gonads, is further corroborated by the occurrence of pelagic larvae of both species year-round in coastal waters off eastern Australia (Gray & Miskiewicz 2000). This extended period of spawning is in contrast to western populations of *S. robusta* occurring at similar latitudes that spawn between December and March (Hyndes & Potter 1996). Our data indicate, however, that a greater proportion of the eastern populations of *S. robusta* spawn between September and March. This predominate spring–summer spawning pattern is also true for the east-Australian estuarine-based sillaginids *S. ciliatia* and *S. maculata* (Morton 1985, Burchmore et al. 1988, Kendall & Gray 2009) as well as a suite of other coastal sillaginids, including *S. burrus*, *S. bassensis*, *S. schomburgkii*, *S. vittata* and *S. aeolus* (Hyndes & Potter 1996, 1997, Hyndes et al. 1996b, Rahman & Tachihara 2005a). Periods of increasing water temperature and photoperiod may potentially trigger reproductive development and spawning in these species (Hyndes & Potter 1996). In contrast, similar proportions of the *S. flindersi* populations sampled here spawned year-round, indicating that changes in water temperature and photoperiod were not the primary cue to trigger reproductive development and spawning in this species.

The extended spawning of eastern populations of *S. robusta* and *S. flindersi* and the potential ability of individuals to spawn multiple times in a spawning season is probably an adaptation to the stochastic and dynamic oceanography of this coastal region. This strategy should enhance the probability of successful survival and recruitment of some eggs and larvae (Lambert & Ware 1984, Lowerre-Barbieri et al. 2011, McBride et al. 2013) and could also be a life history tactic to enhance dispersal of young as well as maintaining genetic connectivity among populations along eastern Australia, as suggested for other coastal boundary current inhabitants (Hare & Cowen 1993, Gray et al. 2012). Further, spawning by species in the depths examined here may be advantageous to maximising alongshore dispersal and subsequent retention of eggs and larvae in shelf waters by the East Australian Current (Roughan et al. 2011).

The observed prolonged reproductive period of both species in eastern Australia compared to other sillaginid populations could also potentially be a response to high levels of fishing-related mortalities in these particular populations. High exploitation rates can impact teleost reproductive strategies, including lowering sizes and ages at maturity, increasing egg production and spawning frequency, and lengthening spawning periods (Rochet 1998, Sharpe & Hendry 2009, Wright & Trippel 2009). The sillaginid populations examined here have been commercially harvested for over 100 yr, with catches over the past 20 yr exceeding 2000 t per annum. These levels of exploitation could potentially be sufficient to have already caused modifications of reproductive strategies of these populations.

Despite the differences outlined above, spawning of S. robusta and S. flindersi overlapped greatly in space and time. This is common among closely related teleosts (Muthiga 2003, Park et al. 2006, Tomaiuolo et al. 2007) and may be related to similar evolutionary histories and environmental requirements (Mercier & Hamel 2010). Thus, there was no specific evidence of large-scale reproductive isolation between these sillaginid species (Wellenreuther & Clements 2007). Reproductive isolation can occur between sympatric species over much finer spatial and temporal scales than examined here (Colin & Clavijo 1988, Colin & Bell 1991, Sancho et al. 2000), which could be further explored in these species.

Maturity and fishery considerations

The estimated L_{50} for both species were mostly smaller for males compared to females, which is in general agreement with other sillaginids (Kendall &

Gray 2009). Length at maturity of S. robusta as determined here was greater by about 1.5 to 2.0 cm than for western populations (Hyndes & Potter 1996). A plethora of biotic and abiotic factors, as well as anthropogenic influences (e.g. fishing) can potentially cause broad-scale variations in length (and age) at maturity among populations (Lassalle et al. 2008). Importantly, these data demonstrate the potential for intraspecific plasticity in life history characteristics among discrete populations of teleosts (Gust et al. 2002, Ruttenberg et al. 2005, Blanck & Lamouroux 2007, Sala-Bozano & Mariani 2011), emphasising the need for regional information of population demographics for fisheries assessment and management.

Significant spatial and temporal differences in the estimated L_{50} of each species were also detected at the smaller regional scale examined here, with differences in parameter estimates between years being equal to between locations. Although such results could be artefacts of variations in sample composition, they demonstrate the potential limitations of such demographic information collected at one place and time (i.e. typical snapshot studies). Indeed, a lack of information of levels of variability in demographic parameters (including rates of growth and mortality) across a species distribution could confound biological-based fisheries assessments and management plans (Morgan & Bowering 1997, Ruttenberg et al. 2005, Jakobsen et al. 2009).

Fisheries managers often set the retained legal lengths of fish at the L_{50} to potentially allow 50% of individuals to spawn at least once prior to harvesting (King & McFarlane 2003). Since eastern populations of S. robusta and S. flindersi are often caught together in large quantities as by-product by commercial trawlers targeting penaeid prawns, a common retained legal length of 14 cm FL (~15 cm total length) could be applied to these species in these fisheries if required. This length corresponds closely to the 50% length selection for these species in 35 mm square mesh cod-ends tested and recommended for use in the east Australian demersal penaeid trawl fisheries in which these species are an important by-product (Broadhurst et al. 2005, Graham et al. 2009, Macbeth et al. 2012). For other fisheries in which these sillaginids are the target species, having gears that specifically select fish > L_{50} (e.g. 20 cm FL) could aid reproductive potential and resource sustainability. In high-volume multispecies trawl fisheries, it is often preferable to manage and regulate the sizes of fish retained by prescribing the selectivity of the fishing gears rather than enforcement of specific legal lengths, which can be logistically problematic for fisheries operators as well as compliance officers. Nevertheless, prior to introducing and mandating any specific fishing gears (or retained legal lengths for these sillaginid species) in any particular fishery, assessments of rates of survival of non-retained individuals in such fishing gears (Broadhurst et al. 2006, Coggins et al. 2007), as well as broader market and economic impacts need to be considered. Moreover, the potential effects on population reproductive output and resource sustainability of alternate management arrangements that protect larger (more fecund) fish, either by harvesting particular slot sizes (Gwinn et al. 2013) or provision of refuge (no-take) areas (Roberts et al. 2005) needs investigating.

Acknowledgements. This research was funded by the NSW Government and done in accordance with the NSW DPI Animal Care and Ethics Permit 2005/05 whilst the authors were located at the (now closed) Cronulla Fisheries Research Centre of Excellence. We thank Don Anderson (*El Margo*) and Bruce Korner (*Little John*) for their fishing expertise and assistance with sampling. Damian Young assisted with fieldwork and Caitlin Young, Justin McKinnon, Martin Jackson and Adam Welfare helped process samples in the laboratory and Jim Craig provided database management. Drs. Chris Walsh, Matt Ives and the journal referees provided constructive reviews of the draft manuscript.

LITERATURE CITED

Barnes LM, Leclerc M, Gray CA, Williamson JE (2011) Dietary niche differentiation of five sympatric species of Platycephalidae. Environ Biol Fish 90:429–441

▶ Barnes LM, van der Meulen DE, Orchard BA, Gray CA (2013) Novel use of an ultrasonic cleaning device for fish reproductive studies. J Sea Res 76:222–226

▶ Berkeley SA, Chapman C, Sogard SM (2004) Maternal age as a determinant of larval growth and survival in a marine fish, *Sebastes melanops*. Ecology 85:1258–1264

▶ Blanck A, Lamouroux N (2007) Large-scale intraspecific variation in life-history traits of European freshwater fish. J Biogeogr 34:862–875

Broadhurst MK, Young DJ, Gray CA, Wooden MEL (2005) Improving selection in south eastern Australian whiting (*Sillago* spp.) trawls: effects of modifying the body, extension and codend. Scientia Mar 69:301–311

▶ Broadhurst MK, Suuronen P, Hulme A (2006) Estimating collateral mortality from towed fishing gear. Fish Fish 7: 180–218

▶ Burchmore JJ, Pollard DA, Middleton MJ, Bell JD, Pease BC (1988) Biology of four species of whiting (Pisces: Sillaginidae) in Botany Bay, NSW. Aust J Mar Freshw Res 39:709–727

▶ Burt A, Kramer DL, Nakatsuru K, Spry C (1988) The tempo of reproduction in *Hyphessobrycon pulchripinnis* (Characidae), with a discussion on the biology of 'multiple spawning' in fishes. Environ Biol Fish 22:15–27

Butcher AR, Hagedoorn WL (2003) Age, growth and mortality estimates of stout whiting, *Sillago robusta* Stead (Sil-

laginidae), from southern Queensland, Australia. Asian Fish Sci 16:215–228

▶ Chen W, Almatar S, Bishop JM (2009) Spatial and temporal variability of fish assemblage in Kuwait Bay. Mar Biol 156:415–424

Cleland KW (1947) Studies on the economic biology of sand whiting (*Sillago ciliata* C. & V.). Proc Linn Soc NSW 72: 215–228

▶ Coggins LG Jr, Catalano MJ, Allen MS, Pine WE III, Walters CJ (2007) Effects of cryptic mortality and the hidden costs of using length limits in fishery management. Fish Fish 8:196–210

▶ Colin PL, Bell LJ (1991) Aspects of the spawning of ladrid and scarid fishes (Pisces, Labroidei) at Enewetak-Atoll, Marshall Islands with notes on other families. Environ Biol Fish 31:229–260

Colin PL, Clavijo IE (1988) Spawning activity of fishes producing pelagic eggs on a shelf edge coral-reef, Southwestern Puerto-Rico. Bull Mar Sci 43:249–279

▶ Colloca F, Carpentieri P, Balestri E, Ardizzone G (2010) Food resource partitioning in a Mediterranean demersal fish assemblage: the effect of body size and niche width. Mar Biol 157:565–574

▶ Coulson PG, Hesp AS, Potter IC, Hall NG (2005) Comparisons between the biology of two co-occurring species of whiting (Sillaginidae) in a large marine embayment. Environ Biol Fish 73:125–139

De Vlaming VL (1983) Oocyte development patterns and hormonal involvements among teleosts. In: Rankin JC, Pitcher TJ, Duggan RT (eds) Control processes in fish physiology. Croom Helm, Beckenham, p 176–199

▶ Duarte cm, Alcaraz M (1989) To produce many small or few large eggs: a size-independent reproductive tactic of fish. Oecologia 80:401–404

▶ Fowler AJ, Short DA (1996) Temporal variation in the early life-history characteristics of the King George whiting (*Sillaginodes punctata*) from analysis of otolith microstructure. Mar Freshw Res 47:809–818

▶ Fowler AJ, McLeay L, Short DA (1999) Reproductive mode and spawning information based on gonad analysis for the King George whiting (Percoidei: Sillaginidae) from South Australia. Mar Freshw Res 50:1–14

▶ Fowler AJ, McLeay L, Short DA (2000) Spatial variation in size and age structures and reproductive characteristics of the King George whiting (Percoidei: Sillaginidae) in South Australian waters. Mar Freshw Res 51:11–22

▶ Genner MJ, Turner GF, Hawkins SJ (1999) Foraging of rocky habitat cichlid fishes in Lake Malawi: coexistence through niche partitioning? Oecologia 121:283–292

▶ Graham KJ, Broadhurst MK, Millar RB (2009) Effects of codend circumference and twine diameter on selection in south-eastern Australian fish trawls. Fish Res 95:341–349

▶ Gray CA, Kennelly SJ (2003) Catch characteristics of the commercial beach-seine fisheries in two Australian barrier estuaries. Fish Res 63:405–422

▶ Gray CA, Miskiewicz AG (2000) Larval fish assemblages in south-east Australian coastal waters: seasonal and spatial structure. Estuar Coast Shelf Sci 50:549–570

▶ Gray CA, Otway NM (1994) Spatial and temporal differences in the assemblages of demersal fishes on the inner continental shelf off Sydney, southeastern Australia. Aust J Mar Freshw Res 45:665–676

▶ Gray CA, Rotherham D, Johnson DJ (2011) Consistency of temporal and habitat-related differences among assemblages of fish in coastal lagoons. Estuar Coast Shelf Sci 95:401–414

Gray CA, Haddy JA, Fearman J, Barnes LM, Macbeth WG,

Kendall BW (2012) Reproduction, growth and connectivity among populations of *Girella tricuspidata* (Pisces: Girellidae). Aquat Biol 16:53–68

Gunderson DR (1977) Population biology of pacific ocean perch, *Sebastes alutus*, stocks in the Washington-Queen Charlotte Sound region, and their response to fishing. Fish Bull 75:369–403

▶ Gunn JS, Milward NE (1985) The food, feeding habits and feeding structures of the whiting species *Sillago sihama* (Forsskal) and *Sillago analis* Whitley from Townsville, North Queensland, Australia. J Fish Biol 26:411–427

▶ Gust N, Choat JH, Ackerman JL (2002) Demographic plasticity in tropical reef fishes. Mar Biol 140:1039–1051

▶ Gwinn DC, Allen MS, Johnston FD, Brown P, Todd CR, Arlinghaus R (2013) Rethinking length-based fisheries regulations: the value of protecting old and large fish with harvest slots. Fish Fish. doi:10.1111/faf.12053

▶ Hajisamae S, Yeesin P, Ibrahim S (2006) Feeding ecology of two sillaginid fishes and trophic interrelations with other co-existing species in the southern part of South China Sea. Environ Biol Fish 76:167–176

▶ Hare JA, Cowen RK (1993) Ecological and evolutionary implications of the larval transport and reproductive strategy of bluefish *Pomatomus saltatrix*. Mar Ecol Prog Ser 98:1–16

Hunter JR, Macewicz BJ (1985) Measurement of spawning frequency in multiple spawning fishes. In: Lasker R (ed) An egg production method for estimating spawning biomass of pelagic fish: application to the northern anchovy, *Engraulis mordax*. US Dep Commerce, NOAA Tech Rep NMFS 36, Springfield, p 79–93

Hunter JR, Lo NCH, Leong RJH (1985) Batch fecundity in multiple spawning fishes. In: Lasker R (ed) An egg production method for estimating spawning biomass of pelagic fish: application to the northern anchovy, *Engraulis mordax*. US Dep Commerce, NOAA Tech Rep NMFS 36, Springfield, p 67–77

▶ Huyer A, Smith RL, Stabeno PJ, Church JA, White NJ (1988) Currents off south-eastern Australia: results from the Australian Coastal Experiment. Aust J Mar Freshw Res 39:245–288

Hyndes GA, Potter IC (1996) Comparisons between the age structures, growth and reproductive biology of two co-occurring sillaginids, *Sillago robusta* and *S. bassensis*, in temperate coastal waters of Australia. J Fish Biol 49:14–32

▶ Hyndes GA, Potter IC (1997) Age, growth and reproduction of *Sillago schomburgkii* in south-western Australian nearshore waters and comparisons of life history styles of a suite of *Sillago* species. Environ Biol Fish 49:435–447

▶ Hyndes GA, Potter IC, Lenanton RCJ (1996a) Habitat partitioning by whiting species (Sillaginidae) in coastal waters. Environ Biol Fish 45:21–40

▶ Hyndes GA, Potter IC, Hesp SA (1996b) Relationships between the movements, growth, age structures, and reproductive biology of the teleosts *Sillago burrus* and *S. vittata* in temperate marine waters. Mar Biol 126:549–558

▶ Hyndes GA, Platell ME, Potter IC (1997) Relationships between diet and body size, mouth morphology, habitat and movements of six sillaginid species in coastal waters: implications for resource partitioning. Mar Biol 128: 585–598

Hyndes GA, Platell ME, Potter IC, Lenanton RCJ (1998) Age composition, growth, reproductive biology, and recruitment of King George whiting, *Sillaginodes punctata*, in coastal waters of south-western Australia. Fish Bull 96: 258–270

▶ Hyndes GA, Platell ME, Potter IC, Lenanton RCJ (1999)

Does the composition of the demersal fish assemblages in temperate coastal waters change with depth and undergo consistent seasonal changes? Mar Biol 134: 335–352

Jakobsen T, Fogarty MJ, Megrey BA, Moksness E (2009) Fish reproductive biology; implications for assessment and management. Blackwell Publishing, Chichester

Kailola PJ, Williams MJ, Stewart PC, Reichelt RE, McNee A, Grieve C (1993). Australian Fisheries Resources. Bureau of Resource Sciences, Department of Primary Industry and Energy, and the Fisheries Research and Development Corporation, Canberra

Kemp J, Lyly J, Rowling K, Ward P (2012) Eastern school whiting *Sillago flindersi*. In: Flood M, Stobutzki I, Andrews J, Begg G and others (eds) Status of key Australian fish stocks reports 2012. Fisheries Research and Development Corporation, Canberra, p 391–395.

▶ Kendall BW, Gray CA (2009) Reproduction, age and growth of *Sillago maculata* (Sillaginidae) in south-eastern Australia. J Appl Ichthyol 25:529–536

▶ Kennelly SJ, Liggins GW, Broadhurst MK (1998) Retained and discarded bycatch from oceanic prawn trawling in New South Wales, Australia. Fish Res 36:217–236

▶ King JR, McFarlane A (2003) Marine fish life history strategies: applications to fishery management. Fish Manag Ecol 10:249–264

▶ Labropoulou M, Damalas D, Papaconstantinou C (2008) Bathymetric trends in distribution and size of demersal fish species in the north Aegean Sea. J Nat Hist 42: 673–686

Lambert TC, Ware DM (1984) Reproductive strategies of demersal and pelagic spawning fish. Can J Fish Aquat Sci 41:1564–1569

▶ Lassalle G, Trancart T, Lambert P, Rochard E (2008) Latitudinal variations in age and size at maturity among allis shad *Alosa alosa* populations. J Fish Biol 73:1799–1809

▶ Lee CS, Hirano R (1985) Effects of water temperature and photoperiod on the spawning cycle of the sand borer, *Sillago sihama*. Prog Fish-Cult 47:225–230

▶ Lowerre-Barbieri SK, Ganias K, Saborido-Rey F, Murua H, Hunter JR (2011) Reproductive Timing in Marine Fishes: Variability, Temporal Scales, and Methods. Mar Coast Fish Dyn Manag Ecosyst Sci 3:71–91

▶ Macbeth WG, Millar RB, Johnson DD, Gray CA, Keech RS, Collins D (2012) Assessment of the relative performance of a square-mesh codend design across multiple vessels in a demersal trawl fishery. Fish Res 134–136:29–41

▶ Macpherson E, Duarte CM (1991) Bathymetric trends in demersal fish size: Is there a general relationship? Mar Ecol Prog Ser 71:103–112

▶ McBride RS, Somarakis S, Fitzhugh GR, Albert A and others (2013) Energy acquisition and allocation to egg production in relation to fish reproductive strategies. Fish Fish, doi:10.1111/faf.12043

McKay RJ (1992) FAO Species Catalogue, Vol 14. Sillaginid fishes of the world (family Sillaginidae). An annotated and illustrated catalogue of the Sillago, smelt or Indo-Pacific whiting species known to date. FAO, Rome

▶ Mercier A, Hamel JF (2010) Synchronized breeding events in sympatric marine invertebrates: role of behavior and fine temporal windows in maintaining reproductive isolation. Behav Ecol Sociobiol 64:1749–1765

▶ Methratta ET, Link JS (2007) Ontogenetic variation in habitat associations for four flatfish species in the Gulf of Maine-Georges Bank region. J Fish Biol 70:1669–1688

Middleton RW, Musick JA (1986) The abundance and distribution of the family Macrouridae (Pisces, Gadiformes) in the Norfolk canyon area. Fish Bull 84:35–62

➤ Morgan MJ, Bowering WR (1997) Temporal and geographic variation in maturity at length and age of Greenland halibut (*Reinhardtius hippoglossoides*) from the Canadian north-west Atlantic with implications for fisheries management. ICES J Mar Sci 54:875–885

Morton RM (1985) The reproductive biology of summer whiting, *Sillago ciliata* C. & V., in northern Moreton Bay, Queensland. Aust Zool 21:491–502

➤ Muthiga NA (2003) Coexistence and reproductive isolation of the sympatric echinoids *Diadema savignyi* Michelin and *Diadema setosum* (Leske) on Kenyan coral reefs. Mar Biol 143:669–677

➤ Nakane Y, Suda Y, Sano M (2013) Responses of fish assemblage structures to sandy beach types in Kyushu Island, southern Japan. Mar Biol 160:1563–1581

➤ Nelder JA, Wedderburn WM (1972) Generalized linear models. J R Stat Soc A 135:370–384

➤ Park YJ, Takemura A, Lee YD (2006) Annual and lunar-synchronized ovarian activity in two rabbitfish species in the Chuuk lagoon, Micronesia. Fish Sci 72:166–172

➤ Parker GA (1992) The evolution of sexual dimorphism in fish. J Fish Biol 41:1–20

Quinn GP, Keough KJ (2002) Experimental Design and Data Analysis for Biologists. Cambridge University Press, Cambridge

Radhakrishnan N (1957) A contribution to the biology of the Indian sand whiting, *Sillago sihama* (Forskal). Indian J Fish 4:254–283

➤ Rahman MH, Tachihara K (2005a) Reproductive biology of *Sillago aeolus* in Okinawa Island, Japan. Fish Sci 71: 122–132

➤ Rahman MH, Tachihara K (2005b) Age and growth of *Sillago aeolus* in Okinawa Island, Japan. J Oceanogr 61: 569–573

Reddy CR, Neelakantan N (1992) Age and growth of Indian whiting, *Sillago sihama* (Forskal) from Karwar waters. Mahasagar 25:61–64

➤ Roberts CM, Hawkins JP, Gell FR (2005) The role of marine reserves in achieving sustainable fisheries. Philos Trans R Soc B 360:123–132

➤ Rochet MJ (1998) Short-term effects of fishing on life history traits of fishes. ICES J Mar Sci 55:371–391

➤ Roff DA (1991) The evolution of life-history variation in fishes, with particular reference to flatfish. Neth J Sea Res 27:197–207

Roff DA (1992) The evolution of life histories: theory and analysis. Chapman & Hall, New York, NY

➤ Ross ST (1977) Patterns of resource partitioning in searobins (Pisces: Triglidae). Copeia 1977:561–571

➤ Ross ST (1986) Resource partitioning in fish assemblages: a review. Copeia 1986:352–388

➤ Roughan M, Middleton JH (2004) On the East Australian Current: variability, encroachment, and upwelling. J Geophys Res Oceans 109:C07003, doi:10.1029/2003JC001833

➤ Roughan M, Macdonald HS, Baird ME, Glasby TM (2011) Modelling coastal connectivity in a Western Boundary Current: Seasonal and inter-annual variability. Deep-Sea Res II 58:628–644

Rowling K, Hegarty A, Ives MC (2010) Status of fisheries resources in NSW 2008/09. Industry and Investment NSW, Cronulla

➤ Ruttenberg BI, Haupt AJ, Chiriboga AI, Warner RR (2005) Patterns, causes and consequences of regional variation in the ecology and life history of a reef fish. Oecologia 145:394–403

➤ Ryer CH, Laurel BJ, Stoner AW (2010) Testing the shallow water refuge hypothesis in flatfish nurseries. Mar Ecol Prog Ser 415:275–282

➤ Sala-Bozano M, Mariani S (2011) Life history variation in a marine teleost across a heterogeneous seascape. Estuar Coast Shelf Sci 92:555–563

➤ Sancho G, Solow AR, Lobel PS (2000) Environmental influences on the diel timing of spawning in coral reef fishes. Mar Ecol Prog Ser 206:193–212

Sarre GA, Potter IC (1999) Comparison between the reproductive biology of black bream *Acanthopagrus butcheri* (Teleosti: Sparidae) in four estuaries with widely differing characteristics. Int J Salt Lake Res 8:179–210

➤ Schoener TW (1974) Resource partitioning in ecological communities. Science 185:27–39

➤ Scott SG, Pankhurst NW (1992) Interannual variations in the reproductive cycle of the New Zealand snapper *Pagrus auratus* (Bloch & Schneider) (Sparidae). J Fish Biol 41: 685–696

Shamsan EF, Ansari ZA (2010) Studies on the reproductive biology of the Indian sand whiting *Sillago sihama* (Forsskal). Indian J Mar Sci 39:280–284

➤ Sharpe DMT, Hendry AP (2009) Life history change in commercially exploited fish stocks: an analysis of trends across studies. Evol Applic 2:260–275

➤ Stefanescu C, Rucabado J, Lloris D (1992) Depth-size trends in western Meditterranean demersal deep-sea fishes. Mar Ecol Prog Ser 81:205–213

➤ Stocks J, Stewart J, Gray CA, West RJ (2011) Using otolith increment widths to infer spatial, temporal and gender variation in the growth of sand whiting *Sillago ciliata*. Fish Manag Ecol 18:121–131

➤ Suthers IM, Young JW, Baird ME, Roughan M and others (2011) The strengthening East Australian Current, its eddies and biological effects—an introduction and overview. Deep-Sea Res II 58:538–546

➤ Tomaiuolo M, Hansen TF, Levitan DR (2007) A theoretical investigation of sympatric evolution of temporal reproductive isolation as illustrated by marine broadcast spawners. Evolution 61:2584–2595

➤ Walsh CT, Gray CA, West RJ, Williams LFG (2011) Reproductive biology and spawning strategy of the catadromous percichthyid, *Macquaria colonorum* (Gunther, 1863). Environ Biol Fish 91:471–486

➤ Wellenreuther M, Clements KD (2007) Reproductive isolation in temperate reef fishes. Mar Biol 152:619–630

➤ Weng HT (1986) Spatial and temporal distribution of whiting (Sillaginidae) in Moreton Bay, Queensland. J Fish Biol 29:755–764

➤ Werner EE, Hall DJ, Laughlin DR, Wagner DJ, Wilsmann LA, Funk FC (1977) Habitat partitioning in a freshwater fish community. J Fish Res Board Can 34:360–370

➤ West G (1990) Methods of assessing ovarian development in fishes: a review. Aust J Mar Freshw Res 41:199–222

➤ Wright PJ, Trippel EA (2009) Fishery-induced demographic changes in the timing of spawning: consequences for reproductive success. Fish Fish 10:283–304

Zeller B, Rowling K, Jebreen E, O'Neill M, Winning M (2012) Stout whiting *Sillago robusta*. In: Flood M, Stobutzki I, Andrews J, Begg G and others (eds) Status of key Australian fish stocks reports 2012. Fisheries Research and Development Corporation, Canberra, p 396–400

5

Harbour seal *Phoca vitulina* movement patterns in the high-Arctic archipelago of Svalbard, Norway

Marie-Anne Blanchet[1,2,*], Christian Lydersen[1], Rolf A. Ims[1,2], Andrew D. Lowther[1], Kit M. Kovacs[1]

[1]Norwegian Polar Institute, Framsentret, 9296 Tromsø, Norway
[2]Department of Arctic and Marine Biology, UiT-Arctic University of Norway, 9037 Tromsø, Norway

ABSTRACT: Harbour seals *Phoca vitulina* are mainly considered a temperate species, but the world's northernmost population resides year-round in the high-Arctic archipelago of Svalbard. In this study we document post-moulting at-sea movements of 30 individuals from this population using satellite relay data loggers deployed in the autumns of 2009 and 2010. All of the seals showed a strong preference for the west side of the archipelago, staying mainly in coastal areas (<50 km over the continental shelf), but seldom entering the fjord systems. Distance swam per day, individual home range size, and trip duration increased throughout the winter to a peak that was reached when drifting sea ice in the region was at a maximum. No effect of age was observed, but sex differences were significant; males occupied larger areas than females. Habitat selection was quantified by modelling time spent in area (TSA) as a function of environmental parameters using Cox proportional hazard models (CPH). The harbour seals avoided heavy ice concentrations (>50%) but did occupy areas with substantial amounts of drifting ice (5 to 25%). Shallow water (<100 m) and steep bathymetric slopes were preferred to deep water or flat-bottom areas. Harbour seal distribution in Svalbard is largely restricted to coastal areas that are heavily influenced by Atlantic water brought northward in the West Spitsbergen Current; both the temperature and influx of this water type are predicted to increase in the future. It is thus likely that environmental conditions in Svalbard in the future will become more favourable for harbour seals.

KEY WORDS: Climate change · Sea ice · Habitat use · Time spent in area · Cox proportional hazard models · Satellite telemetry

INTRODUCTION

Harbour seals *Phoca vitulina* have one of the broadest distributions among the pinnipeds, ranging from temperate areas as far south as southern California and France to Arctic waters of the North Pacific and the North Atlantic (Bigg 1969, Thompson et al. 1989, Bjørge et al. 1995). This species was listed as 'Least Concern' by the IUCN Red List in 2008; at that time the global population was estimated to be 350 000 to 500 000 individuals. However, there have been major declines documented for some harbour seal populations in the last 20 yr (Thompson & Härkönen 2008), with currently ongoing declines in

some regions (Womble et al. 2010, Hanson et al. 2013, Matthiopoulos et al. 2014). Southern harbour seal populations are generally well-studied, protected from exploitation for the most part, and are stable or increasing (e.g. Chavez-Rosales & Gardner 1999, Jeffries et al. 2003, Hassani et al. 2010, Olsen et al. 2010, Reijnders et al. 2010). However, the situation is quite different for northern populations, where up-to-date regional harbour seal abundance information is often lacking (despite local harvesting activities), and some populations have shown marked declines in both the North Atlantic (Lucas & Stobo 2000, Hanson et al. 2013, Matthiopoulos et al. 2014) and the North Pacific (e.g. Frost et al. 1999, Small et

*Corresponding author: marie-anne.blanchet@npolar.no

al. 2008, Womble et al. 2010). The reasons for the observed declines are not clear in all cases, but several hypotheses have been proposed that suggest climate change-related linkages such as shifts in prey distribution, increased predation pressure, vulnerability to pathogens or competitive stress (Trites et al. 2007, Matthiopoulos et al. 2014). The declines in harbour seal numbers in the Aleutian Islands and the Gulf of Alaska over the last 30 yr have been linked to a large-scale oceanographic regime shift that has likely been induced by climate change (Small et al. 2008, Womble et al. 2010). Shark-inflicted mortality seems to be a predominant factor for seal decline in Nova Scotia, where a large proportion of pups and adult females are killed by one or more shark species (Lucas & Stobo 2000, Lucas & Natanson 2010). The substantial decline in the number of harbour seals in the United Kingdom is partly attributable to 2 recent phocine distemper virus epidemics and some colonies have not recovered, particularly in Scotland where severe declines have been documented (Thompson et al. 2005, Lonergan et al. 2007, 2010, 2013, Hanson et al. 2013). Available information from Iceland, Greenland and mainland Norway suggest that harbour seal populations are severely depleted compared to former times, which has been driven in part by overharvesting, but other environmental pressures are likely also contributing to the downward trends (Hauksson 1992, Nilssen et al. 2010, Rosing-Asvid 2010).

The harbour seals in Svalbard constitute the world's northernmost population of this species. In many aspects this population is unique among harbour seals because it experiences a combination of extreme light, sea ice and temperature regimes, exhibits marked sexual dimorphism and individuals are shorter and heavier compared to conspecifics in other populations (Lydersen & Kovacs 2010). No other population in the world encounters such a combination of extreme environmental conditions and only a handful of populations in Canada (Hudson Bay, Bajzak et al. 2013; St Lawrence River Estuary, Lesage et al. 2004) and Greenland (Rosing-Asvid 2010) encounter heavy seasonal sea-ice cover. The core of the harbour seal distribution within the Svalbard Archipelago is off the west coast of Prins Karl Forland. This area contains the only known pupping sites in Svalbard (Lydersen & Kovacs 2010) but observations of hauled-out harbour seals are increasingly reported from other areas within the Svalbard Archipelago. This small, isolated population of ca. 2000 animals (Merkel et al. 2013) has low genetic diversity (Andersen et al. 2011); it is Red-listed in

Norway and hence protected from harvesting. The haul-out behaviour of this population has been studied intensively (Reder et al. 2003), and concomittantly with the current stud (see Hamilton et al. 2014), but the at-sea behaviour of adult harbor seals remains largely undocumented. A few animals in this population were tracked through the fall and early winter period via satellite tag deployments in the early 1990s (Gjertz et al. 2001) and 3 time-depth recorders (TDRs) were deployed for some weeks in the autumn of 1994 (Krafft et al. 2002). However, up-to-date distributional information about areas occupied by this population in a broader seasonal perspective is not available.

The marked declines in Nordic harbour seal populations are likely to be at least partly the result of wide-scale ecosystems changes that are influencing predation levels on, or prey distribution of, harbour seals. However, currently there is insufficient information on the distribution and general ecology of harbour seals in Svalbard to evaluate the various influences of climatic change on this population. Thus, the objectives of this study were to examine present-day movement patterns and at-sea habitat preferences of harbour seals on Svalbard. This will provide a baseline for future studies exploring possible consequences of further climate change on this northernmost harbour seal population.

MATERIALS AND METHODS

Tagging

Thirty harbour seals were captured on the west coast of Prins Karl Forland (PKF; 78° 20 N, 11° 30 E) near a small group of islands (Forlandøyane), in the high-Arctic archipelago of Svalbard (see Fig. 1) in 2009 and 2010 (N = 15 each year). The animals were captured immediately following their annual moult, between 23 August and 13 September using tangle nets set from shore near haul-out sites (see Lydersen & Kovacs 2005 for details). None of the captured adult females was still caring for a pup as the peak of the pupping period occurs during the second half of June (Gjertz & Børset 1992) and pups are weaned in July. All animals were weighed (Salter spring scales ± 0.5 kg) and sex was determined. After being secured in individual holding nets the seals were immobilized with an intramuscular injection of Telazol (1 mg kg^{-1} for juveniles and adult females and 0.75 mg kg^{-1} for adult males) before a lower incisor was extracted for age determination and standard length and girth were measured to the nearest cm (Reder et al. 2003). Seals aged 1 or 2 yr were classified as juveniles while seals aged 6 yr or more were classified as mature. Seals aged 3, 4, or 5 yr were attributed a maturity status based on a combination of their length, girth and mass measurements following Lydersen & Kovacs (2005).

All animals were equipped with conductivity-temperature-depth satellite relay data loggers (CTD-SRDLs) (Sea Mammal Research Unit, University of St Andrews), glued onto the fur mid-dorsally in the neck area using quick-setting epoxy. Data collected by the CTD-SRDLs were transmitted via the Argos satellite system (System Argos). The instruments collected and relayed information on the animals' location, haul-out and dive behaviour, as well as providing CTD up-casts on selected dives (see Boehme et al. 2009 for details on the performance, sampling and transmission protocols of the tags). Only location data are the subject of this study. The CTD-SRDLs weighed 545 g in air, measured 10.5 × 7 × 4 cm and were programmed to send data whenever possible with no duty cycling. All of the seals were also tagged with uniquely numbered plastic tags (Dalton roto-tags) placed through the webbing of each hind flipper for permanent individual identification.

Environmental data extraction

All data processing and analyses were done using the R statistical framework (R Development Core Team 2012). Sea ice classifications were based on daily polygon ice maps from the Norwegian Meteorological Institute; these are produced using a weather-independent SAR radar with a spatial resolution of 50 m, using manual classification of areas into the following concentration categories (http://polarview.met.no/) — (1) open water: 0/10–1/10; (2) very open drift ice: 1/10–4/10; (3) open drift ice: 4/10–7/10; (4) close drift ice: 7/10–9/10; (5) very close drift ice: 9/10–10/10 and; (6) fast ice. Average monthly ice concentrations and sea ice extent were also calculated using the daily ice maps (R package 'Raster'; Hijmans 2014); extent was calculated as the sea surface area covered by an ice concentration of at least 25% (corresponding to the category very open drift ice).

Bathymetry was extracted from a 0.5 × 0.5 km resolution data set from the International Bathymetric Chart of the Arctic Ocean (IBCAO v.3.0, 2012, www.ibcao.org; Jakobsson et al. 2012) and slope was calculated based on the bathymetry map using val-

ues from 8 neighbouring spatial grid cells. Distance to the coast and from the tagging location were calculated in ArcInfo 10 (ESRI), using Norwegian Polar Institute coastline maps (scale 1:250 000) that were produced using aerial photographs of coastlines taken in the period from 1993 to 1998 (with regular updates on the locations of glacier fronts).

Data processing

Satellite-derived location estimates obtained from Service Argos are assigned a location class (LC) based on the number of uplinks received by a passing satellite. LC 3, 2, and 1 have the highest quality with an accuracy ranging from <150 m for LC 3 to 1000 m for LC 1, while LC 0 has unbounded accuracy (>1000 m) and no accuracy information is provided for LC A and B (www.cls.fr/manual). Locations reported for each seal were first filtered using a speed distance and angle filter (SDA filter; Freitas et al. 2008a) that removes all LC Z values and points requiring unrealistic swimming speeds (>2 m s^{-1}) or unlikely turning angles (all spikes with angles smaller than 15 or 25 degrees were removed if their lengths were greater than 2.5 or 5 km, respectively) using the R package 'argosfilter' (Freitas 2013). The remaining location estimates were then processed further using a Kalman filter under a state-space framework using the R package 'crawl' (Johnson 2013) that incorporates a covariate for Argos location error (when available) for each of the 6 LCs (3, 2, 1, 0, A, B). In addition a covariate encompassing the time the animal was hauled out was included allowing the movement along a track line to completely stop during a haul-out event. Processing the raw location estimates in this manner resulted in a model of the most likely track, from which point location estimates could be made for any specific time. Hourly locations were estimated by interpolation using this model; these positions were used for further calculations.

Movement parameters

Movement parameters were calculated for each seal; total distance swum (km), monthly average maximum distance (km) from the tagging location and average trip duration (h). A trip was defined as the time between 2 sequential haul-outs, while a foraging trip was defined as an excursion of at least 3 h duration that took place between 2 successive haul-out events; the 3 h lower limit was set in order to avoid over-emphasizing brief excursions into the water that harbour seals often do close to haul-out locations (Thompson et al. 1998a, Krafft et al. 2002, Blundell et al. 2011). The influence of calendar month, year and sex was tested on each of the movement parameters using linear mixed effects models (LME) ('nlme' package; Pinheiro et al. 2014). Interactions between sex and year, and month and year, were considered whereas interactions between sex and calendar month were not, due the low number of females still transmitting data from February onward in both years (see Table 1). Possible seasonal trends in trip duration were explored using a generalized additive mixed-effects model (GAMM). Trip departure date was entered as a smoothed term based on the number of days since 01 September with a separate smoothed term fitted for each year. Individual animal ID was included as a random effect to deal with pseudoreplication affiliated with multiple points from individual animals.

Home range calculations

Home ranges (HR) were calculated for each animal for each month, for individuals with at least 25 d of tracking data within a given month, by taking 95% of the estimated utilization distribution (UD) (Worton 1989). The UD function was estimated using a kernel method based on Brownian bridge movement (Horne et al. 2007) in the package 'adehabitatHR' (Calenge 2011). This method takes into account the path travelled by the animal between 2 successive locations, whereas classical kernel methods only consider point patterns. Therefore, it avoids identifying areas between habitat patches that were actually not used by the animal (Calenge 2011). The smoothing parameter of the kernel depends on the accuracy and the average number of locations per day after filtering by the state-space model used to calculate the HR. The post-filtering location accuracy was taken from Patterson et al. (2010) based on the mean number of locations per day. HR areas were compared across years, months, ages and sexes and environmental parameters such as sea ice extent using LME. Possible interactions between year and month were tested, but potential influences of sex with sea ice extent and month could not be explored due to the low number of instruments that were still transmitting data from females late in each deployment period. All models were fitted using maximum likelihood methods to ensure they were comparable. Post-

hoc multiple comparison tests were performed using Tukey's pairwise comparisons ('multicomp' package; Hothorn et al. 2014) to test for differences between each month. HR sizes were log-transformed to achieve normality and homogeneity of variances. Animal ID was used as a random effect and 95% confidence intervals were calculated using 500 bootstrap resamples. A first order, moving average, autocorrelation term was chosen to account for temporal autocorrelation between successive months. Model validation was performed by visually inspecting the residuals and model selection was done using Bayesian information criterion (BIC) that penalizes overfitting (Burnham & Anderson 2002).

Habitat use

To quantify space use relative to environmental conditions experienced while at-sea during foraging trips, the time spent at sea (h) was calculated based on a spatial grid with 2.5×2.5 km cells (package 'trip'; Sumner & Luque 2013). Each cell was assigned the geographical location of its centre to facilitate extracting environmental parameters such as bathymetry and slope of the sea floor. Furthermore, a time stamp corresponding to the mean point of the foraging trip was identified in order to be able to associate each value for time spent in a grid cell with data that best described local, dynamic environmental variables such as sea ice. Time spent in area (TSA) per trip was then modelled using mixed-effect Cox proportional-hazards models (CPH) (Cox 1972) in order to see how seal movements were influenced by environmental and intrinsic variables as described in Freitas et al. (2008b). The models were in the form: $h(t) = \exp(\beta_1 X_1 + \beta_2 X_2 + \ldots \beta_p X_p + b) h_0(t)$, where $h(t)$ is the hazard function representing the risk of an individual leaving an area (defined by a 2.5×2.5 km cell) at time t; X represents an environmental variable (explanatory variable) and; β_x denotes the regression coefficient describing the contribution of the variable X. The b term represents the per-subject random effect and $h_0(t)$ is the baseline hazard function at time t, or in other words the risk of leaving an area when all explanatory variables are equal to zero. In this context, the word 'risk' does not imply a negative consequence of the action of leaving an area, but simply depicts the probability of leaving an area given a certain value of the covariates. The hazard ratio $\exp(\beta)$ represents the ratio between the hazard of leaving when the value of the X explanatory variable is increased by 1 unit if X is continuous or, in other words, the relative probability of leaving an area when X is increased by 1 unit. If X is categorical, the hazard ratio (or relative probability) represents the ratio between the hazard of leaving when X assumes a given level and the hazard of leaving when X is equal to a defined reference level. A hazard ratio higher than 1 indicates an increased risk of leaving (or increased probability) and a hazard ratio lower than 1 indicates a lower risk of leaving. Hazard ratios therefore provide a quantitative way to assess how each parameter influences habitat selection: in other words, it is assumed that if an animal remains in an area for an extended period of time, it is likely a favourable habitat. Individual seal ID was included as a frailty term in the model to account for individual variability and to be able to select between a set of candidate models while keeping the individual effect within the model. The predictive map was calculated without the frailty term because the method to predict values from a proportional hazard mixed effect model is not yet implemented in R. However, parameter estimates of models with and without the frailty term were very similar suggesting that the predictions are unbiased. All possible additive combinations of environmental (bathymetry, slope, sea ice concentration, distance-to-coast), time (mo) and individual parameters (sex, age) were tested and optimal models were selected using BIC. In addition the interactions distance to coast × month and distance-to-coast × sea ice concentration were included to see if TSA at different distances to the coast depended on the time of year or on ice concentration, respectively. Similar to the LME analyses described earlier, interactions with sex were not considered here due to the low number of CTD-SRDLs still transmitting data from female seals during the latter stages of the 2 deployment periods. The assumption of proportional hazard ratios (i.e. the hazard ratio of 2 observations being independent of time) was verified for the final model by checking the scaled Schoefeld residuals visually and by testing if their slope was zero (Collett 2003). All models were fitted in R using 'survival' and 'coxme' (Therneau 2012).

RESULTS

The CTD-SRDLs deployed on adult and juvenile harbour seals (N = 30) in this study provided data for periods ranging between 54 and 298 d, with an average individual data record of 200 ± 79 d (Table 1). Twelve animals had records that extended into the month of June in the year following tag deployment;

but all tags ceased to function before July. Year of tag deployment (ANOVA $F_{1,58}$ = 2.14, p = 0.15) and seal age class (ANOVA $F_{1,58}$ = 0.92, p = 0.40) had no effect on the duration of the tracking records; but males had longer data records than females in both years (ANOVA $F_{1,58}$ = 6.62, p < 0.01). There was no difference in body length within age groups (ANOVA$_{ad}$ $F_{1,12}$ = 1.80, p = 0.30; ANOVA$_{juv}$ $F_{1,14}$ = 1.12, p = 0.30) or sexes between the 2 different study years (ANOVA$_M$ $F_{1,13}$ = 0.06, p = 0.81; ANOVA$_F$ $F_{1,13}$ = 0.22, p = 0.65).

At-sea movements

Seal locations were concentrated in an area that was on average within 82 km (± 58 km) north or south of the tagging location. All of the animals stayed close to the coast, within 50 km west of PKF and none of the instrumented animals travelled to the east coast of Spitsbergen (Fig. 1). Similarly, most of the animals did not go south of 77.5° N or north of 80° N, except in 2010 when 4 males spent some time outside these boundaries. The total distance swum per month ranged between 70 and 1362 km, with clear differences being displayed between sexes and months. The top 3 LMEs with a combined AIC$_c$ weight of 0.7 did not include the interactions between year and month and between sex and year. The most parsimonious model included month and sex (AIC$_c$ = 160.9). Males travelled more than females (mean dist$_M$ = 648 ± 295 km; dist$_F$ = 383 ± 186 km) and all animals covered more distance during the winter and spring months (December through May) compared to the fall (September through November) (mean dist$_{Dec-May}$ = 705 ± 291 km; mean dist$_{Sep-Nov}$ = 368 ± 143 km). The most parsimonious LME model explaining the monthly maximum distance from the tagging location included year and month (AIC$_c$ = 278; AIC$_c$ weight = 0.371). Animals travelled further away from the tagging location in 2009 compared to 2010 (mean dist$_{2009}$ = 53 ± 38 km; mean dist$_{2010}$ = 114 ± 59 km) and all animals were further away from PKF during the winter and spring months compared to the fall (mean dist$_{Sep-Nov}$ = 69 ± 55 km; mean dist$_{Dec-May}$ = 95 ± 58 km). Monthly tracks were longer and the animals were progressively further away from the tagging location as winter set in (Fig. 2). Four males travelled more than 200 km from the tagging location (Table 1). The best GAMM model describing the variation in trip duration included date as a smooth curve, fitted separately for each tagging year. Sex and age were not included in the best model (Table 2). Trip duration varied with the season in both years, though markedly

Table 1. *Phoca vitulina*. Summary of the locations data for 30 harbour seals equipped with CTD satellite relay data loggers (CTD-SRDLs) in Svalbard in 2009 and 2010. Ad: adult; Juv: juvenile; M: male; F: female; dist: distance; loc(s): location(s); tag: tagging. Deployment date is shown as dd/mm/yyyy

Seal ID	Sex	Mass (kg)	Girth (cm)	Length (cm)	Age class	Age (yr)	Deployment date	Tagging duration (d)	Mean no. of filtered locs d^{-1}	Total dist travelled (km)	Max. dist from tag loc (km)
ct52-F41-09	F	41.0	90	104	Juv	2	02/09/2009	278	4.8	2866.7	76.5
ct52-F44-09	F	44.0	89	116	Juv	2	02/09/2009	147	7.7	1491.2	70.1
ct52-F47-09	F	47.0	92	118	Juv	2	02/09/2009	129	11.1	1627.9	67.0
ct52-F48-09	F	48.0	93	120	Juv	3	01/09/2009	101	10.5	1052.6	63.9
ct52-F60-09	F	60.0	98	126	Ad	4	02/09/2009	201	8.4	3111.2	185.3
ct52-F66-09	F	66.0	100	133	Ad	5	02/09/2009	116	9.1	1092.5	22.0
ct52-F74-09	F	74.0	108	139	Ad	11	01/09/2009	160	4.6	1096.8	70.8
ct52-F76-09	F	76.0	107	139	Ad	9	04/09/2009	280	11.0	4082.6	41.8
ct52-M43-09	M	43.0	86	113	Juv	2	04/09/2009	177	9.3	3239.5	150.7
ct52-M51-09	M	51.0	96	120	Juv	2	04/09/2009	279	12.8	6402.8	64.5
ct52-M52-09	M	52.0	98	122	Juv	2	05/09/2009	277	12.5	6040.3	81.5
ct52-M56-09	M	56.0	101	128	Ad	4	04/09/2009	284	11.0	5653.4	49.7
ct52-M64-09	M	64.0	94	130	Ad	3	05/09/2009	256	10.8	4919.5	71.6
ct52-M65-09	M	65.0	109	128	Ad	3	13/09/2009	268	8.0	4396.7	72.4
ct52-M77-09	M	77.0	111	142	Ad	5	10/09/2009	107	6.8	1211.8	73.2
ct69-F42-10	F	42.0	88	109	Juv	2	26/08/2010	151	3.5	563.9	71.1
ct69-F44-10	F	44.0	88	116	Juv	3	28/08/2010	144	6.2	1791.8	76.8
ct69-F50-10	F	50.0	87	127	Juv	3	23/08/2010	54	10.6	649.6	16.2
ct69-F53-10	F	53.0	92	120	Ad	4	03/09/2010	157	6.2	2861.8	180.2
ct69-F58a-10	F	58.0	95	124	Ad	5	25/08/2010	86	5.8	1492.3	61.6
ct69-F58b-10	F	58.0	94	132	Ad	5	28/08/2010	297	9.5	5675.9	86.0
ct69-F59-10	F	59.0	93	125	Juv	3	03/09/2010	103	2.3	993.2	85.0
ct69-M41-10	M	41.0	89	105	Juv	1	25/08/2010	284	9.0	6954.3	180.2
ct69-M45-10	M	45.0	86	122	Juv	2	31/08/2010	151	3.5	1585.9	103.7
ct69-M48-10	M	48.0	86	128	Juv	3	24/08/2010	288	6.4	6157.5	222.7
ct69-M53a-10	M	53.0	97	126	Juv	3	30/08/2010	293	7.3	7401.2	178.6
ct69-M53b-10	M	53.0	93	123	Juv	2	03/09/2010	176	9.3	3369.7	74.9
ct69-M57-10	M	57.0	96	132	Ad	5	24/08/2010	299	8.6	6953.3	228.1
ct69-M64-10	M	64.0	104	127	Ad	4	26/08/2010	289	11.0	8194.3	207.8
ct69-M65-10	M	65.0	95	137	Ad	5	24/08/2010	235	11.1	6279.1	209.0

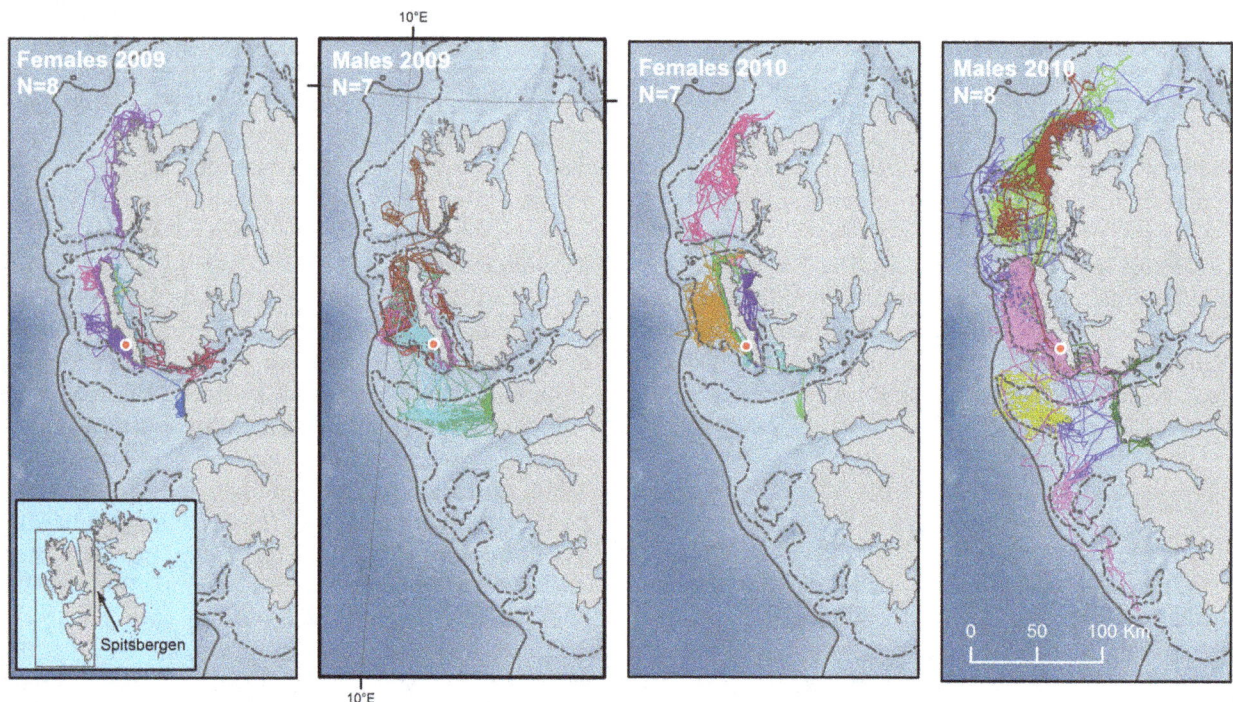

Fig. 1. *Phoca vitulina*. Filtered tracks of the 30 adult or juvenile harbour seals equipped with CTD satellite relay data loggers in Svalbard, Norway during 2009 and 2010, overlaid on bathymetry (darker shades indicate deeper water). The 500 m (solid grey line) and 200 m (dashed grey line) isobaths are shown. The red dot represents the tagging location on Forlandøyane. The inset shows the Svalbard Archipelago and the island of Spitsbergen

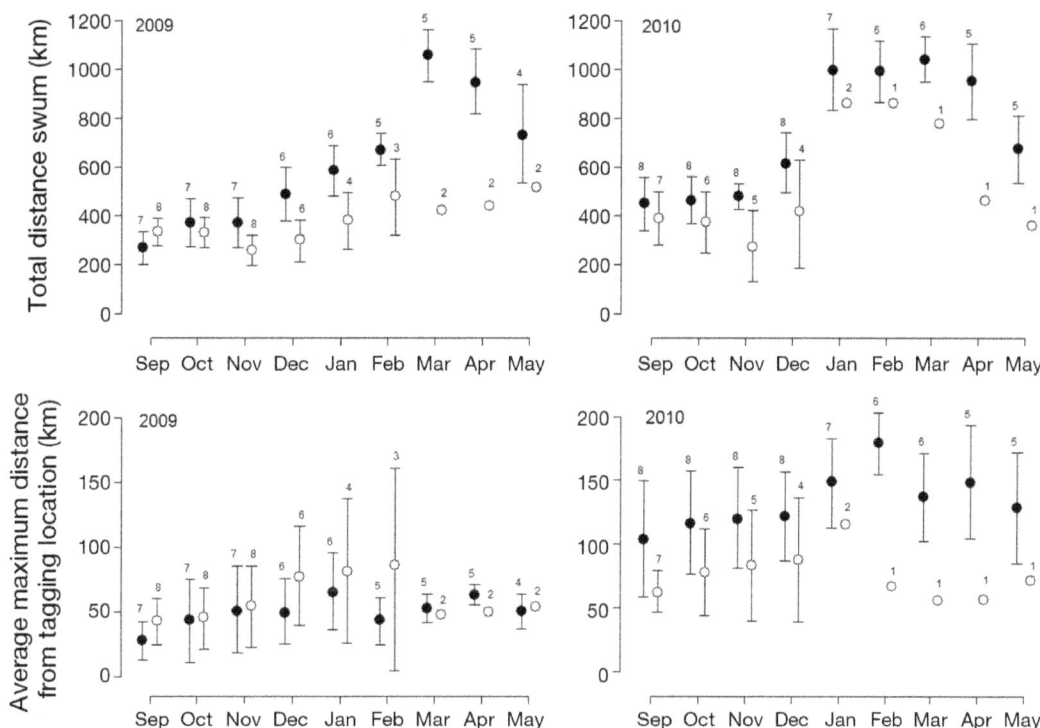

Fig. 2. *Phoca vitulina*. Mean and bootstrapped 95 % confidence interval (CI) of the total distance swum per month and the average maximum distance from the tagging location for 30 harbour seals instrumented with CTD satellite relay data loggers (CTD-SRDLs) in Svalbard, Norway in 2009 and 2010. Males are displayed with black circles and females with white circles. Only months with complete records are presented (September through May). The number of individuals in each month is displayed above each CI

more so in 2010. Trips were shorter in the immediate post-moult period (mean$_{Sep2009}$ = 19 ± 17 h trip^{-1}; mean$_{Sep2010}$ = 18 ± 15 h trip^{-1}), right after the tags were deployed, and gradually increased to a maximum around April 1 (mean$_{Apr2009}$ = 52 ± 56 h trip^{-1}) in the first year, peaking earlier and more markedly around February 1 (mean$_{Feb2010}$ 72 ± 63 h trip^{-1}) in the second year (Fig. S1 in the Supplement at www.int-res.com/articles/suppl/b021p167_suppl.pdf). Trip duration was not influenced by sex.

Home range

Monthly HR varied greatly between individuals (Fig. 3), ranging from 4 km^2 for a juvenile female in September 2010 to 1258 km^2 for an adult male in February 2010. The LME explaining the greatest amount of variation in HR sizes included sea ice extent (monthly), sex and month. Age, year and the interactions between year and month did not contribute to the fit of the model (Table 3). Sea ice extent positively influenced HR size ($t_{1,145}$ = 2.3, p < 0.05) and males had significantly larger HRs than

females ($t_{1,28}$ = 3.6, p < 0.001). There was a distinct influence of month; seals used larger HRs from December through May and smaller HRs from September to November (Fig. S2 in the Supplement, Table 3). There was no significant difference between the months within each of the 2 periods (September to November and December to May) with the exception of December and March (z = −4.06, p < 0.01). There was a moderate amount of co-

Table 2. *Phoca vitulina*. Different GAMM models for trip duration for 30 harbor seals equipped with CDT satellite relay data loggers (CDT-SRDLs) in Svalbard, Norway in 2009 and 2010. For each model the Akaike Information Criteria corrected for small samples (AIC$_c$), the difference with the top-ranked model and its weight are presented. Date: starting date of a trip; year: year of tagging; age: level of maturity (adult or juvenile); ID: individual identification

Model structure	AIC$_c$	ΔAIC$_c$	AIC$_c$ weight
f(date,by=year) + (1|ID)	43185.86	0	0.764
f(date,by=year) + sex + age + (1|ID)	43188.24	2.374	0.233
f(date) + sex + age + year + (1|ID)	43198.14	12.278	0.002
f(date) + sex + (1|ID)	43199.86	13.999	<0.001
f(date) + sex + age + (1|ID)	43201.29	15.431	<0.001
f(date,by=sex) + age + year + (1|ID)	43208.74	22.883	<0.001
f(date,by=age) + sex + year + (1|ID)	43211.16	25.301	<0.001

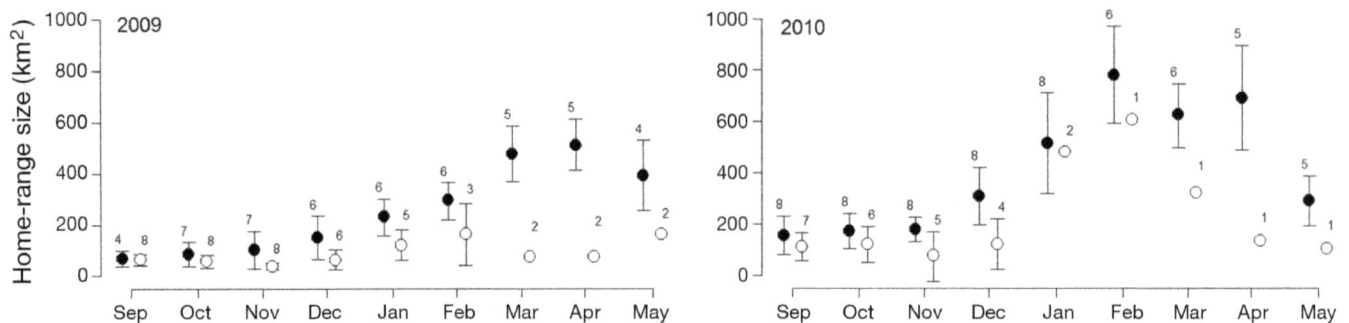

Fig. 3. *Phoca vitulina*. Mean (with bootstrapped 95% CI) home range size as a function of month for 30 harbour seals instrumented with CTD satellite relay data loggers (CTD-SRDLs) in Svalbard, Norway in 2009 and 2010. Males are displayed with black circles and females with white. Only animals with data for at least 25 d in each month are presented (September through May). The number of individuals in each month is displayed above each CI

linearity between month and sea ice extent (variance inflation factor > 3), but both variables had important partial effects given that they were both included in the best fit model.

Table 3. *Phoca vitulina*. Parameter estimates, standard errors, degrees of freedom, *t*-values and p-values for the model that best explained the variation in home range size of harbour seals instrumented with CTD satellite relay data loggers (CTD-SRDLs) in Svalbard, Norway in 2009 and 2010. The reference level for the variables sex and month are female (F) and September respectively. Significant values are in **bold**

Parameters	Estimates	SE	df	t	p
Intercept	3.815	0.199	145	19.149	**<0.001**
Sex (male)	0.872	0.242	28	3.601	**<0.001**
Month (Oct)	−0.075	0.150	145	−0.499	0.618
Month (Nov)	−0.417	0.218	145	−1.915	0.057
Month (Dec)	0.189	0.204	145	0.924	0.357
Month (Jan)	0.626	0.240	145	2.609	**<0.01**
Month (Feb)	0.989	0.247	145	4.013	**<0.001**
Month (Mar)	1.025	0.252	145	4.060	**<0.001**
Month (Apr)	0.959	0.257	145	3.733	**<0.001**
Month (May)	0.678	0.230	145	2.943	**<0.01**
Sea ice extent	<0.001	<0.001	145	2.227	**<0.05**

Habitat use

Several variables influenced TSA and the most appropriate CPH model (based on the lowest BIC; see Table 4) included sea ice concentration, slope, distance to the coast, sex and the interaction between distance to the coast and month. Coefficients for ice, month, sex, bathymetry, distance and distance × month are presented in Table 5. TSA ranged from 1 to 94 h. Females were generally more stationary (TSA = 4.1 ± 6 h) than males (TSA = 2.4 ± 3.9 h). This was confirmed by the hazard ratio associated with sex, which showed that the risk of leaving an area for a male was on average 35% higher than for a female (Table 5). TSA was also significantly influenced by sea ice concentration; animals spent twice as much time on average in nearly ice-free waters (ca. 5% ice concentration) compared to waters containing close drift ice (ca. 95% ice concentration) (Fig. 4). CPH modelling suggested that there was no significant change in the risk of leaving an area between the lowest ice concentrations (25%) and nearly ice-free waters but that heavier ice concentrations (>55%) were associated with an increased risk of leaving

Table 4. *Phoca vitulina*. Model ranking (most parsimonious model at the top in **bold**) based on Bayesian Information Criterion (BIC) for time spent in area (TSA) for harbour seals instrumented with CTD satellite relay data loggers (CTD-SRDLs) in Svalbard, Norway in 2009 and 2010. Log(L): log likelihood; age: level of maturity (adult or juvenile); ice: a categorical variable representing the ice concentration with 5 levels; slope: a continuous variable representing the computed slope from bathymetric values from 8 neighbour cells; bat: the bathymetry as a categorized variable (0–100 m, 100–200 m or >200 m depth); dist: the distance to the nearest coast as a categorized variable (0–10 km, 10–50 km, >50 km)

Model	BIC	log(L)	df	BIC$_c$ weight
Bat + Dist + Ice + Month + Sex + Slope + Dist × Month	624887.7	−312282	31	0.995
Age + Bat + Dist + Ice + Month + Sex + Slope + Dist × Month	624898.1	−312282	32	0.005
Bat + Dist + Month + Sex + Slope + Dist × Month	624914.5	−312317	27	0
Bat + Ice + Month + Sex + Slope	624915.1	−312369	17	0
Bat + Dist + Ice + Month + Sex + Slope	624922.2	−312362	19	0
Age + Bat + Dist + Month + Sex + Slope + Dist × Month	624924.4	−312316	28	0
Age + Bat + Ice + Month + Sex + Slope	624925.5	−312369	18	0

Table 5. *Phoca vitulina*. Effects of ice concentration, slope, month, sex, bathymetry and distance to the nearest coast, obtained by modeling the time spent in an area (TSA) (2.5 × 2.5 km) for harbour seals instrumented with CTD satellite relay data loggers (CTD-SRDLs) in Svalbard, Norway in 2009 and 2010. β: Cox proportional hazard coefficients; exp(β): hazard ratios; CI: 95% confidence intervals. Coefficients are given in relation to the ice concentration Category I, September, for females, for a bathymetry comprised between 0 and 100 m and for a distance to the coast between 0 and 10 km. The number of observations (N) in each category is also given. Significant values are in **bold**

Variable	N	β	exp(β)	—— CI —— Lower	Upper	p
Ice (I)	18047	–	–	–	–	–
Ice (II)	6128	−0.018	0.982	0.950	1.015	0.27
Ice (III)	5062	0.054	1.055	1.018	1.094	**<0.001**
Ice (IV)	3609	0.153	1.165	1.116	1.216	**<0.001**
Ice (V)	481	0.151	1.163	1.058	1.279	**<0.001**
Slope	33327	−0.042	0.958	0.949	0.969	**<0.001**
Month (Sep)	2740	–	–	–	–	–
Month (Oct)	2839	−0.139	0.871	0.826	0.918	**<0.001**
Month (Nov)	2725	−0.265	0.767	0.726	0.811	**<0.001**
Month (Dec)	3274	−0.258	0.772	0.731	0.816	**<0.001**
Month (Jan)	4896	−0.155	0.856	0.807	0.909	**<0.001**
Month (Feb)	4845	0.004	1.004	0.940	1.072	0.91
Month (Mar)	4705	0.110	1.116	1.049	1.189	**<0.001**
Month (Apr)	3961	0.248	1.282	1.204	1.365	**<0.001**
Month (May)	2687	0.065	1.0673	1.005	1.134	**<0.05**
Month (Jun)	655	0.193	1.213	1.103	1.333	**<0.001**
Sex (female)	7999	–	–	–	–	–
Sex (male)	25328	0.303	1.353	1.318	1.390	**<0.001**
Bat (0–100 m)	21437	–	–	–	–	–
Bat (100–200 m)	8821	0.242	1.274	1.235	1.314	**<0.001**
Bat (<200 m)	3069	0.531	1.701	1.627	1.780	**<0.001**
Dist (0–10 km)	20782	–	–	–	–	–
Dist (10–50 km)	12099	0.457	1.580	1.278	1.953	**<0.001**
Dist(>50km)	446	−0.430	0.650	0.309	1.368	0.26
Dist (0–10 km):Sep		–	–	–	–	–
Dist (10–50 km):Oct		−0.261	0.770	0.529	1.123	0.18
Dist (10–50 km):Nov		−0.033	0.968	0.745	1.256	0.80
Dist (10–50 km):Dec		−0.123	0.884	0.702	1.115	0.30
Dist (10–50 km):Jan		−0.345	0.708	0.569	0.882	**<0.001**
Dist (>50 km):Jan		0.528	1.696	0.793	3.628	0.17
Dist (10–50 km):Feb		−0.454	0.634	0.509	0.791	**<0.001**
Dist (>50 km):Feb		0.761	2.141	1.007	4.552	0.05
Dist (10–50 km):Mar		−0.485	0.615	0.494	0.766	**<0.001**
Dist (>50 km):Ma		0.078	1.081	0.362	3.226	0.89
Dist (10–50 km):Apr		−0.625	0.535	0.429	0.667	**<0.001**
Dist (10–50 km):May		−0.603	0.547	0.436	0.686	**<0.001**
Dist (10–50 km):Jun		−0.239	0.787	0.594	1.042	0.09

(by up to 16% compared to nearly ice-free waters; Table 5, Fig. S3 in the Supplement). Bathymetry was an influential parameter, with the risk of leaving increasing with depth (71% higher for water depths of >200 m compared to water depths between 0 and 100 m). The seals also appeared to prefer particular benthic habitats; CPH showed a decreased risk of leaving areas associated with steeper slopes.

Seasonal differences in the temporal and spatial use of available habitat were also displayed by the seals. For example, months from October through to January were associated with a lower risk of leaving relative to September, while the tendency to leave an area increased from February through June. The influence of distance to the coast was complex to interpret due to its interaction with month, but generally the risk of leaving an area increased with distance to the coast (increased risk of leaving by 58% for distances 10 to 50 km) except in January, February, March, April and May when drift ice is more likely to be present close to shore (Fig. S3). Fig. 5 shows the predicted risk of leaving based on water depth, distance to coast, slope and sea ice for males in September in the absence of sea ice. Female patterns were similar (not shown). The variables used to produce the predictions were bound to the range of values in the original dataset to avoid extrapolation beyond the span of the model. The seals had lower risks of leaving an area that was shallow and close to the coast; deep water is clearly avoided by most seals (Fig. 5).

DISCUSSION

This study is one of few that have explored harbour seal behaviour at the edge of their distributional range in seasonally ice-covered waters (Gjertz et al. 2001, Lesage et al. 2004, Bajzak et al. 2013). It is by far the most extensive in terms of number of animals instrumented (30 ind. over 2 yr) and average durations of data streams (200 d), which spanned from the immediate post-moulting period to late spring of the following year, taking them through most of the ice-covered period in the Svalbard region. Previous studies on harbour seals movement patterns indicate that this species is relatively sedentary throughout the year, although seasonal shifts between favoured sites is common and in some areas the seals make offshore trips in the vicinity of the main haul-out sites to distances between 20 and 30 km (Thompson & Miller 1990, Thompson et al. 1991, Thompson et al. 1998b, Vincent et al. 2010, Sharples et al. 2012). However, long-distance movements have also been recorded in some areas, linked mainly to seasonal migrations to avoid ice or dispersal by immature animals (Lesage et al. 2004, Peterson et al. 2012, Sharples et al. 2012,

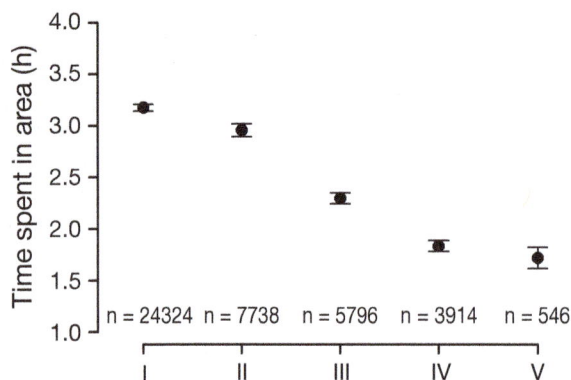

Fig. 4. *Phoca vitulina*. Mean ±SE of time spent in an area (2.5 × 2.5 km), by 30 adult or juvenile harbour seals instrumented with CTD satellite relay data loggers (CTD-SRDLs) in Svalbard, Norway in 2009 and 2010 as a function of sea ice concentration for 2009 and 2010. The ice concentration categories correspond to I = open water, II = very open drift ice, III = open drift ice, IV = close drift ice, V = very close drift ice. The number of 2.5 × 2.5 km cells considered in each category is indicated

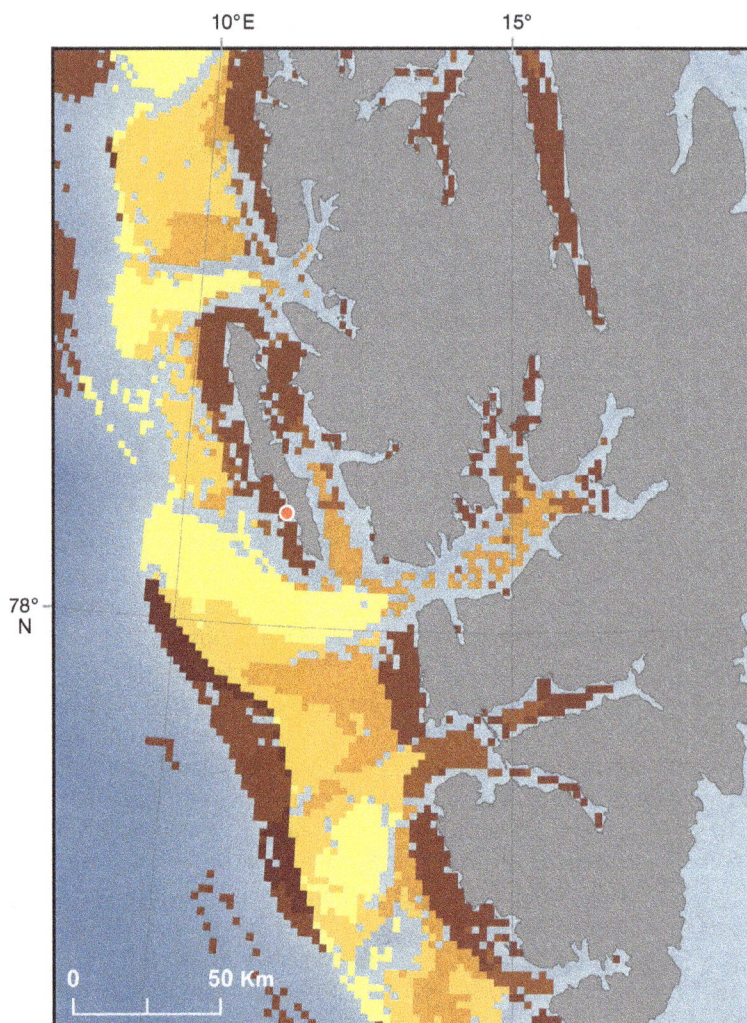

Bajzak et al. 2013, Womble & Gende 2013). In the present study, all of the animals stayed within 100 km of the coast of Spitsbergen throughout the study period, though some few individuals dispersed away from the tagging location at PKF, travelling up to ~230 km. PKF is the main haul-out area and the only recorded pupping site on Svalbard (Lydersen & Kovacs 2010). Gjertz et al. (2001) suggested that harbour seals that disperse to other areas of Svalbard likely return to PKF for pupping, mating and possibly moulting. Identified terrestrial haul-out sites are certainly not limited to PKF; animals haul out along much of the west coast of Spitsbergen (Hamilton et al. 2014), demonstrating that at least some individuals are not resident around PKF throughout the year. The large individual variations observed in movement metrics in this study suggest the existence of individual strategies, with part of the population staying close to PKF while other individuals disperse much more broadly, though members of this population remain tied to the west coast of Spitsbergen. The reason for their residency in this area is likely a combination of several features including water depth and water temperature as well as biotic factors such as predation risk levels and availability of suitable prey.

Bathymetry was an influential parameter in the habitat selection model (Table 5) in this study; individuals showed a marked preference for water depths shallower than 100 m. The seals also clearly avoided deep channels located in front of fjord mouths and did not go into deep waters off the shelf either (Figs. 1 & 5). These results are consistent with previous observations from Svalbard (Gjertz et al. 2001, Krafft et al. 2002) where all dives were shallower than 250 m

Fig. 5. *Phoca vitulina*. A composite grid showing predicted habitat use intensities for harbour seals instrumented with CTD satellite relay data loggers (CTD-SRDLs) in Svalbard, Norway in 2009 and 2010, displaying the estimated risk of leaving a 2.5 × 2.5 km area. Predictions were based on bathymetry, slope and ice concentration extracted at the centre of each grid cell for males in September. The variables used to produce the predictions were bound to the range of values in the data to avoid extrapolating beyond the span of the model. Warmer colours indicate lower risks of leaving an area (i.e. preferred areas). Red dot indicates tagging location

and half of the diving performed was shallower than 40 m. Edges along steep slopes close to the coast were favoured by the seals in this study, with a decreased risk of leaving an area of 5% for each additional degree of steepness (Table 5). However, this finding might be heavily influenced by a few individuals that spent large amounts of time foraging along the edge of the continental slope in slightly deeper waters (Fig. 5).

Areas within 50 km of the coast were more attractive to the seals regardless of the time of the year. Shallow coastal waters and Arctic shelves are generally areas of high productivity where nutrients are easily mixed in the water column (Reigstad et al. 2011). This is particularly true for the west coast of Spitsbergen where cold, relatively fresh Arctic water meets and mixes at the Polar Front with warmer, more saline Atlantic water carried northward by the West Spitsbergen Current (Boyd & D'Asaro 1994, Eriksen et al. 2012). This mixing is enhanced by steep bathymetric features like the continental shelf break, which is located 35 km west of PKF, and the strong surface winds that occur on the west coast of Spitsbergen, rendering this area very productive (Cottier et al. 2007, Ingvaldsen & Loeng 2009). Indeed, primary productivity derived from model simulations show that the highest values are found tied with regions dominated by inflows of Atlantic waters in the western and southeastern parts of Spitsbergen. However, the latter areas in the southeast are found at large distances from the coast (Reigstad et al. 2011). This explain why none of the harbour seals in this study went to the east coast; this resource-rich area is likely too far from the coast to be profitable (Fig. 1), despite the absence of sea ice during the summer and bathymetric and slope conditions similar to the areas in the west that they do use.

Polar bear *Ursus maritimus* densities on the west coast are lower than in the north or east of Svalbard, which likely reinforces the harbour seals' preference for the west coast of Spitsbergen (Mauritzen et al. 2001, Mauritzen et al. 2002, Freitas et al. 2012). Interestingly, seals seem to avoid entering nearby deep fjord systems on the west coast even though these areas have features identified as favourable in our model. This might be linked to the fact that the fjords in this region are often dominated by Arctic water, with warm Atlantic water entering the fjord systems only under unusual wind and current conditions (Berge et al. 2005). Additionally, ringed seals *Pusa hispida*—which overlap in terms of diet with harbour seals—occupy the fjords (Lydersen 1998, Freitas et al. 2008c), and competitive exclusion might occur.

Whether all of the animals in this study returned to PKF for the breeding season following their tracking period is uncertain because many of the data records terminated too early, but the few animals (mainly males) still transmitting data in June were all close to the island. Studies of harbour seals in Alaska have shown that although some individuals undertake extensive migratory movements (up to 900 km away from pupping sites), there is a substantial degree of site fidelity, with 93% of the females returning to their original tagging site for the following breeding season (Womble & Gende 2013).

Svalbard harbour seals increased their daily travel distance, trip duration and HR size towards the winter months. Changes in distribution in other harbour seal populations have been linked to changes in prey availability (Thompson et al. 1991, 1996), which might also be the case here. Gadoid species (the cod family) were the most abundant prey consumed by harbour seals on Svalbard in the study period (Colominas 2012) with Atlantic cod *Gadus morhua* and polar cod *Boreogadus saida* being particularly important in the diet. The polar cod is a dominant fish species on Arctic shelves and is tightly linked to Arctic water masses, while Atlantic cod is associated with Atlantic water in cold, north temperate areas (Renaud et al. 2012, Hop & Gjøsæter 2013). The west coast of Spitsbergen is strongly influenced by both of these water masses and their relative dominance changes both seasonally and inter-annually in this region (Willis et al. 2008). In such an environment, seasonal dietary changes would be expected and have previously been suggested for harbour seals in the Svalbard population (Gjertz et al. 2001, Andersen et al. 2004). In recent years, increased influxes of warm Atlantic water have taken place on the west coast of Svalbard and a northward expansion of Atlantic cod has been observed (Renaud et al. 2012, Hop & Gjøsæter 2013).

Other environmental features in Svalbard that likely limit harbour seal distribution are strongly seasonal. Hamilton et al. (2014) have shown that harbour seals from Svalbard spend less time hauled out (and consequently more time in the water) during the winter compared to summer and suggested that it could be a strategy to avoid thermal stress when hauling out on ice in winter when air temperatures can drop below −30°C. However, the potential thermal challenges imposed by low winter temperatures are unlikely to be the only reason for the increased aquatic behaviour, since increased time spent in the water and increasing distances travelled per day occurred concomitantly. On the west coast of Spitsbergen the winter months are usually associated with the arrival of drifting pack ice along the outer shores and also some land-fast sea ice in the fjords (Vinje 2009). A strong coastal Arctic current brings drift ice from around the southern tip of the Spitsbergen northward (Vinje 2009) along the west coast; the distribution of this drifting ice is then determined by wind direction and the action of other currents. High sea ice concentrations and land-fast sea ice are limiting factors for harbour seals because they cannot maintain breathing holes like ringed seals do and rapidly changing ice conditions might force the seals to travel more to avoid being caught in the ice (Mansfield 1967, Lesage et al. 2004, Bajzak et al. 2013). This general pattern is supported by the CPH habitat use model in this study that shows that high ice concentrations increased the risk of leaving an area. There was also a positive relationship between sea ice extent and HR size. However, the seals did use moderately ice-covered areas and they travelled through ice concentrations up to 55%. Sea ice extent was more important in the second year of this study, when ice cover was heaviest and HR sizes were also larger during the second winter.

Ice cover might also have some positive aspects for harbour seals. In the St Lawrence River Estuary in Canada, the presence of ice cover is known to concentrate food resources in areas of open water and along ice edges, increasing the ease of access to prey for this seal species (Lesage et al. 2004). This could also be the case in Svalbard, at least during periods when there is sufficient light to trigger phytoplankton blooms (Engelsen et al. 2002, Hop & Gjøsæter 2013). It is however difficult to test this hypothesis given the low resolution of ice maps for the region, which do not permit fine-scale spatial analyses for precisely where the seals were within ice-covered areas.

Age, size and sex have also been shown to influence movement patterns of harbour seals (Thompson et al. 1998b, Peterson et al. 2012). Thompson et al. (1998b) reported that sex and body size were correlated with trip duration in harbour seals in Moray Firth (Scotland) and that males had larger foraging ranges than females. The present study is in agreement with these observations; males generally had larger HRs than females, travelled greater distances per day and had a higher risk of leaving areas, suggesting that they were generally less sedentary. Possible reasons for this gender difference in at-sea movements are different reproduction costs for the 2 sexes and different levels of intra-specific competition. No age effects were found for any of the movement metrics, which suggests that body size differences were not responsible for the observed differences in at-sea movements, leaving a more intrinsic sex difference as the likely explanation. Males and females might have somewhat different diets for parts of the year, but no sex-related dietary information exists for Svalbard's harbour seals to explore this suggestion.

Climate change has already had major influences on air and water temperatures in Svalbard and dramatically changed the sea-ice cover in this region (Pavlov et al. 2013, Nordli et al. 2014). Forward-looking climate scenarios predict that the Barents Sea ecosystem will continue to experience rapid unidirectional change that will include higher sea surface temperature, increased frequency of strong winds, decreases in sea-ice cover and extension of the ice-free period (see Vinje 2009 and Kovacs et al. 2011 for reviews). This physical environmental change is expected to cause a shift in species distributions and abundance and to affect all trophic levels within the Arctic ecosystem (e.g. Gilg et al. 2012). Already, exceptionally warm years have occurred and some boreal fish species have been recorded at the northernmost edges of their distribution range (Wienerroither et al. 2011). With the predicted increase of Atlantic water inflow through the Fram Strait this situation is likely to become more common in the future, affecting the type and distribution of prey for the harbour seals in this region. The recent increases in water temperature and reduction in sea ice seen in Svalbard are predicted to intensify in the future, which are likely to make conditions in the archipelago more favourable for harbour seals by increasing the available habitat.

In summary, we present the first year-round observations of at-sea movements of juvenile and adult harbour seals from the world's northernmost population, residing in the high-Arctic archipelago of Svalbard. Tracking data provides a solid baseline for understanding how individuals from this population use their environment currently and provide a basis for predicting how they might react to environmental change in the future (New et al. 2014). It revealed that animals show a strong preference for the west coast of Spitsbergen through all seasons. None of the animals travelled to the east side of Spitsbergen, suggesting that they do not find sufficiently favourable conditions there, despite the absence of sea ice during part of the year and the presence of available terrestrial haul-out platforms. The influx of warm saline Atlantic water that occurs close to the coast on the west side of the islands is likely essential for maintaining a high productivity area close to the coast via mixing of water masses and shelf upwelling, along with tolerable amounts of sea ice in the winter. The recent increases in water temperature and reduction in sea ice seen in Svalbard are predicted to intensify in the future, thus making conditions in the archipelago more favourable for harbour seals.

Acknowledgements. We thank Lisa LeClerc, Benjamin Merkel, Morten Tryland and Bjørn Waalberg for their assistance in the field.

LITERATURE CITED

Andersen SM, Lydersen C, Grahl-Nielsen O, Kovacs KM (2004) Autumn diet of harbour seals (*Phoca vitulina*) at Prins Karls Forland, Svalbard, assessed via scat and fatty-acid analyses. Can J Zool 82:1230–1245

Andersen LW, Lydersen C, Frie AK, Rosing-Asvid A, Hauksson E, Kovacs KM (2011) A population on the edge: genetic diversity and population structure of the world's northernmost harbour seals (*Phoca vitulina*). Biol J Linn Soc 102:420–439

Bajzak C, Bernhardt W, Mosnier M, Hammill M, Stirling I (2013) Habitat use by harbour seals (*Phoca vitulina*) in a seasonally ice-covered region, the western Hudson Bay. Polar Biol 36:477–491

Berge J, Johnsen G, Nilsen F, Gulliksen B, Slagstad D (2005) Ocean temperature oscillations enable reappearance of blue mussels *Mytilus edulis* in Svalbard after a 1000 year absence. Mar Ecol Prog Ser 303:167–175

Bigg MA (1969) The harbour seal in British Columbia. Bull Fish Res Board Can 172:1–33

Bjørge A, Thompson D, Hammond P, Fedak M and others (1995) Habitat use and diving behaviour of harbour seals in a coastal archipelago in Norway. In: Blix AS, Walløe L, Ulltang Ø (eds) Whales, seals, fish and man. Proc Int Symp Biol Mar Mamm NE Atl, Tromsø, Norway, 29 Nov to 1 Dec 1994. Elsevier, Amsterdam, p 211–223

Blundell GM, Womble JN, Pendleton GW, Karpovich SA, Gende SM, Herreman JK (2011) Use of glacial and terrestrial habitats by harbor seals in Glacier Bay, Alaska: costs and benefits. Mar Ecol Prog Ser 429:277–290

Boehme L, Lovell P, Biuw M, Roquet F and others (2009) Animal-bourne CTD-Satellite relay data loggers for real-time oceanographic data collection. Ocean Sci 5:685–695

Boyd TJ, D'Asaro EA (1994) Cooling of the west Spitsbergen current: wintertime observations west of Svalbard. J Geophys Res 99(C11):22597–22618

Burnham KP, Anderson DR (eds) (2002) Model selection and multimodel inference: a practical information-theoretic approach. Springer, New York, NY

Calenge C (2011) Home range estimation in R: the adehabitat HR package. http://cran.at.r-project.org/web/packages/adehabitatHR/vignettes/adehabitatHR.pdf

Chavez-Rosales S, Gardner SC (1999) Recent harbour seal (*Phoca vitulina richardsi*) pup sightings in Magdalena Bay, Baja California Sur, Mexico. Aquat Mamm 25:169–171

Collett D (ed) (2003) Modelling survival data in medical research, 2nd edn. Chapman & Hall, CRC, Boca Raton, FL

Colominas R (2012) Harbour seal diet in a changing Arctic (Svalbard, Norway). MSc thesis, University of Bergen

Cottier FR, Nilsen F, Inall ME, Gerland S, Tverberg V, Svendsen H (2007) Wintertime warming of an Arctic shelf in response to large-scale atmospheric circulation. Geophys Res Lett 34:L10607, doi:10.1029/2007GL029948

Cox DR (1972) Regression models and life tables (with discussion). J R Stat Soc B 34:187–120

Engelsen O, Hegseth EN, Hoop H, Hansen E, Falk-Petersen S (2002) Spatial variability of chlorophyll-*a* in the Marginal Ice Zone of the Barents Sea with relations to sea ice and oceanographic conditions. J Mar Syst 35:79–97

Eriksen E, Ingvaldsen R, Stiansen JE, Johansen GO (2012) Thermal habitat for 0-group fish in the Barents Sea; how climate variability impacts their density, length, and geographic distribution. ICES J Mar Sci 69:870–879

Freitas C (2013) Package 'Argosfilter'. http://cran.at.r-project.org/web/packages/argosfilter/argosfilter.pdf

Freitas C, Kovacs KM, Ims RA, Lydersen C (2008a) Predicting habitat use by ringed seals (*Phoca hispida*) in a warming Arctic. Ecol Modell 217:19–32

Freitas C, Kovacs KM, Lydersen C, Ims RA (2008b) A novel method for quantifying habitat selection and predicting space use. J Appl Ecol 45:1213–1220

Freitas C, Lydersen C, Fedak MA, Kovacs KM (2008c) A simple new algorithm to filter marine mammal Argos locations. Mar Mamm Sci 24:315–325

Freitas C, Kovacs KM, Andersen M, Aars J and others (2012) Importance of fast ice and glacier fronts for female polar bears and their cubs during spring in Svalbard, Norway. Mar Ecol Prog Ser 447:289–304

Frost KJ, Lowry LF, Ver Hoef JM (1999) Monitoring the trend of harbor seals in Prince William Sound, Alaska, after the Exxon Valdez oil spill. Mar Mamm Sci 15:494–506

Gilg O, Kovacs KM, Aars J, Fort J and others (2012) Climate change and the ecology and evolution of Arctic vertebrates. Ann N Y Acad Sci 1249:166–190

Gjertz I, Børset A (1992) Pupping in the most northerly harbour seals (*Phoca vitulina*). Mar Mamm Sci 8:103–109

Gjertz I, Lydersen C, Wiig Ø (2001) Distribution and diving of harbour seals (*Phoca vitulina*) in Svalbard. Polar Biol 24:209–214

Hamilton CD, Lydersen C, Ims RA, Kovacs KM (2014) Haulout behaviour of the world's northermost population of harbour seals (*Phoca vitulina*) throughout the year. PLoS ONE 9:e86055

Hanson N, Thompson D, Duck C, Moss S, Lonergan M (2013) Pup mortality in a rapidly declining harbour seal (*Phoca vitulina*) population. PLoS ONE 8:e80727

Hassani S, Dupuis L, Elder JF, Caillot E and others (2010) A note on harbour seal (*Phoca vitulina*) distribution and abundance in France and Belgium. NAMMCO Sci Publ 8:107–116

Hauksson E (1992) Counting of common seals (*Phoca vitulina* L.) and grey seals (*Halichoerus grypus* Fabr.) in 1980–1990, and the state of the seal population at the coast of Iceland. Hafrannsoknir 43:5–22 (in Icelandic with a summary in English)

Hijmans RJ, van Etten J, Mattiuzzi M, Summer M and others (2014) Raster: geographic data analysis and modeling. http://cran.at.r-project.org/web/packages/raster/raster.pdf

Hop H, Gjøsæter H (2013) Polar cod (*Boreogadus saida*) and capelin (*Mallotus villosus*) as key species in marine food webs of the Arctic and the Barents Sea. Mar Biol Res 9:878–894

▶ Horne JS, Garton EO, Krone SM, Lewis JS (2007) Analysing animal movements using brownian bridges. Ecology 88: 2354–2363

Hothorn T, Bretz F, Westfall P, Heiberger R, Schuetzenmeister A (2014) multcomp: simultaneous inference in general parametric models. http://cran.r-project.org/web/packages/multcomp/index.html

Ingvaldsen R, Loeng H (2009) Physical oceanography. In: Sakshaug E, Johnsen G, Kovacs K (eds) Ecosystem Barents Sea. Tapir Academic Press, Trondheim, p 33–64

▶ Jakobsson M, Mayer LA, Coakley B, Dowdeswell JA and others (2012) The international bathymetric chart of the Arctic Ocean (IBCAO) v. 3.0. Geophys Res Lett 39: L12609, doi:10.1029/2012GL052219

▶ Jeffries S, Huber H, Calambokidis J, Laake J (2003) Trends and status of harbor seals in Washington State: 1978-1999. J Wildl Manag 67:207–218

Johnson DS (2013) Crawl: fit continuous-time correlated random walk models to animal movement data. http://cran.r-project.org/web/packages/crawl/index.html

▶ Kovacs KM, Lydersen C, Overland JE, Moore SE (2011) Impacts of changing sea-ice conditions on Arctic marine mammals. Mar Biodivers 41:181–194

Krafft BA, Lydersen C, Gjertz I, Kovacs KM (2002) Diving behaviour of sub-adult habour seals (*Phoca vitulina*) at Prins Karls Forland, Svalbard. Polar Biol 25:230–234

▶ Lesage V, Hammill MO, Kovacs KM (2004) Long-distance movements of harbour seals (*Phoca vitulina*) from a seasonally ice-covered area, the St. Lawrence River estuary, Canada. Can J Zool 82:1070–1081

▶ Lonergan M, Duck CD, Thompson D, Mackey BL, Cunningham L, Boyd IL (2007) Using sparse survey data to investigate the declining abundance of British harbour seals. J Zool 271:261–269

▶ Lonergan M, Hall A, Thompson H, Thompson PM, Pomeroy P, Harwood J (2010) Comparison of the 1988 and 2002 phocine distemper epizootics in British harbour seal *Phoca vitulina* populations. Dis Aquat Org 88:183–188

▶ Lonergan M, Duck C, Moss S, Morris C, Thompson D (2013) Rescaling of aerial survey data with information from small numbers of telemetry tags to estimate the size of a declining harbour seal population. Aquat Conserv 23: 135–144

Lucas ZN, Natanson LJ (2010) Two shark species involved in predation on seals at Sable Island, Nova Scotia, Canada. Proc Nova Scotia Inst Sci 45:64–88

▶ Lucas ZN, Stobo WT (2000) Shark-inflicted mortality on a population of harbour seals (*Phoca vitulina*) at Sable Island, Nova Scotia. J Zool 252:405–414

Lydersen C (1998) Status and biology of ringed seals (*Phoca hispida*) in Svalbard. NAMMCO Sci Publ 1:46–62

▶ Lydersen C, Kovacs KM (2005) Growth and population parameters of the world's northernmost harbour seals *Phoca vitulina* residing in Svalbard, Norway. Polar Biol 28:156–163

Lydersen C, Kovacs KM (2010) Status and biology of harbour seals (*Phoca vitulina*) in Svalbard. NAMMCO Sci Publ 8:47–60

Mansfield AW (1967) Distribution of the harbor seal, *Phoca vitulina* Linnaeus, in Canadian Arctic waters. J Mamm 48:249–257

▶ Matthiopoulos J, Cordes L, Mackey B, Thompson D and others (2014) State-space modelling reveals proximate causes of harbour seal population declines. Oecologia 174:151–162

▶ Mauritzen M, Derocher AE, Wiig Ø (2001) Space-use strategies of female polar bears in a dynamic sea ice habitat. Can J Zool 79:1704–1713

▶ Mauritzen M, Derocher AE, Wiig Ø, Belikov SE, Boltunov AN, Hansen E, Garner GW (2002) Using satellite telemetry to define spatial population structure in polar bears in Norwegian and western Russian Arctic. J Appl Ecol 39:79–90

▶ Merkel B, Lydersen C, Yoccoz NG, Kovacs KM (2013) The world's northernmost harbour seals population—how many are they? PLoS ONE 8:e67576

▶ New LF, Clark JS, Costa DP, Fleishman E and others (2014) Using short-term measures of behaviour to estimate long-term fitness of southern elephant seals. Mar Ecol Prog Ser 496:99–108

Nilssen KT, Skavberg NE, Poltermann M, Haug T, Härkönen T, Henriksen G (2010) Status of harbour seals (*Phoca vitulina*) in mainland Norway. NAMMCO Sci Publ 8: 61–69

▶ Nordli Ø, Przybylak R, Ogilvie AEJ, Isaksen K (2014) Long-term temperature trends and variability on Spitsbergen: the extended Svalbard Airport temperature series, 1898-2012. Polar Res 33:21349

Olsen MT, Andersen SM, Teilmann J, Dietz R, Endren SMC, Linnet A, Härkönen T (2010) Status of harbour seals (*Phoca vitulina*) in the southern Scandinavia. NAMMCO Sci Publ 8:77–94

▶ Patterson TA, McConnell BJ, Fedak MA, Bravington MV, Hindell MA (2010) Using GPS data to evaluate the accuracy of state-space methods for correction of Argos satellite telemetry errors. Ecology 91:273–285

▶ Pavlov AK, Tverberg V, Ivanov BV, Nilsen F, Falk-Petersen S, Granskog MA (2013) Warming of Atlantic water in two west Spitsbergen fjords over the last century (1912-2009). Polar Res 32:11206

▶ Peterson SH, Lance MM, Jeffries SJ, Acevedo-Gutierrez A (2012) Long distance movements and disjunct spatial use of harbor seals (*Phoca vitulina*) in the inland waters of the Pacific northwest. PLoS ONE 7:e39046

Pinheiro J, Bates D, DebRoy S, Sarkar D (2014) nlme: Linear and nonlinear mixed effects models. http://cran.us.r-project.org/web/packages/nlme/nlme.pdf

R Development Core Team (2012) R: a language and environment for statistical computing. R Foundation for Statistical Computing, Vienna

▶ Reder S, Lydersen C, Arnold W, Kovacs KM (2003) Haulout behaviour of High Artcic harbour seals (*Phoca vitulina vitulina*) in Svalbard, Norway. Polar Biol 27:6–16

▶ Reigstad M, Carroll J, Slagstad D, Ellingsen I, Wassmann P (2011) Intra-regional comparison of productivity, carbon flux and ecosystem composition within the northern Barents Sea. Prog Oceanogr 90:33–46

Reijnders PJH, Brasseur SMJM, Tougaard S, Siebert U, Borchardt T, Stede M (2010) Population development and status of harbour seals (*Phoca vitulina*) in the Wadden Sea. NAMMCO Sci Publ 8:95–105

▶ Renaud PE, Berge J, Varpe Ø, Lønne OJ, Nahrgang J, Ottesen C, Hallanger I (2012) Is the poleward expansion by Atlantic cod and haddock threatening native polar cod, *Boreogadus saida*? Polar Biol 35:401–412

Rosing-Asvid A (2010) Catch history and status of the harbour seal (*Phoca vitulina*) in Greenland. NAMMCO Sci Publ 8:161–174

▶ Sharples RJ, Moss SE, Patterson TA, Hammond PS (2012) Spatial variation in foraging behaviour of a marine top predator (*Phoca vitulina*) determined by a large-scale satellite tagging program. PLoS ONE 7:e37216

Small RJ, Boveng PL, Byrd GV, Withrow DE (2008) Harbor seal population decline in the Aleutian Archipelago. Mar Mamm Sci 24:845–863

Sumner MD, Luque S (2013) Spatial analysis of animal track data in R: the trip package. http://cran.r-project.org/web/packages/trip/trip.pdf

Therneau TM (2012) A package for survival analysis in S. R package v. 2.36-14. http://cran.r-project.org/web/packages/survival/survival.pdf

Thompson D, Härkönen T (2008) *Phoca vitulina* IUCN Red List of Threatened Species v.20132

Thompson PM, Miller D (1990) Summer foraging activity and movements of radio-tagged common seals (*Phoca vitulina* L.) in the Moray Firth, Scotland. J Appl Ecol 27:492–501

Thompson PM, Fedak MA, McConnell BJ, Nicholas KS (1989) Seasonal and sex related variation in the activity patterns of common seals (*Phoca vitulina*). J Appl Ecol 26:521–535

Thompson PM, Pierce GJ, Hislop JRG, Miller D, Diack JSW (1991) Winter foraging by common seals (*Phoca vitulina*) in relation to food availability in the Inner Moray Firth, N. E. Scotland. J Anim Ecol 60:283–294

Thompson PM, McConnell BJ, Tollit DJ, MacKay A, Hunter C, Racey PA (1996) Comparative distribution, movements and diet of harbour and grey seals from the Morey Firth, N. E. Scotland. J Appl Ecol 33:1572–1584

Thompson D, Duck CD, McConnell BJ, Garrett J (1998a) Foraging behaviour and diet of lactating female southern sea lions (*Otaria flavescens*) in the Falkland Islands. J Zool (Lond) 246:135–146

Thompson PM, Mackay A, Tollit DJ, Enderby S, Hammond PS (1998b) The influence of body size and sex on the characteristics of harbour seal foraging trips. Can J Zool 76:1044–1053

Thompson D, Lonergan M, Duck C (2005) Population dynamics of harbour seals *Phoca vitulina* in England: monitoring growth and catastrophic declines. J Appl Ecol 42:638–648

Trites AW, Deecke VB, Gregr EJ, Ford JKB, Olesiuk PF (2007) Killer whales, whales, and sequential megafaunal collapse in the North Pacific: a comparative analysis of the dynamics of marine mammals in Alaska and British Columbia following commercial whaling. Mar Mamm Sci 23:751–765

Vincent C, McConnell BJ, Delayat S, Elder JF, Gautier G, Ridoux V (2010) Winter habitat use by harbour seals (*Phoca vitulina*) fitted with Fastloc GPS/GSM tags in two tidal bays in France. NAMMCO Sci Publ 8:285–302

Vinje T (2009) Sea-ice. In: Sakshaug E, Johnsen G, Kovacs K (eds) Ecosystem Barents Sea. Tapir Academic Press, Trondheim, p 65–82

Wienerroither R, Johannesen E, Dolgov A, Byrkjedal I and others (2011) Atlas of the Barents Sea fishes. IMR/PINRO Joint Report Series 1-2011

Willis KJ, Cottier FR, Kwasnieswski S (2008) Impact of warm water advection on the winter zooplankton community in an Arctic fjord. Polar Biol 31:475–481

Womble JN, Gende SM (2013) Post-breeding season migrations of a top predator, the harbor seal (*Phoca vitulina richardii*), from a marine protected area in Alaska. PLoS ONE 8:e55386

Womble JN, Pendleton GW, Mathews EA, Blundell GM, Bool NM, Gende SM (2010) Harbor seal (*Phoca vitulina richardii*) decline continues in the rapidly changing landscape of Glacier Bay National Park, Alaska 1992–2008. Mar Mamm Sci 26:686–697

Worton BJ (1989) Kernel methods for estimating the utilization distribution in home-range studies. Ecology 70:164–168

6

Effects of ocean acidification on the larvae of a high-value pelagic fisheries species, mahi-mahi *Coryphaena hippurus*

Sean Bignami[1,4,*], Su Sponaugle[1,2,3], Robert K. Cowen[1,3]

[1]Division of Marine Biology and Fisheries, Rosenstiel School of Marine and Atmospheric Science, University of Miami, Miami, FL 33149, USA

[2]Department of Integrative Biology, Oregon State University, Corvallis, OR 97331, USA

[3]Hatfield Marine Science Center, Oregon State University, Newport, OR 97365, USA

[4]*Present address:* Concordia University Irvine, Irvine, CA 92612, USA

ABSTRACT: Negative impacts of CO_2-induced ocean acidification on marine organisms have proven to be variable both among and within taxa. For fishes, inconsistency confounds our ability to draw conclusions that apply across taxonomic groups and highlights the limitations of a nascent field with a narrow scope of study species. Here, we present data from a series of 3 experiments on the larvae of mahi-mahi *Coryphaena hippurus*, a large pelagic tropical fish species of high economic value. Mahi-mahi larvae were raised for up to 21 d under either ambient seawater conditions (350 to 490 µatm pCO_2) or projected scenarios of ocean acidification (770 to 2170 µatm pCO_2). Evaluation of hatch rate, larval size, development, swimming activity, swimming ability (U_{crit}), and otolith (ear stone) formation produced few significant effects. However, larvae unexpectedly exhibited significantly larger size-at-age and faster developmental rate during 1 out of 3 experiments, possibly driven by metabolic compensation to elevated pCO_2 via a corresponding decrease in routine swimming velocity. Furthermore, larvae had significantly larger otoliths at 2170 µatm pCO_2, and a similar but non-significant trend also occurred at 1200 µatm pCO_2, suggesting potential implications for hearing sensitivity. The lack of effect on most variables measured in this study provides an optimistic indication that this large tropical species, which inhabits the offshore pelagic environment, may not be overly susceptible to ocean acidification. However, the presence of some treatment effects on growth, swimming activity, and otolith formation suggests the presence of subtle, but possibly widespread, effects of acidification on larval mahi-mahi, the cumulative consequences of which are still unknown.

KEY WORDS: Ocean acidification · Larval fish · Otolith · Mahi-mahi · CO_2 · U_{crit} · Behavior

INTRODUCTION

A diversity of marine organisms are negatively affected by CO_2-induced ocean acidification, yet the occurrence and severity of such effects are highly variable among taxa and life history stages (Kroeker et al. 2013). While much research has focused on calcifying marine invertebrates, until recently there has been relatively little consideration of acidifica-tion impacts on fishes. This is due, in part, to the fact that adult fishes and other organisms with high metabolic rates are relatively efficient at regulating their internal acid–base balance in response to metabolic or environmental disturbances, and thus are considered to be more capable of resisting the effects of ocean acidification (Melzner et al. 2009a). Unfortunately, there are few comparable data on the ability of early life history stages of fishes to

*Corresponding author: sean.bignami@cui.edu

cope with increased environmental partial pressure of carbon dioxide (pCO_2). Embryonic and larval fishes are not equipped with the same physiological mechanisms as juveniles and adults, but are capable of some internal pH regulation (Brauner 2008). Nonetheless, existing literature indicates that early life stages of fishes are more susceptible to elevated pCO_2 relative to adults (reviewed by Pörtner et al. 2005). As the primary period of dispersal for many marine fishes, the larval stage is critical to population replenishment and connectivity (Cowen & Sponaugle 2009); therefore, our understanding of the broader population- and ecosystem-level impacts of ocean acidification requires consideration of this life stage.

The impact of ocean acidification on larval fishes is highly variable among studies and taxa. The range of effects include reduced growth and survival (Baumann et al. 2011), skeletal deformation (Pimentel et al. 2014a), altered neurological function (Nilsson et al. 2012), and disrupted behavior (Munday et al. 2010, Ferrari et al. 2012, Hamilton et al. 2014). Some species have experienced more discrete effects, such as altered otolith (ear stone) development (Checkley et al. 2009, Munday et al. 2011a, Hurst et al. 2012, Bignami et al. 2013a) and impaired tissue health (Frommel et al. 2011). Several other studies report no pCO_2 effects on fish larvae (e.g. Munday et al. 2011b, Frommel et al. 2013). Though intriguing, this inter- and intra-specific variability in response to ocean acidification confounds our ability to formulate general conclusions regarding the fate of fish populations under future environmental scenarios. To better clarify the underlying factors driving patterns of susceptibility or resistance to ocean acidification, it is necessary to examine a diversity of species with distinct differences in life history strategies, physiological capabilities, and habitat use. To date, most studies of ocean acidification impacts on fish larvae have focused on demersal, reef-associated species, with few large pelagic species represented in the literature (except see Checkley et al. 2009, Bignami et al. 2013a,b, Pimentel et al. 2014b).

Fishes with a strictly offshore pelagic life history spawn pelagic eggs, undergo planktonic larval development, and live as juveniles and adults in an offshore environment that lacks the extreme diurnal cycles in pH and temperature observed in shallow coastal environments (Hofmann et al. 2011, Frieder et al. 2012, Price et al. 2012). It has been suggested that pelagic species that have adapted to such a stable environment may be more susceptible to future environmental changes, such as ocean acidification,

compared to demersal species that regularly experience fluctuating environmental conditions (Munday et al. 2008, Pörtner 2008). Prior studies on larval cobia *Rachycentron canadum* and mahi-mahi *Coryphaena hippurus* have shown variable impacts of ocean acidification on these tropical pelagic fishes. In particular, cobia larvae have shown some resistance to direct impacts of ocean acidification on growth, development, swimming activity, and swimming ability (Bignami et al. 2013a), although larvae exhibited altered otolith formation, which has implications for auditory sensation (Bignami et al. 2013b). However, cobia also use nearshore environments, and the evolution of adaptations to withstand such variable environmental conditions may reduce their susceptibility to acidification (Bignami et al. 2013a). Conversely, very young mahi-mahi larvae exhibit metabolic depression and behavioral changes under elevated pCO_2 conditions (Pimentel et al. 2014b). Mahi-mahi, a large tropical pelagic species of high economic value, has an entirely pelagic life history and their response to ocean acidification may be more representative of other strictly pelagic species.

In this study, we examined the response of larval mahi-mahi to predicted levels of ocean acidification. We present data from a series of experiments that investigated a suite of factors that are comparable to those reported for larval cobia and other species, including hatch rate, growth, development, swimming activity, critical swimming speed (U_{crit}), and otolith formation (Munday et al. 2009, Baumann et al. 2011, Bignami et al. 2013a). We hypothesized that mahi-mahi would be more susceptible to acidification than species adapted to more variable natural environments. Thus, we expected larvae exposed to elevated pCO_2 treatments to exhibit reduced growth and development, altered swimming activity, diminished swimming ability, and enhanced otolith growth.

MATERIALS AND METHODS

Study species

Mahi-mahi *Coryphaena hippurus* is a highly migratory, epipelagic, predatory marine fish with a circumglobal distribution in tropical to subtropical waters (Palko et al. 1982, FAO 2013). Spawning of pelagic eggs occurs throughout the year, primarily during warmer months (>24°C, April to November), with planktonic larvae that hatch at ~3.5 mm standard length (SL), undergo flexion at ~7 to 9 mm SL, and develop via a gradual transition into the pelagic juvenile stage (Ditty et al. 1994). Mahi-mahi grow rapidly and are capable of hatching, developing, and reaching reproductive maturity in as little as 4 mo (Beardsley 1967, Oxenford 1999). Maximum size is as large as 2 m and over 30 kg, however, longevity is short compared with other pelagic species, with an average of 2 yr and maximum of 4 to 5 yr (Beardsley 1967, FAO 2013). The species is highly targeted by commercial, artisanal, and recreational fisheries throughout its distribution (Potoschi et al. 1999, Folpp & Lowry 2006, Zúñiga Flores et al. 2008) and is one of the top 7 harvested pelagic species in the western central Atlantic (Oxenford 1999). It has high commercial value, with global fishery landings of over 57 000 tons in 2008 (FAO 2013).

Experimental design and water chemistry

Three experiments were conducted to assess the effects of elevated pCO_2 on mahi-mahi larvae. In each experiment, ambient seawater was utilized as a control (350 to 490 µatm pCO_2, Table 1). One (during Expt 1) or 2 (during Expts 2 and 3) elevated pCO_2 treatments ranging from 770 to 2170 µatm pCO_2 were applied to simulate projected scenarios of ocean acidification over the next 2 centuries (Caldeira & Wickett 2003, 2005, Meehl et al. 2007), as well as present-day conditions in some nearshore marine environments (Feely et al. 2008, Thomsen et al. 2010, Melzner et al. 2013). Elevated pCO_2 treatments between 770 and 1600 µatm correspond to RCP8.5 projections that may be reached within the next 75 to 150 yr, while the 2170 µatm pCO_2 treat-

ment slightly exceeds RCP8.5 projections through year 2300 (Meinshausen et al. 2011).

Seawater carbonate chemistry was manipulated by the addition of equimolar HCl and $NaHCO_3$ to seawater prior to introduction into tanks (Gattuso 2010, Bignami et al. 2013a; see Table 1 for water parameter summary). Tank pH was monitored daily using a handheld pH meter (pH11, Oakton) and Ross Electrode (Orion 9102BWNP, Thermo Scientific) calibrated daily with TRIS buffer. Water samples were collected every 5 d in 250 ml PET bottles, fixed with 100 µl of saturated mercuric chloride, and the total alkalinity (TA) and pH_T were measured using automated Gran titration checked for accuracy with Dickson standards (Scripps Institution of Oceanography; Langdon et al. 2000). CO2SYS was used to solve the carbonate system using the 2 measured parameters (pH_T and TA; Lewis & Wallace 1998). Temperature and dissolved oxygen were measured with a combination meter (550A, YSI) twice and once per day, respectively, and salinity was measured once per day using a refractometer (RHS-10 ATC, Premium Aquatics). During Expts 1 and 3, ambient pCO_2 of seawater was 50 to 200 µatm higher than expected, likely due to respiration in the source seawater (Biscayne Bay, Florida) and within the facility plumbing, but was well below elevated pCO_2 treatment levels. This did not occur during Expt 2. Although temperature was stable within each experiment, there were differences among experiments due to seasonal changes in source seawater and limited heating capability in the rearing system. Due to differences in pCO_2 and temperature between experiments, direct comparison of results between Expts 1, 2, and 3 is not intended.

Larval rearing

All experiments were conducted using adapted methodology for the rearing of larval cobia (Benetti et al. 2008, Bignami et al. 2013a), described briefly here. Mahi-mahi eggs and larvae for each experiment were produced at the University of Miami Experimental Hatchery (UMEH) from a population of 3 wild-caught broodstock (2 females, 1 male). Mahi-mahi eggs were collected within 12 h of evening/nighttime spawning, gently transferred into an aerated 300 l fiberglass tank, and provided with a 1 h

Table 1. Water chemistry conditions during Expts 1, 2, and 3. Temperature, pH, and total alkalinity (TA) were measured and mean pCO_2 calculated with the software CO2SYS (Lewis & Wallace 1998). Values are means ± SE

Treatment level	Temp. (°C)	pH_T	TA (µmol kg^{-1})	pCO_2 (µatm)
Expt 1				
Ambient/control	23.9 ± 0.2	8.00 ± 0.01	2457 ± 22	487 ± 24
1600 µatm pCO_2	23.9 ± 0.2	7.54 ± 0.01	2462 ± 22	1595 ± 85
Expt 2				
Ambient/control	25.3 ± 0.2	8.11 ± 0.005	2387 ± 24	348 ± 17
770 µatm pCO_2	25.3 ± 0.2	7.82 ± 0.01	2384 ± 23	767 ± 52
1460 µatm pCO_2	25.3 ± 0.2	7.56 ± 0.01	2394 ± 23	1461 ± 56
Expt 3				
Ambient/control	28.1 ± 0.2	7.99 ± 0.01	2326 ± 30	454 ± 17
1190 µatm pCO_2	28.1 ± 0.2	7.62 ± 0.01	2320 ± 28	1189 ± 65
2170 µatm pCO_2	28.1 ± 0.2	7.38 ± 0.02	2332 ± 43	2172 ± 159

formalin sterilization treatment (100 ppm) followed by rapid flushing of seawater, and transferred to either an incubator tank or an experimental tank. Experimental tanks received filtered, UV-sterilized seawater at an exchange rate of ~400% d^{-1} and were aerated with a small amount (<1 l min^{-1}) of bubbled air. Tanks were partially submerged in a large water bath to maintain replicate tank temperatures within ±0.2°C of each other, positioned under 95% shade cloth to reduce light intensity, and fitted with translucent white polyethylene lids to prevent CO_2 offgassing, intrusion of rain water, or introduction of contaminants.

During Expt 1, eggs were hatched within ~36 h in a 1000 l incubator tank where they were allowed to develop until 2 days post hatch (dph). Then, larvae were stocked into 6 replicated 400 l flow-through experimental tanks per treatment, at a density of 9 to 10 eggs l^{-1}. During Expts 2 and 3, eggs were stocked directly into 4 replicated experimental tanks per treatment immediately following sterilization treatment, to expose them to elevated pCO_2 conditions during a possible window of susceptibility in early development (Baumann et al. 2011). Acidification treatments were initiated just prior to stocking, reaching full strength within 12 h. In all experiments, feeding began at 2 dph with the addition of enriched rotifers, *Brachionus plicatilis*, 4 or 5 times d^{-1}, providing total densities of 3 to 8 rotifers ml^{-1} d^{-1} for larvae 2 dph to 10 or 12 dph (depending on larval size), gradually increasing with age (Benetti et al. 2008). Beginning at 5 to 8 dph (depending on larval size), larvae were also provided enriched *Artemia* nauplii at total densities of 0.25 to 1.5 ml^{-1} d^{-1}, gradually increasing with age (Benetti et al. 2008). Greenwater rearing techniques were utilized to provide supplemental nutrients and increase tank turbidity, which has been shown to improve larval feeding and survival in other species (Naas et al. 1992, Faulk & Holt 2005). Between 1 and 12 dph, we added 15 to 20 ml d^{-1} of a concentrated blend of whole cell suspended microalgae (RotiGreen Omega, Reed Mariculture) to rearing tanks throughout the day along with rotifer feedings, maintaining a level of turbidity in which visibility of the tank bottom (~75 cm depth) was obscured.

Throughout all experiments, mortality rates were estimated visually but were not measured quantitatively due to practical limitations. Daily mortality rates were highly variable throughout ontogeny, which is a common scenario during intensive rearing of mahi-mahi larvae at UMEH. Cumulative mortality rates were estimated to reach as high as the mid-90th percentile, regardless of treatment. During Expts 1

and 2, equipment failure (air flow and water supply) resulted in complete mortality of larvae in some tanks, forcing premature termination (prior to 21 dph) of Expts 1 and 2, and analysis was limited to those data collected prior to the failures. During Expts 1, 2, and 3, rearing continued until 17, 13, and 21 dph, respectively.

Hatch rate, size, and development

While stocking experimental tanks during Expts 2 and 3, 50 eggs from each replicate tank were reserved and placed into a closed 1 l high-density polyethylene jar with a 300 µm nylon mesh window and flow-through seawater. During Expt 2, these chambers were suspended in each tank until hatching in experimental tanks was observed to be complete, at which point chambers were retrieved and eggs/larvae preserved for later analysis of hatch rate and size. To determine if embryonic duration varied with CO_2 treatment, during Expt 3, hatching chambers were collected at a set time (sundown), while hatching in experimental tanks was still in progress. Throughout each experiment, subsamples of 5 to 20 larvae were collected from each replicate tank throughout ontogeny, specifically targeting the period of transition from pre-flexion through postflexion. During Expt 3, fewer sampling days were selected to ensure sufficient sample sizes for swimming tests at 20 dph. Larvae were stored in 95% ethanol, a digital image of each larva was captured, and SL was measured to the nearest 0.1 mm using the software Image-Pro Plus (v7.0, MediaCybernetics). We qualitatively assessed progression through flexion according to notochord position (i.e. flexing upwards) and development of caudal fin rays, scoring each larva as 'pre-flexion' or 'undergoing/post-flexion'.

Swimming activity and ability

Swimming activity tests were performed at 8 dph during Expts 1 and 2 according to previous techniques (Bignami et al. 2013a), described briefly here. Prior to the first daily feeding, 3 subsamples of 5 larvae each were removed from each replicate tank for swimming activity observations. Larvae were allowed to acclimate to a 15 cm diameter observation container for at least 1 h. Following acclimation, each container of 5 larvae was placed individually on a clear shelf inside an enclosed PVC observation chamber with a translucent lid and allowed to recover for 2 min. Larval routine swimming activity was recorded for 2 min using a low-light video camera (Hi-Res EXvision, Super Circuits) set 40 cm beneath the observation chamber. Directly following routine swimming observation, a pipette was used to gently add 1 ml of an olfactory stimulant to the center of the observation dish, and swimming activity in response to olfactory stimulation was recorded for an additional 2 min. During Expt 1, food-scented water (30 μm filtered rotifer culture water) was used as an olfactory stimulant. A different olfactory stimulant was used during Expt 2: chemical alarm cues, which are known to be released by a diversity of fish taxa upon damage to the skin (Mathis et al. 1995, Brown 2003, Holmes & McCormick 2010). Prior to behavioral observations, a single 2.5 cm juvenile mahi-mahi was anesthetized with 10% quinaldine, euthanized by immersion in MS-222, and rinsed with seawater. Both sides of the fish were scored with a blade and it was submerged in 100 ml of seawater for 10 min. This suspension was subsequently used as the olfactory stimulant.

Observation videos were converted to digital files, reduced to 2 frames s^{-1}, and analyzed for mean and maximum swimming speed, average angle change between 0.5 s observation points, and net-to-gross displacement (del Carmen Alvarez & Fuiman 2005) with the software ImageJ (v1.46p, National Institutes of Health) using the MTrackJ plug-in (Meijering et al. 2012). Individual larvae were tracked over the entire 2 min routine swimming and subsequent 2 min olfactory stimulation observation periods. To capture periods of active swimming, and because larvae were sometimes stationary for extended periods of time, analyses of average angle change and net-to-gross displacement were performed on 1 randomly selected 20 s period of active swimming per larva (defined as a minimum of 5 cm gross distance traveled in 20 s).

U_{crit} was measured during Expt 3 for 20 dph post-flexion fish using a 6 lane swimming flume similar to that described by Munday et al. (2009). Larvae were tested in their respective CO_2 treatment groups by adding equimolar HCl and NaHCO$_3$ to the large, recirculating seawater reservoir connected to the flume. This required larvae to be tested in order of increasing pCO$_2$ treatment, beginning with ambient seawater and ending with the highest treatment. U_{crit} was measured using the same methodology as Bignami et al. (2013a), modified from Stobutzki & Bellwood (1997) and described briefly here.

Prior to U_{crit} testing, subsamples of 6 larvae were collected in 1 l containers from each replicate tank be-

fore the first daily feeding. Larvae were allowed to acclimate to flume water conditions for at least 1 h before being randomly assigned lanes and allowed to acclimate to the flume for 5 min at a current speed of 2 cm s^{-1}. During U_{crit} testing, water current speed was increased by approximately 3 cm s^{-1} at 2 min intervals until larvae failed to maintain their position and were swept against a mesh barrier at the end of each flume lane. Flow calibration was verified using dye immediately before and after the day of testing, and U_{crit} speeds were calculated in body lengths s^{-1} using the equation $U_{crit} = V_p + [(t_f / t_i) \times V_i)]$, where V_p is the penultimate velocity increment successfully completed for the full time interval (t_i), prior to the velocity increment at which failure occurred (V_i) at a failure time (t_f) less than the full time interval (Hammer 1995).

Otolith analysis

During Expt 3, left and right sagittal and lapillar otoliths were dissected from four 21 dph larvae per replicate tank (total of 16 treatment^{-1}), stored in medium viscosity immersion oil, and imaged sulcus side down under a total magnification of 400×. Using Image-Pro Plus (v7.0) software, we digitally outlined each otolith according to pixel contrast and collected measurement data for dimensions including area, length, width, rectangularity, and roundness (otolith area divided by the area of the smallest rectangle able to contain it, and otolith perimeter divided by the circumference of the smallest circle able to contain it, respectively; Munday et al. 2011a,b). Otolith dimensions were scaled according to the SL of each larva.

Data analysis

All statistical analyses were conducted in R (v2.15.1) using tank mean data, and results were considered significant at p < 0.05. SL, proportion of fish having entered flexion, U_{crit}, swimming activity metrics, and left sagittal and left lapillar otolith shape (roundness and rectangularity) data were analyzed using 1-way ANOVA with CO_2 as a fixed factor and the respective measurement as the response variable. Otolith size metrics were tested using AN-COVA procedures, with each metric as a response variable, pCO$_2$ as a fixed factor, and SL as a continuous covariate. In the event that one treatment exhibited a significantly different regression slope, that treatment was removed from further analysis. Arcsine transformations were applied to the proportion of fish having entered flexion. Normality and homo-scedasticity were verified using Shapiro-Wilk and Bartlett's tests prior to all statistical procedures. Although the majority of data met all assumptions for ANOVA, during Expt 2, tank mean SL data from control treatments were not normally distributed on 13 dph and tank mean data of net-to-gross displacement were heteroscedastic across treatments. However, underlying distributions of subsampled fish were normal and homoscedastic. ANOVA procedures were conducted because of this underlying normality and because F-values are robust to departures from normality and homoscedasticity, with negligible to modest inflation of Type I error (Harwell et al. 1992, Underwood 1997).

RESULTS

Expt 1: The SL of larvae during Expt 1 was not significantly affected by elevated pCO_2 treatments for any sampling day (5, 8–12 dph, all p > 0.05; Fig. 1a). Larvae collected at the termination of Expt 1 (17 dph) showed a similar trend, but statistical analyses were not conducted due to the aforementioned high mortality in some replicate tanks. Larval progression through flexion was not affected by elevated pCO_2 treatments on any targeted sampling day, ranging from 9 to 12 dph (all p > 0.05; Fig. 1b). Similarly, no metric of routine swimming activity was affected by CO_2 treatment, and larvae did not exhibit any treatment-related change in swimming activity following olfactory stimulation with food-scented water (all p > 0.05). See Table 2 for summary statistics from all Expt 1 analyses.

Expt 2: Hatch rate (>95% in all treatments) and size-at-hatch were not significantly affected by CO_2 treatment during Expt 2 (both p > 0.05; data not shown), but larvae demonstrated significantly increased SL under both elevated pCO_2 treatments (770 and 1460 µatm pCO_2) compared to controls at 5 dph (p = 0.017; Fig. 2a). A similar result was found at the highest treatment level at 8 dph (p = 0.032), but not at 9 or 13 dph. Correspondingly, a significantly higher proportion of larvae were undergoing flexion at 8 dph in the 1460 µatm pCO_2 treatment compared to controls (p = 0.035; Fig. 2b). By the second day of sampling for flexion, nearly all larvae had begun to flex and no significant differences were evident. Routine swimming activity was significantly impacted by CO_2 treatment for 1 metric: larvae from the 1460 µatm pCO_2 treatment level exhibited significantly lower maximum swimming velocity compared

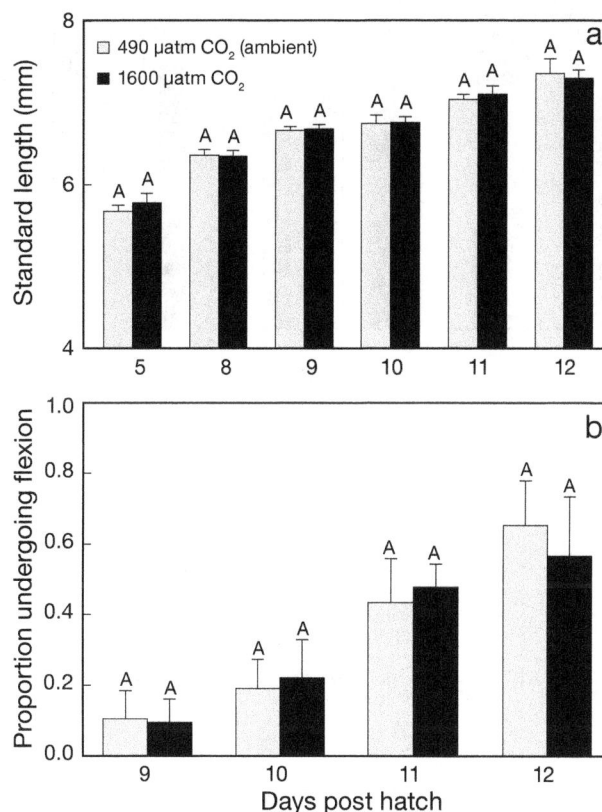

Fig. 1. Effects of ocean acidification on larval mahi-mahi (a) size-at-age and (b) proportion of larvae undergoing flexion during Expt 1. Within each day post hatch (dph), bars that share a letter are statistically similar (ANOVA, p > 0.05). Values are tank means (a) +SE and (b) +SD; n = 6 treatment[-1]

Table 2. Summary of statistical results from ANOVA of standard length (SL), flexion, and routine swimming activity of mahi-mahi larvae across multiple days post hatch (dph) during Expt 1

Metric	F	p	df
SL (dph)			
5	0.557	0.473	1,10
8	0.014	0.909	1,10
9	0.058	0.814	1,10
10	0.012	0.915	1,10
11	0.286	0.605	1,10
12	0.089	0.772	1,10
Flexion (dph)			
9	0.061	0.810	1,10
10	0.293	0.601	1,10
11	0.527	0.484	1,10
12	0.966	0.349	1,10
Swimming activity			
Mean velocity	2.535	0.142	1,10
Max velocity	0.117	0.740	1,10
Net-to-gross displacement	0.892	0.367	1,10
Angle change	0.574	0.466	1,10

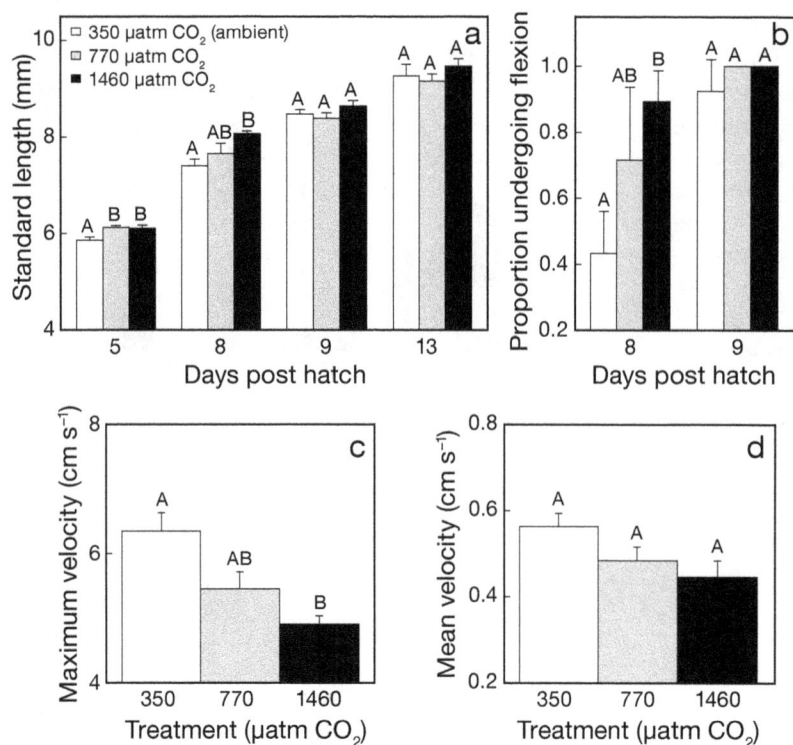

Fig. 2. Effects of ocean acidification on larval mahi-mahi (a) size-at-age, (b) proportion of larvae undergoing flexion, (c) maximum routine swimming velocity, and (d) mean routine swimming velocity during Expt 2. Bars not sharing a letter (within each day post hatch in panels a and b) are significantly different (Tukey's test, p < 0.05). Values are tank means (a,c,d) +SE and (b) +SD; n = 4 treatment^{-1}

1190 µatm pCO_2 treatment were of intermediate size and not significantly different from either treatment or controls (all p > 0.05), with the exception of the left sagitta being significantly wider compared to controls (p < 0.001). The regression of lapillus length and area against SL from the 1190 µatm pCO_2 treatment did not satisfy the assumption of equal slopes compared to ambient and 2170 µatm pCO_2 treatments, and was removed from further analysis. There were no significant differences in otolith roundness or rectangularity between any treatments (p > 0.05). See Table 4 for summary statistics from all Expt 3 analyses.

DISCUSSION

Contrary to our original hypotheses, mahi-mahi larvae were not particularly susceptible to ocean acidification, exhibiting no negative effects on hatch rate, size, development, or swimming ability. However, young

to controls (p = 0.006; Fig. 2c). A similar trend was observed for mean swimming velocity, but this pattern was not significant (p = 0.093; Fig. 2d). Angle change and net-to-gross-displacement were not significantly affected by CO_2 treatment, and there were no significant treatment effects on swimming activity following olfactory stimulation with conspecific chemical alarm cue (all p > 0.05). See Table 3 for summary statistics from all Expt 2 analyses.

Expt 3: Both elevated pCO_2 treatments (1190 and 2170 µatm pCO_2) exhibited a trend of lower proportion of eggs hatched compared to controls, but this trend was not significant (p = 0.076; Fig. 3a). There were no significant differences in the SL of larvae on any sampling day (2, 6, and 21 dph, all p > 0.05; Fig. 3b) and fish were not sampled to analyze progression through flexion. Larvae tested for U_{crit} at 20 dph demonstrated a trend of lower U_{crit} with increasing pCO_2, but this pattern was not significant (p = 0.179; Fig. 3c).

Left sagittal and lapillar otolith length, width, and area were all significantly larger in larvae from the 2170 µatm pCO_2 treatment compared to controls (all p < 0.05; Fig. 4a–c). Otoliths from larvae in the

Table 3. Summary of statistical results from ANOVA of proportion of mahi-mahi larvae hatched (Hatch), standard length (SL), flexion, and routine swimming activity across multiple days post hatch (dph) during Expt 2. Significant results (p < 0.05) are shown in **bold**. Tukey's post-hoc test results compare ambient pCO_2 (C), 770 µatm pCO_2 (1), and 1460 µatm pCO_2 (2) treatments

Metric	F	p	df	Tukey's
Hatch	0.685	0.528	2,9	C = 1 = 2
SL (dph)				
0	0.598	0.570	2,9	C = 1 = 2
5	6.635	**0.017**	2,9	C < 1 = 2
8	5.191	**0.032**	2,9	1 = C < 2 = 1
9	1.512	0.272	2,9	C = 1 = 2
13	0.866	0.457	2,8	C = 1 = 2
Flexion (dph)				
8	4.988	**0.035**	2,9	C < 1 = 2
9	2.835	0.112	2,9	C = 1 = 2
Swimming activity				
Mean velocity	3.122	0.093	2,9	C = 1 = 2
Max velocity	9.454	**0.006**	2,9	C < 1 = 2
Net-to-gross displacement	0.321	0.734	2,9	C = 1 = 2
Angle change	0.021	0.980	2,9	C = 1 = 2

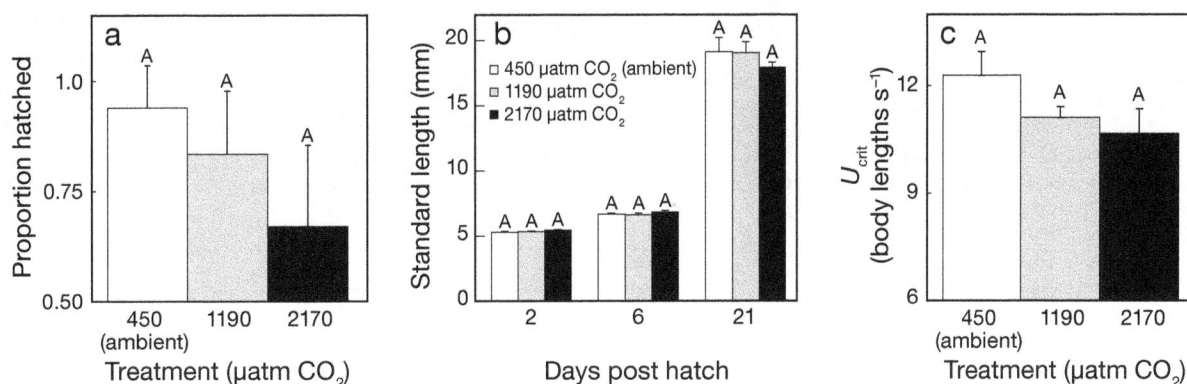

Fig. 3. Effects of ocean acidification on (a) the proportion of mahi-mahi eggs hatched, (b) size-at-age, and (c) critical swimming speed (U_{crit}) of 20 day post hatch (dph) larvae during Expt 3. Bars sharing a letter (within each dph in panel b) are statistically similar (Tukey's test, p > 0.05). Values are tank means (a) +SD and (b,c) +SE; n = 4 treatment^{-1}

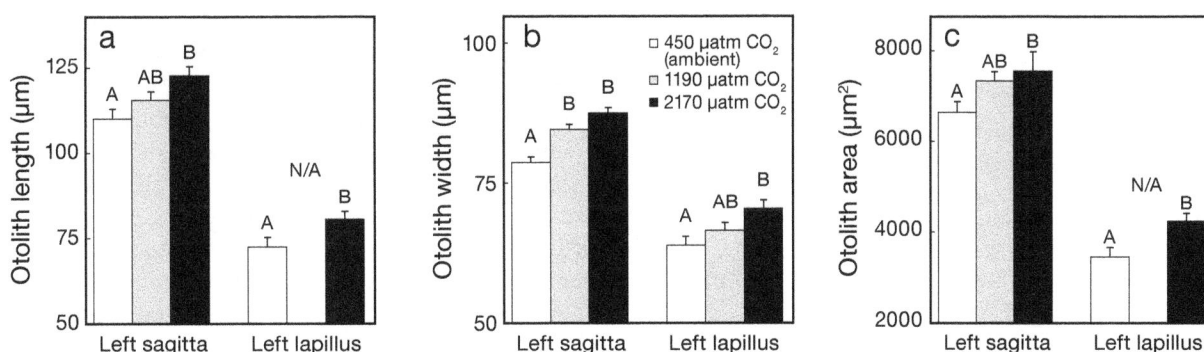

Fig. 4. Effects of ocean acidification on sagittal and lapillar otolith (a) length, (b) width, and (c) area for 20 day post hatch (dph) larval mahi-mahi during Expt 3. Results are from ANCOVA of left lapillus and left sagitta. In (a) and (c), the slope of the regression of left lapillus length and area against fish standard length differed from other treatments; therefore it was removed from analysis. Within each otolith type, bars not sharing a letter are significantly different (Tukey's p < 0.05). Values are adjusted tank means +SE (n = 4 treatment^{-1})

larvae exposed to acidification were of greater size-at-age and had decreased routine swimming activity than controls during Expt 2, and acidification also resulted in significant otolith overgrowth compared to controls during Expt 3. The absence of major impacts of acidification on mahi-mahi larvae is consistent with the concept that some resistance to acidification may be possible in species with the ability to maintain tight physiological regulation of their internal environment, a general trait of organisms with high metabolic rates (Melzner et al. 2009a). By maintaining tight control over internal pH, mahi-mahi may be able to avoid reduced protein biosynthesis that occurs at lower pH (Langenbuch & Pörtner 2003), thus preventing negative impacts on size and development. Alternatively, more extensive effects of increased pCO_2 may have been too subtle to be detected during the present study.

During Expt 1, it is possible that an effect of acidification on growth and development was masked by temperature-driven depression of metabolic and growth rates across controls and treatments. Due to seasonal variation in source seawater, Expt 1 was conducted at the lowest average temperature (~24°C) and larvae were observed to grow and develop at the slowest rates of any experiment. If acidification impacted the metabolism of larvae, which could be manifested as either a depression (Pörtner et al. 2005) or increase in metabolic rate (Miller et al. 2012), the magnitude of this effect could have been reduced to the point of non-detection due to the overall lower metabolic rates and slower growth.

During early development (5 and 8 dph), mahi-mahi larvae in Expt 2 exhibited an increase in size and developmental rate in high-CO_2 seawater. The growth–mortality hypothesis suggests that such results could be beneficial to larval survival in the wild, allowing them to pass through gape-limited predation windows more quickly than smaller larvae (Anderson 1988).

Table 4. Summary of statistical results from ANOVA of proportion of mahi-mahi larvae hatched (Hatch), standard length (SL), critical swimming speed (U_{crit}), and otolith size/shape across multiple days post hatch (dph) during Expt 3. ANCOVA results presented for otolith length, width, and area are from left lapillus (LL) and left sagitta (LS). Significant results (p < 0.05) are shown in **bold**. Regression slopes for LL length and area of 1190 µatm pCO_2 treatment otoliths differed from ambient and 2170 µatm pCO_2 treatments, and were therefore removed from analysis. Tukey's post-hoc test results compare ambient pCO_2 (C), 1190 µatm pCO_2 (1), and 2170 µatm pCO_2 (2) treatments

Metric	F	p	df	Tukey's
Hatch	3.475	0.076	2,9	C = 1 = 2
SL (dph)				
0	0.817	0.472	2,9	C = 1 = 2
2	1.546	0.265	2,9	C = 1 = 2
6	0.909	0.437	2,9	C = 1 = 2
21	0.802	0.482	2,9	C = 1 = 2
U_{crit} **(dph)**				
20	2.091	0.179	2,9	C = 1 = 2
Otolith size/shape				
LS length	6.797	**0.023**	2,7	1 = C < 2 = 1
LS width	28.444	**<0.001**	2,7	C < 1 = 2
LS area	11.321	**0.006**	2,7	1 = C < 2 = 1
LS rectangularity	0.321	0.735	2,8	C = 1 = 2
LS roundness	1.795	0.227	2,8	C = 1 = 2
LL length	8.529	**0.043**	1,4	C < 2
LL width	5.806	**0.033**	2,7	1 = C < 2 = 1
LL area	13.948	**0.020**	1,4	C < 2
LL rectangularity	0.381	0.695	2,8	C = 1 = 2
LL roundness	0.939	0.430	2,8	C = 1 = 2

This is not the first report of an increase in the size-at-age of larval fish under acidified conditions. Chambers et al. (2014) reported a temporary increase in the length-at-age of larval flounder under elevated pCO_2 conditions (1808 and 4714 ppm), during the first 2 wk of development. Additionally, damselfish larvae reared at elevated pCO_2 (550 to 1030 ppm) exhibited up to an 18% increase in length-at-age, which the authors suggested could be accomplished either through increased energy intake or decreased energy expenditure (Munday et al. 2009). In a captive rearing environment, prey are consistently provided at higher quantities than what is expected in nature; thus, larvae in the present study may have been able to increase their energy intake to compensate, or overcompensate, for an increase in the metabolic cost of acid–base balance under acidified conditions (Deigweiher et al. 2008, Miller et al. 2012). This has been demonstrated with mussels in both laboratory and natural marine environments, where food availability predominantly influences growth compared to increased pCO_2 (Thomsen et al. 2013). Reduced energy expenditure is also possible through either metabolic depression or behavioral acclimation. Young mahi-mahi larvae (3 dph) reduce their metabolism under ocean acidification conditions (Pimentel et al. 2014b), a phenomenon that occurs in other marine organisms as well (Langenbuch & Pörtner 2003). However, metabolic depression would likely result in reduced growth (Pörtner et al. 2004); therefore, a more likely explanation of increased size-at-age may be a reduction in energy expenditure via altered behavioral activity, as was observed for swimming velocity at the same age and during the same experiment (Expt 2) when increased size was detected. A reduction in swimming velocity could help compensate for possible CO_2-driven increases in the metabolic cost of acid-base regulation, allowing for re-allocation of metabolic resources to somatic growth. Pimentel et al. (2014b) also reported altered swimming behavior in mahi-mahi larvae approximately 3 to 5 d younger than those in our study, under slightly higher pCO_2 conditions (1670 µatm), though there was no corresponding change in larval size. In the present study, differences in larval size among treatments decreased to a point of non-significance after 8 dph, suggesting that this behavioral response or its effectiveness may be transient.

Reduced routine swimming velocity may have important ecological implications for the feeding and survival of larvae in the wild. Prey encounter rate is directly related to predator swimming speed (Gerritsen & Strickler 1977), thus larvae under elevated pCO_2 conditions may exhibit reduced feeding success due to lower encounter rates. This would not be evident in a laboratory scenario with abundant availability of prey, but could be critical in nature, where starvation is thought to be an important process influencing larval survival and recruitment magnitude (Houde 1989). Reduced food intake also has the potential to interact with larval susceptibility to elevated pCO_2: for example, cobia exhibit reduced starvation resistance under acidified conditions (Bignami 2013). This combination of effects could negatively influence cumulative survival through the larval stages and result in lower recruitment magnitude into the adult population.

During Expt 3, we observed no significant change in hatch rate or U_{crit} under elevated pCO_2 treatments. These U_{crit} results are consistent with other species at various ontogenetic stages (Melzner et al. 2009b, Munday et al. 2009, Bignami et al. 2013a). Not surprisingly, otolith growth was higher in larvae raised under acidified conditions. This has been observed in a number of species (e.g. Checkley et al. 2009, Munday et al. 2011a, Bignami et al. 2013b) and is thought to be caused by the establishment of a more favorable calcifying environment through the retention of bicarbonate ions in the extracellular fluid, a physiological mechanism used by fishes to compensate for increased blood pCO_2 during hypercapnic events (Esbaugh et al. 2012). The presence, absence, or severity of otolith overgrowth may be directly related to the magnitude of this physiological response as well as the non-carbonate blood buffer capacity of different species. The lowest pCO_2 known to affect otolith development is 800 µatm, for 21 dph cobia (Bignami et al. 2013a), while in other species much higher treatments (~4000 µatm) have failed to induce a significant change (Frommel et al. 2013). Otolith size and density influences the function of otoliths as auditory sensory organs: modeling of larval cobia otoliths with up to 58% greater mass indicated increased auditory sensitivity and a ~50% extension of hearing range (Bignami et al. 2013b). In the present study, we found an increase in otolith size; thus mass is likely to be increased as well, which could similarly impact the auditory sensitivity and hearing range of mahi-mahi larvae. Although mahi-mahi do not navigate and recruit to a noisy benthic environment like many reef fishes, auditory sensation likely plays an important role in prey detection, communication, and navigation in the pelagic environment.

This study provides perspective on the impact of acidification on the larvae of a species with a similar life history strategy and physiological characteristics as other offshore pelagic fishes. The impacts on growth, behavior, and otolith development reported here, along with metabolic and behavioral effects reported by Pimentel et al. (2014b), indicate that mahi-mahi are not immune to the impacts of ocean acidification. However, neither mahi-mahi nor other pelagic species such as cobia have demonstrated substantially greater susceptibility to ocean acidification than the nearshore demersal species that are presumably better adapted to variable environmental conditions (Munday et al. 2008, Pörtner 2008). Although these findings offer a potentially optimistic outlook for the future of large pelagic species in the face of ocean acidification, research on entirely pelagic tropical species remains limited. Furthermore, the present study did not address several important factors, such as more subtle impacts of acidification, the cumulative effect of long-term exposure to ocean acidification and its potential effects on growth and reproduction (Pörtner et al. 2005), or the combined effect of multiple stressors, which fishes will experience as a result of climate change. Continued study of the impacts of ocean acidification on species with distinct life history characteristics will strengthen our knowledge of the diversity of ecological responses to acidification and enable informed management and conservation decisions.

Acknowledgements. All live animal use was conducted with the approval of the University of Miami Institutional Animal Care and Use Committee (protocol #12-075). This study was supported by grants from the National Science Foundation *GK-12* program, University of Miami Maytag Ichthyology Chair, Guy Harvey Ocean Foundation, Florida Sea Grant, International Light Tackle Tournament Association, International Women's Fishing Association, Manasquan River Marlin and Tuna Club, and the Yamaha Contender/Miami Billfish Tournament. We thank UMEH, T. Capo, D. Benetti, C. Langdon, K. Ternus, C. Li, M. Iwane, D. Martin, M. Huebner, and A. Rivard for facility use and assistance, as well as M. Grosell and J. Dallman for discussions. We also thank P. Munday for helpful information and discussion on swimming flume design.

LITERATURE CITED

Anderson J (1988) A review of size dependent survival during pre-recruit stages of fishes in relation to recruitment. J Northwest Atl Fish Sci 8:1–12

Baumann H, Talmage SC, Gobler CJ (2011) Reduced early life growth and survival in a fish in direct response to increased carbon dioxide. Nat Clim Change 2:38–41

► Beardsley GL Jr (1967) Age, growth, and reproduction of the dolphin, *Coryphaena hippurus*, in the Straits of Florida. Copeia 1967:441–451

► Benetti DD, Sardenberg B, Welch A, Hoenig R, Orhun MR, Zink I (2008) Intensive larval husbandry and fingerling production of cobia *Rachycentron canadum*. Aquaculture 281:22–27

Bignami S (2013) Effects of ocean acidification on the early life history of two pelagic tropical fish species, cobia (*Rachycentron canadum*) and mahi-mahi (*Coryphaena hippurus*). PhD dissertation, University of Miami, Coral Gables, FL

► Bignami S, Sponaugle S, Cowen RK (2013a) Response to ocean acidification in larvae of a large tropical marine fish, *Rachycentron canadum*. Glob Change Biol 19: 996–1006

► Bignami S, Enochs IC, Manzello DP, Sponaugle S, Cowen RK (2013b) Ocean acidification alters the otoliths of a pantropical fish species with implications for sensory function. Proc Natl Acad Sci USA 110:7366–7370

Brauner C (2008) Acid–base balance. In: Finn RN, Kapoor BG (eds) Fish larval physiology. Science Publishers, Enfield, NH, p 185–198

► Brown GE (2003) Learning about danger: chemical alarm cues and local risk assessment in prey fishes. Fish Fish 4: 227–234

► Caldeira K, Wickett ME (2003) Anthropogenic carbon and ocean pH. Nature 425:365

► Caldeira K, Wickett ME (2005) Ocean model predictions of chemistry changes from carbon dioxide emissions to the atmosphere and ocean. J Geophys Res 110:C09S04, doi: 1029/2004JC002671

► Chambers RC, Candelmo AC, Habeck EA, Poach ME and others (2014) Effects of elevated CO_2 in the early life stages of summer flounder, *Paralichthys dentatus*, and potential consequences of ocean acidification. Biogeosciences 11:1613–1626

► Checkley DM, Dickson AG, Takahashi M, Radich JA, Eisenkolb N, Asch R (2009) Elevated CO_2 enhances otolith growth in young fish. Science 324:1683

► Cowen RK, Sponaugle S (2009) Larval dispersal and marine population connectivity. Annu Rev Mar Sci 1:443–466

Deigweiher K, Koschnick N, Pörtner HO, Lucassen M (2008) Acclimation of ion regulatory capacities in gills of marine fish under environmental hypercapnia. Am J Physiol 295:R1660–R1670

► del Carmen Alvarez M, Fuiman L (2005) Environmental levels of atrazine and its degradation products impair survival skills and growth of red drum larvae. Aquat Toxicol 74:229–241

Ditty JG, Shaw RF, Grimes CB, Cope JS (1994) Larval development, distribution, and abundance of common dolphin, *Coryphaena hippurus*, and pompano dolphin, *C. equiselis* (family: Coryphaenidae), in the northern Gulf of Mexico. Fish Bull 92:275–291

► Esbaugh AJ, Heuer R, Grosell M (2012) Impacts of ocean acidification on respiratory gas exchange and acid–base balance in a marine teleost, *Opsanus beta*. J Comp Physiol B 182:921–934

FAO (2013) Species fact sheets, *Coryphaena hippurus* (Linnaeus, 1758). www.fao.org/fishery/species/3130/en (accessed on 11 Mar 2013)

► Faulk CK, Holt GJ (2005) Advances in rearing cobia *Rachycentron canadum* larvae in recirculating aquaculture systems: live prey enrichment and greenwater culture. Aquaculture 249:231–243

► Feely RA, Sabine CL, Hernandez-Ayon JM, Ianson D, Hales B (2008) Evidence for upwelling of corrosive 'acidified' water onto the continental shelf. Science 320:1490–1492

► Ferrari MCO, McCormick M, Munday PL, Meekan MG, Dixson DL, Lönnstedt O, Chivers DP (2012) Effects of ocean acidification on visual risk assessment in coral reef fishes. Funct Ecol 26:553–558

Folpp H, Lowry M (2006) Factors affecting recreational catch rates associated with a fish aggregating device (FAD) off the NSW coast, Australia. Bull Mar Sci 78: 185–193

► Frieder CA, Nam SH, Martz TR (2012) High temporal and spatial variability of dissolved oxygen and pH in a nearshore California kelp forest. Biogeosciences 9:3917–3930

Frommel AY, Maneja R, Lowe D, Malzahn AM and others (2011) Severe tissue damage in Atlantic cod larvae under increasing ocean acidification. Nat Clim Change 2:42–46

► Frommel AY, Schubert A, Piatkowski U, Clemmesen C (2013) Egg and early larval stages of Baltic cod, *Gadus morhua*, are robust to high levels of ocean acidification. Mar Biol 160:1825–1834

Gattuso JP (2010) Approaches and tools to manipulate the carbonate chemistry. In: Riebesell U, Fabry V, Hannson L, Gattuso JP (eds) Guide to best practices for ocean acidification research and data reporting. Publications office of the European Union, Luxembourg, p 44–52

Gerritsen J, Strickler JR (1977) Encounter probabilities and community structure in zooplankton: a mathematical model. Can J Fish Aquat Sci 34:73–82

► Hamilton TJ, Holcombe A, Tresguerres M (2014) CO_2-induced ocean acidification increases anxiety in Rockfish via alteration of $GABA_A$ receptor functioning. Proc R Soc Lond B Biol Sci 281:20132509

► Hammer C (1995) Fatigue and exercise tests with fish. Comp Biochem Physiol A 112:1–20

Harwell MR, Rubinstein EN, Hayes WS, Olds CC (1992) Summarizing Monte Carlo results in methodological research: the one-and two-factor fixed effects ANOVA cases. J Educ Behav Stat 17:315–339

► Hofmann GE, Smith JE, Johnson KS, Send U and others (2011) High-frequency dynamics of ocean pH: a multi-ecosystem comparison. PLoS ONE 6:e28983

► Holmes TH, McCormick M (2010) Smell, learn and live: the role of chemical alarm cues in predator learning during early life history in a marine fish. Behav Processes 83: 299–305

Houde ED (1989) Comparative growth, mortality, and energetics of marine fish larvae: temperature and implied latitudinal effects. Fish Bull 87:471–495

► Hurst TP, Fernandez ER, Mathis JT, Miller JA (2012) Resiliency of juvenile walleye pollock to projected levels of ocean acidification. Aquat Biol 17:247–259

► Kroeker KJ, Kordas RL, Crim R, Hendriks IE and others (2013) Impacts of ocean acidification on marine organisms: quantifying sensitivities and interaction with warming. Glob Change Biol 19:1884–1896

► Langdon C, Takahashi T, Sweeney C, Chipman D and others (2000) Effect of calcium carbonate saturation state on the calcification rate of an experimental coral reef. Global Biogeochem Cycles 14:639–654

► Langenbuch M, Pörtner HO (2003) Energy budget of hepatocytes from Antarctic fish (*Pachycara brachycephalum* and *Lepidonotothen kempi*) as a function of ambient CO_2: pH-dependent limitations of cellular protein biosynthesis? J Exp Biol 206:3895–3903

Lewis E, Wallace D (1998) CO2SYS—Program developed for the CO_2 system calculations. Carbon Dioxide Information Analysis Center, Oak Ridge National Laboratory, US Dept of Energy, Oak Ridge, TN

► Mathis A, Chivers DP, Smith RJF (1995) Chemical alarm signals: Predator deterrents or predator attractants? Am Nat 145:994–1005

Meehl G, Stocker TF, Collins W, Friedlingstein P and others (2007) Global climate projections. In: Solomon S, Qin D, Manning M, Chen Z and others (eds) Climate change 2007: the physical science basis. Contribution of Working Group I to the Fourth Assessment Report of the Intergovernmental Panel on Climate Change. Cambridge University Press, Cambridge, p 748–845

Meijering E, Dzyubachyk O, Smal I (2012) Methods for cell and particle tracking. In: Conn PM (ed) Methods in enzymology, Vol 504: imaging and spectroscopic analysis of living cells. Elsevier, London, p 183–200

► Meinshausen M, Smith SJ, Calvin K, Daniel JS and others (2011) The RCP greenhouse gas concentrations and their extensions from 1765 to 2300. Clim Change 109:213–241

Melzner F, Goebel S, Langenbuch M, Gutowska MA, Pörtner HO, Lucassen M (2009b) Swimming performance in Atlantic Cod (*Gadus morhua*) following long-term (4–12 months) acclimation to elevated seawater pCO_2. Aquat Toxicol 92:30–37

► Melzner F, Gutowska MA, Langenbuch M, Dupont S and others (2009a) Physiological basis for high CO_2 tolerance in marine ectothermic animals: Pre-adaptation through lifestyle and ontogeny? Biogeosciences 6: 2313–2331

► Melzner F, Thomsen J, Koeve W, Oschlies A and others (2013) Future ocean acidification will be amplified by hypoxia in coastal habitats. Mar Biol 160:1875–1888

Miller GM, Watson SA, Donelson JM, McCormick M, Munday PL (2012) Parental environment mediates impacts of increased carbon dioxide on a coral reef fish. Nat Clim Change 2:858–861

► Munday PL, Jones GP, Pratchett MS, Williams AJ (2008) Climate change and the future for coral reef fishes. Fish Fish 9:261–285

► Munday PL, Donelson JM, Dixson DL, Endo GGK (2009) Effects of ocean acidification on the early life history of a tropical marine fish. Proc R Soc Lond B Biol Sci 276: 3275–3283

► Munday PL, Dixson DL, McCormick M, Meekan M, Ferrari MCO, Chivers DP (2010) Replenishment of fish populations is threatened by ocean acidification. Proc Natl Acad Sci USA 107:12930–12934

► Munday PL, Hernaman V, Dixson DL, Thorrold SR (2011a) Effect of ocean acidification on otolith development in larvae of a tropical marine fish. Biogeosciences 8: 1631–1641

► Munday PL, Gagliano M, Donelson JM, Dixson DL, Thorrold SR (2011b) Ocean acidification does not affect the early life history development of a tropical marine fish. Mar Ecol Prog Ser 423:211–221

► Naas KE, Naess T, Harboe T (1992) Enhanced first feeding of halibut larvae (*Hippoglossus hippoglossus* L.) in green water. Aquaculture 105:143–156

Nilsson GE, Dixson DL, Domenici P, McCormick M, Sørensen C, Watson SA, Munday PL (2012) Near-future carbon dioxide levels alter fish behaviour by interfering with neurotransmitter function. Nat Clim Change 2:201–204

► Oxenford HA (1999) Biology of the dolphinfish (*Coryphaena hippurus*) in the western central Atlantic: a review. Sci Mar 63:303–315

Palko BJ, Beardsley GL Jr, Richards WJ (1982) Synopsis of the biological data on dolphin-fishes, *Coryphaena hippurus* Linnaeus and *Coryphaena equiselis* Linnaeus. NOAA Tech Rep, NMFS Circ 443, US Dept of Commerce, Washington, DC

► Pimentel MS, Faleiro F, Dionísio G, Repolho T, Pousão-

Ferreira P, Machado J, Rosa R (2014a) Defective skeletogenesis and oversized otoliths in fish early stages in a changing ocean. J Exp Biol 217:2062–2070

► Pimentel M, Pegado M, Repolho T, Rosa R (2014b) Impact of ocean acidification in the metabolism and swimming behavior of the dolphinfish (*Coryphaena hippurus*) early larvae. Mar Biol 161:725–729

► Pörtner HO (2008) Ecosystem effects of ocean acidification in times of ocean warming: a physiologist's view. Mar Ecol Prog Ser 373:203–217

► Pörtner HO, Langenbuch M, Reipschläger A (2004) Biological impact of elevated ocean CO_2 concentrations: lessons from animal physiology and earth history. J Oceanogr 60: 705–718

► Pörtner HO, Langenbuch M, Michaelidis B (2005) Synergistic effects of temperature extremes, hypoxia, and increases in CO_2 on marine animals: from Earth history to global change. J Geophys Res 110:C09S10, doi:1029/2004 JC002561

► Potoschi A, Cannizzaro L, Milazzo A, Scalisi M, Bono G (1999) Sicilian dolphinfish (*Coryphaena hippurus*) fishery. Sci Mar 63:439–445

► Price NN, Martz TR, Brainard RE, Smith JE (2012) Diel variability in seawater pH relates to calcification and benthic community structure on coral reefs. PLoS ONE 7:e43843

► Stobutzki I, Bellwood D (1997) Sustained swimming abilities of the late pelagic stages of coral reef fishes. Mar Ecol Prog Ser 149:35–41

► Thomsen J, Gutowska MA, Saphörster J, Heinemann A and others (2010) Calcifying invertebrates succeed in a naturally CO_2 enriched coastal habitat but are threatened by high levels of future acidification. Biogeosciences Discuss 7:5119–5156

► Thomsen J, Casties I, Pansch C, Körtzinger A, Melzner F (2013) Food availability outweighs ocean acidification effects in juvenile *Mytilus edulis*: laboratory and field experiments. Glob Change Biol 19:1017–1027

Underwood AJ (1997) Experiments in ecology: their logical design and interpretation using analysis of variance. Cambridge University Press, Cambridge

► Zúñiga Flores MS, Ortega-García S, Klett-Traulsen A (2008) Interannual and seasonal variation of dolphinfish (*Coryphaena hippurus*) catch rates in the southern Gulf of California, Mexico. Fish Res 94:13–17

7

Distribution of ampullary pores on three catshark species (*Apristurus* spp.) suggest a vertical-ambush predatory behaviour

D. M. Moore*, I. D. McCarthy

School of Ocean Sciences, College of Natural Sciences, Bangor University, Menai Bridge, Anglesey LL59 5AB, UK

ABSTRACT: *Apristurus* is a genus of typically small sharks that inhabit deep waters around the globe. Relatively little is known about the feeding behaviour of these species. Here, the electrosensory biology of 3 species, *A. aphyodes*, *A. melanoasper* and *A. microps*, was investigated. Intra-specific variation in ampullary pore abundance was high in all species, highlighting the need for studies to examine multiple individuals. Abundance and distribution of ampullary pores on the head indicate that all 3 species are vertical ambush predators.

KEY WORDS: *Apristurus aphyodes* · *A. melanoasper* · *A. microps* · Rockall Trough · Ampullae of Lorenzini

INTRODUCTION

Electroreception in elasmobranchs is thought to facilitate various types of behaviour, including social interaction (Sisneros et al. 1998) and navigation (Paulin 1995, Montgomery & Walker 2001). However, by far the most important use of electroreception in elasmobranchs is believed to be predator/prey detection (Kalmijn 1971, Raschi et al. 2001). The distribution of ampullary pores in elasmobranchs has been studied in numerous species (Fishelson & Baranes 1998, Kajiura 2001, Atkinson & Battaro 2006), and there is an increasing interest in mapping ampullary pore distribution as a tool to infer behaviour and ecology (Fishelson & Baranes 1998, Kajiura 2001, Atkinson & Bottaro 2006, Kempster & Collin 2011a,b). Some published works of this field adopt a traditional meristical approach, whereby ampullary pores are counted and mapped from a single individual and taken as typical for the species (Kempster & Collin 2011a,b), particularly in large or rarely sighted shark species. As most studies take data from multiple individuals, this study will attempt to examine intra-specific variation in ampullary pore counts and furthermore elucidate the reliability of single individual representations.

The majority of studies on elasmobranch electroreception to date have focussed on coastal or upper pelagic species, and few on species from deep water environments, owing to limited accessibility. However, there is a growing need to better understand the ecology of deeper-dwelling species such as *Apristurus* spp., which are under increasing pressure from both targeted and non-targeted fishing. The genus *Apristurus* Garman, 1913 comprises 36 recognised species, which are poorly understood. We suggest that the ecology of *Apristurus* spp. may be inferred from ampullary pore distribution, by comparison with species in which their ecology is better studied.

At least 5 species of *Apristurus* are found in the Rockall Trough, Northeast Atlantic, an area under intense fishing pressure from Europe (Neat et al. 2008): *A. aphyodes* Nakaya & Stehmann, 1998; *A. laurussonii* (Synonym *A. maderensis*) Saemundsson, 1922; *A. manis* Springer, 1979; *A. melanoasper* Iglésias,

*Corresponding author: dm_moore@hotmail.co.uk

Nakaya & Stehmann, 2004, and *A. microps* Gilchrist, 1922. Three of these species (selected due to availability of useable sample numbers) shall be the focus herein.

A. aphyodes lives in depths of 1014 to 1800 m, from 49° 1.9' N to 60° 49.7' N (Nakaya & Stehmann 1998). It preys upon crustaceans, cephalopods and small teleost fishes (Duffy & Huveneers 2004). It is classed as Data Deficient on the IUCN Red List of Threatened Species.

A. melanoasper inhabits waters from 512 to at least 1520 m depth and is distributed across the North Atlantic between 39° 17' N and 61° 06' N (Iglésias et al. 2002). It is classed as Data Deficient on the IUCN Red List and there are currently no studies on its diet.

A. microps inhabits waters between 700 and 1200 m depth in both the North and South Atlantic Ocean. It feeds on small teleost fishes, crustaceas and cephalopods (Compagno et al. 1989) and is classed as Least Concern on the IUCN Red List.

All 3 species are regularly caught as by-catch in the deep-water fisheries of the Rockall Trough region (F. C. Neat, pers. comm.), but there are no detailed assessments.

MATERIALS AND METHODS

Specimens of *Apristurus* spp. were collected during a Marine Scotland Science deep-water trawl survey of the Rockall Trough area during September 2009 (see Moore et al. 2013). A maximum equal-sex and -length stratified subsample was taken, consisting of 20 *A. aphyodes* (10♂ and 10♀), 10 *A. melanoasper* (5♂ and 5♀) and 18 *A. microps* (9♂ and 9♀). Heads of specimens were removed by a single dorsoventral cut just posterior of the 5th gill slit, as all ampullary pores were found anterior of this point upon visual inspection. Heads were frozen for transportation.

After defrosting in a warm water bath, heads were examined using a Schott KL 1500 fibre optic cold light source (Fig. 1). Digital images were taken using a Ricoh Caplio R6 7.2 megapixel camera and then subsequently used for drawing of pore maps using Paint v6.1. Pores were counted *in situ* by direct visual inspection with counted pores marked in ink to limit observer error. Pore density estimations were established following Raschi (1986), utilising ImageJ picture analysis software.

Data were tested for normality (Anderson-Darling) and homoscedasticity (Hartlets) to ensure that assumptions made by statistical tests were met.

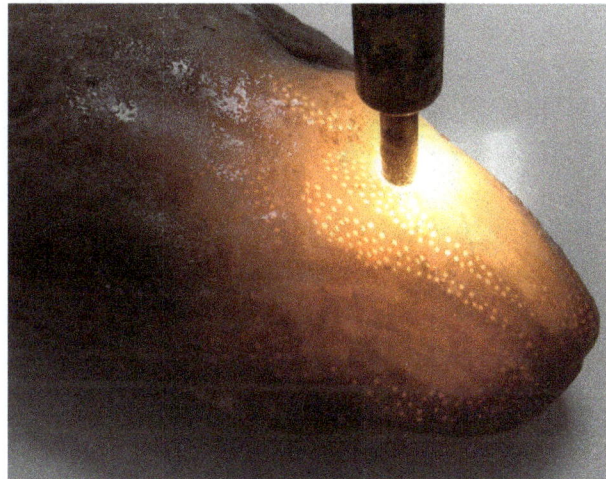

Fig. 1. Use of a fibre optic cold light source to highlight electrosensory pores in *Apristurus microps*

RESULTS

All 3 *Apristurus* species showed similarity in their pore grouping distribution pattern (Fig. 2). Most pores were grouped around the anterior snout region of the head, with only 12.1 ± 1.4 %, 9.0 ± 0.7 % and 11.3 ± 1.5 % (mean ± SD) of pores being positioned posterior to the eye in *A. aphyodes*, *A. melanoasper* and *A. microps*, respectively. There was large intra-specific variation from the mean total pore count for all species. Variation from the mean was found to be highest in *A. aphyodes* at up to 20.6 % (177 pores). Intra-specific variation in mean total pore count was 18.5 % and 13.0 % for *A. melanoasper* and *A. microps* respectively. All 3 species displayed variation in pore aperture diameter.

Total pore counts in *A. aphyodes* ranged from 761 to 1037. Female *A. aphyodes* had significantly more pores than males with a mean (±SD) pore count of 898 and 822, respectively (Table 1; Kruskal-Wallis test, $H_{20} = 7.41$, p = 0.007). There was no significant difference in total length (TL) between the sexes (ANOVA, $F_{(1,18)} = 0.60$, p = 0.448). Both males and females displayed significantly higher pore counts and pore density on the dorsal surface than on the ventral surface (Table 1; ANOVA, ♂ $F_{(1,19)} = 91.97$, p < 0.001, ♀ $F_{(1,19)} = 108.19$, p < 0.001). There was no significant relationship between pore count and TL (GLM, $r^2 = 0.91$, $F_{(1,15)} = 1.75$, p = 0.357).

Total pore counts in *A. melanoasper* ranged from 1355 to 1781. There was no significant difference between males and females (ANOVA, $F_{(1,9)} = 0.03$,

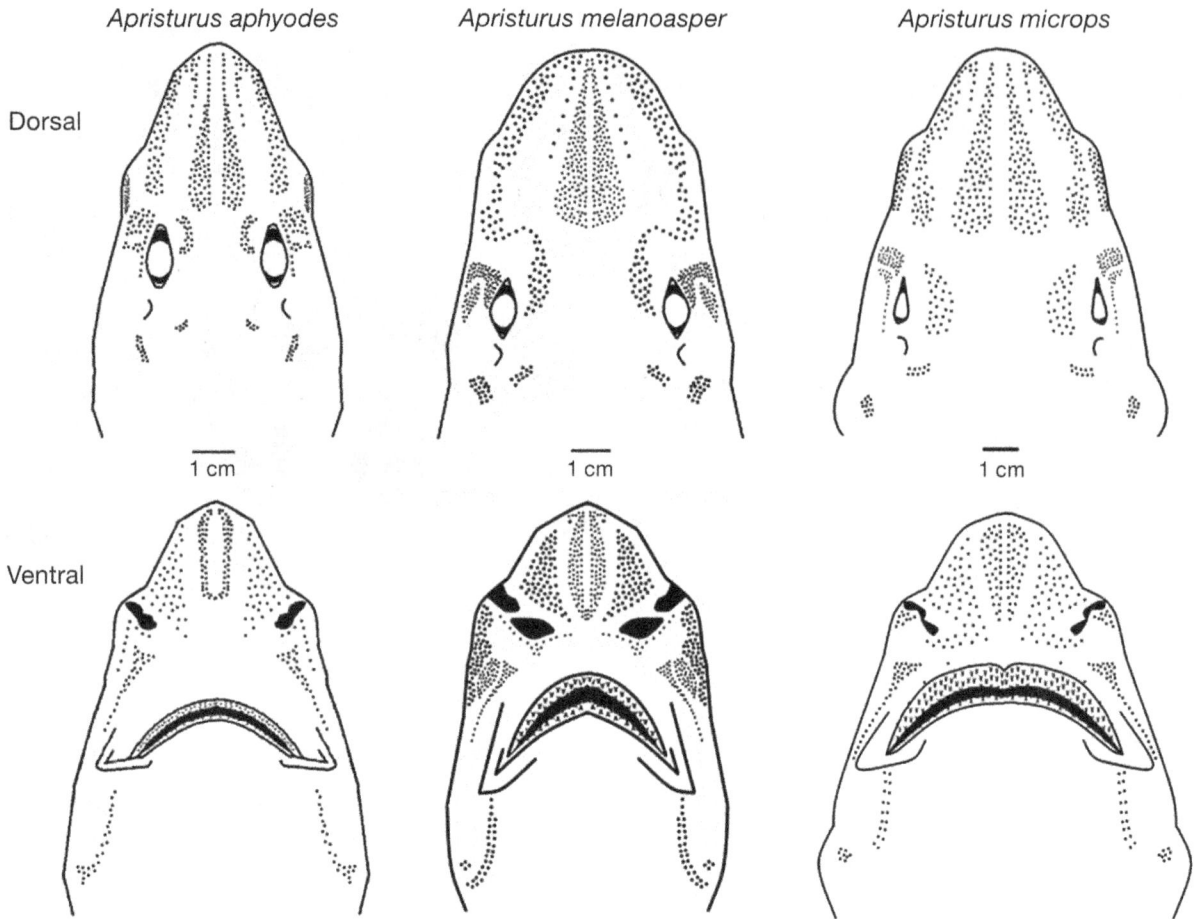

Fig. 2. Comparison of ampullary pore regions between *Apristurus aphyodes*, *A. melanoasper* and *A. microps* showing dorsal and ventral perspectives. Representations are not to scale

p = 0.869). The difference in TL between sexes was significant (Kruskal-Wallis test, H_{10} = 5.77, p = 0.016). There were no significant differences between dorsal and ventral pore counts (ANOVA, ♂ $F_{(1,9)}$ = 1.41, p = 0.269, ♀ $F_{(1,9)}$ = 1.34, p = 0.281), nor between pore count and TL (GLM, r^2 = 0.07, $F_{(1,9)}$ = 0.52, p = 0.496). *A. melanoasper* displayed a higher mean pore den-

sity (Table 1) on the ventral surface than on the dorsal surface, and the highest pore densities of the 3 species.

Total pore counts in *A. microps* ranged from 820 to 1035, with no significant difference between males and females (ANOVA, $F_{(1,18)}$ = 0.21, p = 0.653). There was no significant difference in TL between the sexes (ANOVA, $F_{(1,18)}$ = 0.70, p = 0.415). Both males and females displayed significantly higher pore counts and pore density on the dorsal surface than on the ventral surface (ANOVA, ♂ $F_{(1,19)}$ = 204.99, p < 0.001, ♀ $F_{(1,17)}$ = 302.08, p < 0.001). There was no significant relationship between pore count and TL (GLM, r^2 = 0.01, $F_{(1,18)}$ = 0.01, p = 0.914).

Table 1. Total pore count (mean ± SD) and mean total length (TL) for *Apristurus aphyodes*, *A. melanoasper* and *A. microps* by sex

Species	Sex	TL (cm)	Mean	±SD	Dorsal	±SD	Ventral	±SD	Dorsal	Ventral
					Pore number				Pore density (pores/cm²)	
Apristurus aphyodes	M	44.8	822	70	495	49	326	25	10.35	6.65
	F	46.9	898	81	559	62	339	26	11.69	6.91
Apristurus melanoasper	M	60.2	1496	165	716	83	780	86	13.90	14.58
	F	45.2	1511	121	732	44	779	80	14.21	14.56
Apristurus microps	M	56.0	910	69	585	54	325	21	10.91	5.66
	F	52.1	923	56	596	41	327	22	11.12	5.69

DISCUSSION

All species of *Apristurus* displayed similar characteristics in their ampullary pore distribution pattern, including bi-lateral symmetry and concurrency with previously generated pore maps for other species within this genus (Cornett 2006). The dorsal surface displayed distinct longitudinally arranged central and peripheral pore regions to the anterior of the eye, a crescent region above the eye and two smaller regions posterior to the eye. Similarities in ventral surface distribution included a double line of pores running longitudinally posterior of the mouth, a roughly triangular pore assemblage anterior of the nares and a longitudinal region in the mesial zone of the snout. This distribution pattern was also found in *A. brunneus* (Cornett 2006) and is likely common to the whole genus. Only *A. microps* had pores in the labial area. Total pore counts for *Apristurus* spp. were high, compared to other scyliorhinids (Kajiura 2010). Maximum pore density was typically in the pore assemblages below the eyes, although in *A. melanoasper* pore density in this region was also equalled by that of the mesial snout assemblages.

Only *A. aphyodes* showed sexual dimorphism in total pore count, females having significantly higher numbers of ampullary pores. However, Fishelson & Baranes (1998) also found the highest number of pores on female individuals of another carcharhiniform, the Oman shark *Iago omanensis*, which suggests sexual dimorphism in total pore counts could be interesting area of future research. Neither *A. aphyodes* nor *A. microps* showed significant differences in TL between sexes and it is likely the positive result for *A. melanoasper* is due to low sample size. With this in mind it is likely that intra-specific variation in total pore number is not linked to sexual dimorphism and is more likely of genetic origin.

The higher pore counts on dorsal surfaces than ventral surfaces in *A. aphyodes* and *A. microps* are not concurrent with findings for other deep water habitat species such as *Galeus melastomus* or *Etmopterus spinax* where ventral pores were more numerous (Atkinson & Bottaro 2006). Both *G. melastomus* and *E. spinax* hunt fast-moving pelagic prey (Wurtz & Vacchi 1981) and a higher ventral pore number may suggest that these species hunt by ambushing prey found below them in the water column. The higher dorsal pore number in *A. aphyodes* and *A. microps* suggests a more benthic existence; this higher dorsal pore number may facilitate detection of passing prey or predators above. This suggestion of vertical ambush is further reinforced by the depressed morphology of the snout and head within all study species. However this can be countered by the sub-terminal position of the mouth in all 3 target species which would only be concurrent with targeting prey from above; a strategy made more likely in *A. melanoasper* by its higher pore count being on the ventral surface. The labial pores in *A. microps* may suggest some in-sediment feeding and would therefore suggest that the increased mean pore number on the dorsal surface is more for detection of predators than prey. The mean pore densities in all 3 target species (Table 1) match the relationship displayed by the mean pore number; density however provides a much better indication of electroreception resolution (Raschi 1986). Following this, both *A. aphyodes* and *A. microps* have greater electroreception resolution on the dorsal surface, whereas it is on the ventral surface that *A. melanoasper* exhibits the greater electroreception resolution.

Mauchline & Gordon (1983) outline stomach contents for unidentified *Apristurus* spp. as being mainly composed of decapod crustaceans with an ontogenetic shift in diet to include more cephalopods and crustaceans with increasing total length. In addition, Ebert et al. (1996) also showed a mixed diet of teleosts and crustaceans for *Apristurus* spp., but in the South Atlantic. Such a varied diet over the sharks' lifetimes is likely to suggest that *Apristurus* spp. are opportunistic generalist feeders and the conflicting sensory/morphological evidence supports this. Such a feeding strategy is also very agreeable in a deepwater environment where feeding opportunities are likely intermittent.

There was no significant relationship between total length and pore count in any species, suggesting *Apristurus* spp. individuals are born with a set number of electrosensory pores which remain active throughout their lives, as found in other shark species (Kajiura 2001, Raschi & Gerry 2003, Cornett 2006, Mello 2009). However, the increase in TL with age necessarily results in a decrease in pore density as pore number remains constant. High pore densities are correlated with less mobile prey (like epifaunal and infaunal crustaceans) whereas lower pore densities are correlated with more mobile prey (like cephalopods and teleosts), due to a reduction in resolution as pore density decreases (with age-related growth), leaving the animal less able to detect immobile prey items (Jordan 2008, Bedore et al. 2014). It is therefore likely that the ontogenetic diet shift as described by Mauchline & Gordon (1983) is driven by both a reduced ability to detect more immobile prey and an increase in mouth gape size that allows it to target larger prey items.

Intra-specific variation in total pore count was high (>10%) for all target species, as found in other studies (Jordan 2008, Bedore et al. 2014), but is unlikely due to observer error owing to the marking methodology employed. Future research should aim to understand to what extent, if any, such variation has on electrosensory capability. In addition, identifying the species' prey would assist in understanding the significance of pore distribution patterns in these deep-water elasmobranch species.

Acknowledgements. The Marine Scotland Science deep-water trawl survey (Cruise 1209S) was funded by the Scottish Government (Grant MF0763). Thanks go to Francis Neat (Marine Scotland) and the skipper and crew of the FRV 'Scotia' for assistance in obtaining samples. Additional thanks go to E. G. Cunningham and M. Kurr as well as 3 anonymous referees for constructive comments that have improved this manuscript.

LITERATURE CITED

▶ Atkinson CJL, Bottaro W (2006) Ampullary pore distribution of *Galeus melastomus* and *Etmopterus spinax*: possible relations with predatory lifestyle and habitat. J Mar Biol Assoc UK 86:447–448

▶ Bedore CN, Harris LL, Kajiura SM (2014) Behavioral responses of batoid elasmobranchs to prey-simulating electric fields are correlated to peripheral sensory morphology and ecology. Zoology 117:95–103

Compagno LJV, Ebert DA, Smale MJ (1989) Guide to the sharks and rays of southern Africa. New Holland, London

Cornett AD (2006) Ecomorphology of shark electroreceptors. MS thesis, Florida Atlantic University, Boca Raton, FL

Duffy C, Huveneers C (2004) *Apristurus aphyodes*. IUCN Red List of Threatened Species.Version 2010.4. www.iucnredlist.org (accessed 26 January 2013)

▶ Ebert DA, Cowley PD, Compagno LJV (1996) A preliminary investigation of the feeding ecology of catsharks (Scyliorhinidae) off the west coast of southern Africa. S Afr J Mar Sci 17:233–240

▶ Fishelson L, Baranes A (1998) Distribution, morphology, and cytology of ampullae of Lorenzini in the Oman shark, *Iago omanensis* (Triakidae), from the Gulf of Aqaba, Red Sea. Anat Rec 251:417–430

Iglésias SP, Du Buit MH, Nakaya K (2002) Egg capsules of deep-sea catsharks from eastern North Atlantic, with first description of the capsule of *Galeus murinus* and *Apristurus aphyodes*. Cybium 26:59–63

▶ Jordan LK (2008) Comparative morphology of stingray lateral line canal and electrosensory systems. J Morphol 269:1325–1339

▶ Kajiura SM (2001) Head morphology and electrosensory pore distribution of carcharhinid and sphyrnid sharks. Environ Biol Fishes 61:125–133

Kajiura SM (2010) Sensory adaptations to the environment: electroreceptors as a case study. In: Carrier JC, Musick JA, Heithaus MR (eds) Sharks and their relatives. II. Biodiversity, adaptive physiology and conservation. CRC Press, Boca Raton, p 393–433

▶ Kalmijn AJ (1971) The electric sense of sharks and rays. J Exp Biol 55:371–383

▶ Kempster RM, Collin SP (2011a) Electrosensory pore distribution and feeding in the megamouth shark *Megachasma pelagios* (Lamniformes: Megachasmidae). Aquat Biol 11:225–228, doi:10.3354/ab00311

▶ Kempster RM, Collin SP (2011b) Electrosensory pore distribution and feeding in the basking shark *Cetorhinus maximus* (Lamniformes: Cetorhinidae). Aquat Biol 12: 33–36, doi:10.3354/ab00328

▶ Mauchline J, Gordon JDM (1983) Diets of the sharks and chimaeroids of the Rockall Trough, North Eastern Atlantic Ocean. Mar Biol 75:269–278

▶ Mello W (2009) The electrosensorial pore system of the cephalofoil in the four most common species of hammerhead shark from the southwestern Atlantic. C R Biol 332: 404–412

▶ Montgomery J, Walker M (2001) Orientation and navigation in elasmobranchs: which way forward? Environ Biol Fishes 60:109–116

▶ Moore DM, Neat FC, McCarthy ID (2013) Population biology and ageing of the deep water sharks *Galeus melastomus*, *Centroselachus crepidater* and *Apristurus aphyodes* from the Rockall Trough, north-east Atlantic. J Mar Biol Assoc UK 93:1941–1950

Nakaya K, Stehmann M (1998) A new species of deep-water catshark, *Apristurus aphyodes* n. sp., from the eastern North Atlantic (Chondrichthyes: Carchariniformes: Scyliorhinidae). Arch Fish Mar Res 46:77–90

Neat F, Burns F, Drewery J (2008) The deepwater ecosystem of the continental shelf slope and seamounts of the Rockall Trough: a report on the ecology and biodiversity based on FRS scientific surveys. Fish Res Serv Int Rep 02/08

Paulin MG (1995) Electroreception and the compass sense of sharks. J Theor Biol 174:325–339

Raschi W (1986) A morphological analysis of the Ampullae of Lorenzini in selected skates (Pisces, Rajoidei). J Morphol 189:225–247

Raschi WG, Gerry S (2003) Adaptations in the elasmobranch electroreceptive system. In: Val AL, Kapoor BG (eds) Fish adaptations. Science Publishers, Plymouth, p 233–258

Raschi W, Aadlond C, Keithar ED (2001) A morphological and functional analysis of the ampullae of Lorenzini in selected galeoid sharks. In: Kapoor BG, Hara TJ (eds) Sensory biology of jawed fishes. Science Publishers, Plymouth, p 297–316

▶ Sisneros JA, Tricas TC, Luer CA (1998) Response properties and biological function of the skate electrosensory system during ontogeny. J Comp Physiol A 183:87–99

Wurtz M, Vacchi M (1981) Ricerca di cicli nittemerali nell'alimentazione di selaci batiali. Quad Lab Tecnol Pesca 3(Suppl 1):155–164

8

Trophic ecology of common elasmobranchs exploited by artisanal shark fisheries off south-western Madagascar

Jeremy J. Kiszka[1,2,*], Kevin Charlot[1], Nigel E. Hussey[3], Michael R. Heithaus[2], Benoit Simon-Bouhet[1], Frances Humber[4,5], Florence Caurant[1], Paco Bustamante[1]

[1]Littoral Environnement et Sociétés (LIENSs), UMR 7266 CNRS-Université de la Rochelle, 2 rue Olympe de Gouges, 17000 La Rochelle, France

[2]Marine Sciences Program, Department of Biological Sciences, Florida International University, 3000 NE 151st Street, North Miami, Florida 33181, USA

[3]Great Lakes Institute for Environmental Research, University of Windsor, 401 Sunset Avenue, Ontario N9B 3P4, Canada

[4]Blue Ventures, Level 2 Annex, Omnibus Business Centre, 39-41 North Road, London N7 9DP, UK

[5]Centre for Ecology and Conservation, College of Life and Environmental Sciences, University of Exeter, Penryn TR10 9FE, UK

ABSTRACT: Knowledge of the trophic ecology and interactions of marine top predators is fundamental for understanding community structure and dynamics as well as ecosystem function. We examined the feeding relationships of 4 heavily exploited elasmobranchs caught in coastal artisanal shark fisheries in south-western Madagascar in 2009 and 2010—*Sphyrna lewini, Loxodon macrorhinus, Carcharhinus falciformis* and *Rhynchobatus djiddensis*—using stable isotope (δ^{15}N and δ^{13}C) analysis. Relative trophic position (indicated by δ^{15}N) and foraging location (indicated by δ^{13}C) differed among species. Isotopic niche width was highly variable: more pelagic species, such as *S. lewini* and *C. falciformis*, had the broadest isotopic niches while the benthic *R. djiddensis* had the narrowest. A high percentage of niche overlap occurred between *R. djiddensis* and 2 of the species, *C. falciformis* (93.2%) and *L. macrorhinus* (73.2%), and to a lesser extent *S. lewini* (13.3%). Relative trophic position of *S. lewini* significantly increased with size, suggesting a dietary shift with age. Sex differences in δ^{15}N values were observed in *L. macrorhinus*, suggesting intraspecific niche partitioning. Variation in stable isotope values among these 4 highly exploited elasmobranch species indicates trophic structuring, likely driven by differences in diet and habitat use as well as by size and sex. This study provides the first baseline information on the trophic ecology of elasmobranchs caught in artisanal fisheries from south-western Madagascar.

KEY WORDS: Artisanal fisheries · Sharks · Trophic ecology · δ^{15}N · δ^{13}C · Ontogenetic shift · Sex differences

INTRODUCTION

Predicting the community-level consequences of changes in the abundance of a particular species, due to natural or anthropogenic factors, requires an understanding of its trophic interactions and trophic similarity amongst species (i.e. the level of trophic redundancy). In both terrestrial and marine environments, the removal of predators across multiple trophic levels has been shown to disrupt ecosystem function (reviewed in Estes et al. 2011). Declines of top marine predators, such as sharks, have been widely documented (Ferretti et al. 2010) and have raised international concern (Dulvy et al. 2008, Worm

*Corresponding author: jeremy.kiszka@gmail.com

et al. 2013), but there is a paucity of data regarding trophic relationships among species within this predatory guild. For coastal ecosystems, it has been proposed that predator declines could initiate trophic cascades, whereby their removal disrupts the natural population abundances, or behaviours, of consecutive lower trophic level species (Myers et al. 2007, Heithaus et al. 2008, Burkholder et al. 2013). Elasmobranchs are upper trophic level predators in many marine ecosystems, but there is considerable variation in diets and relative trophic position among species (Wetherbee & Cortés 2004, Hussey et al. 2014). The presence of a diverse marine top predator community, such as elasmobranchs, including an abundance of sympatric and ecologically interacting species, may be important in structuring some marine communities (Heithaus et al. 2010). It is unclear, however, whether diverse marine top predator faunas represent trophically redundant species (Myers et al. 2007) or inhabit unique foraging niches with differential impacts on overall community structure (Kinney et al. 2011, Heithaus et al. 2013).

Although 95 % of global fishers are artisanal (Pauly 2006), detailed information on fisheries catch composition is limited due to a lack of monitoring and reporting as a result of restricted financial and logistical capacity. In developing countries, artisanal fisheries are the principal fishing practice and are consequently of considerable social and economic importance to regional human populations. These fisheries, however, can negatively impact the abundance and species composition of vulnerable species such as elasmobranchs (Pinnegar & Engelhard 2008). Continued unregulated exploitation can lead to declines of key species with consequences extending to the broader food web, including commercial species that are critical to the livelihoods of local populations. Along the coast of Madagascar, elasmobranchs are heavily exploited both for subsistence (meat) and commercial (fins) purposes with an active and developed export market (McVean et al. 2006, Robinson & Sauer 2013). Between October 2001 and October 2002, 13 species were identified from a total of 1164 individual elasmobranchs caught off the south-western region of Madagascar. Hammerhead sharks (mostly *Sphyrna lewini*, and to a lesser extent *S. mokarran* and *S. zygaena*) represented 29 % of the catch by number and 24 % of the total wet weight with an estimate of over 123 metric tons landed (McVean et al. 2006). Hammerhead sharks (*Sphyrna* spp.) are globally threatened, with *S. lewini* currently classified as Endangered (IUCN Red List). Currently, no baseline ecological data exists for elasmobranchs

from this region, but a decline in elasmobranch abundance has been observed for the most exploited species (McVean et al. 2006). This decline of elasmobranchs from the coastal waters of south-western Madagascar could have adverse effects on both fishing communities and marine ecosystems. To assess this possibility, there is a need for information on these species including their trophic interactions and levels of trophic redundancy in the elasmobranch community.

Nitrogen and carbon stable isotopes provide chemical tracers of the diets and foraging habitats of organisms in a given ecosystem. Because offshore or pelagic-derived food webs tend to be ^{13}C-depleted compared to inshore or benthic food webs (Hobson 1999), carbon isotope values (δ^{13}C) can be used to indicate the foraging habitat of a species. In addition, the relative position of the consumer in the food web can be estimated from nitrogen isotopes values (δ^{15}N) because of the enrichment of ^{15}N through successive trophic transfers (Hobson 1999, Caut et al. 2009, Hussey et al. 2014). To date, carbon and nitrogen stable isotopes have been used to elucidate aspects of elasmobranch trophic and foraging ecology, including niche breadth and separation (Kinney et al. 2011, Speed et al. 2011, Vaudo & Heithaus 2011, Heithaus et al. 2013), individual foraging specialization (Matich et al. 2011), and ontogenetic and sex variation in trophic interactions and habitat use (e.g. Hussey et al. 2011, Carlisle et al. 2012). Long-term integrated stable isotope values also provide information on the role of elasmobranchs within a food web (McMeans et al. 2010, Vaudo & Heithaus 2011). In this study, we used stable isotopes to assess trophic relationships, isotopic niche breadth and overlap, as well as ontogenetic variation in trophic interactions among the most commonly caught elasmobranch species in artisanal fisheries operating in the coastal waters off south-western Madagascar (Table 1), including the scalloped hammerhead shark *S. lewini*, the slit-eye shark *Loxodon macrorhinus*, the giant guitarfish *Rhynchobatus djiddensis* and the silky shark *Carcharhinus falciformis*.

MATERIALS AND METHODS

Study sites and sample collection

Shark samples were collected at 5 fish villages (landing sites) in south-western Madagascar: Andavadoaka, Nosy Be, Nosy Hao, Lamboara and Nosy Andriamitaroka (Fig. 1). Southern areas (Andavado-

Table 1. Habitat, minimum depth range ('min depth') and major prey type ('diet') of elasmobranch species investigated in this study

Species	Habitat	Min depth (m)	Diet	Reference
Rhynchobatus djiddensis	Benthic	0–200	Small fishes, molluscs and crustaceans	Darracott (1977), Compagno (1986)
Loxodon macrorhinus	Demersal	0–80	Small demersal fishes and cephalopods, crustaceans	Compagno (1984)
Carcharhinus falciformis	Epipelagic, oceanic	0–500	Large squids, pelagic fishes, pelagic crabs	Compagno (1984), Bonfil (2008), Cabrera-Chávez-Costa et al. (2010)
Sphyrna lewini	Coastal, oceanic	0–275	Pelagic to demersal fishes, squids and elasmobranchs	Compagno (1984), Baum et al. (2007), Hussey et al. (2011)

aka, Nosy Be, Nosy Hao, Lamboara) are characterised by 2 distinct fringing and barrier reef systems separated by a 5 km wide passage or channel in which several patch reefs are situated. The northern area (Nosy Andriamitaroka) is characterised by a shallow underwater shelf, approx. 30 km in width and generally less than 20 m deep. A scattered coral bank lies at the seaward periphery adjacent to the

continental shelf drop-off. Sharks were caught by traditional fishers using longlines and gillnets from traditional non-motorised sailing pirogues (6 to 8 m long). Longlines consisted of an anchor line 50 to 100 m long and a buoyed surface line, 50 to 100 m long, with three 12 m long snoods approx. 25 m apart which are attached to the surface line (McVean et al. 2006). Gillnets, the most commonly used gear, are approx. 50 m long and 4.5 m deep with a mesh size of 20 to 25 cm. The nets are typically set on the bottom, in water approx. 30 m deep, and are generally baited with fish. Trained local data collectors surveyed the 5 fish landing sites year round from April 2009 to May 2010 to collect elasmobranch muscle samples for stable isotope analysis. White muscle tissue samples from the dorsal region of freshly landed sharks were collected from the most commonly recorded species, and were frozen at −20°C until further processing. Basic morphometric measurements, including total and fork length (TL and FL, respectively) and sex were recorded for each individual sampled.

Stable isotope analyses

Elasmobranch white muscle tissue was freeze-dried, ground into a homogeneous powder and lipids were removed by 2 successive extractions (1 h shaking in cyclohexane at room temperature and subsequent centrifugation) prior to analysis to standardize data among individuals and across species within the food web (Hussey et al. 2012a). This process also removes urea and trimethyl-amine oxide (TMAO) present in shark tissues, which can potentially affect $\delta^{15}N$ values (Hussey et al. 2012b). A small sub-sample of tissue (0.35 to 0.45 mg) was weighed and stable isotope measurements performed with a continuous-flow isotope-ratio mass spectrometer (Delta V Advan-

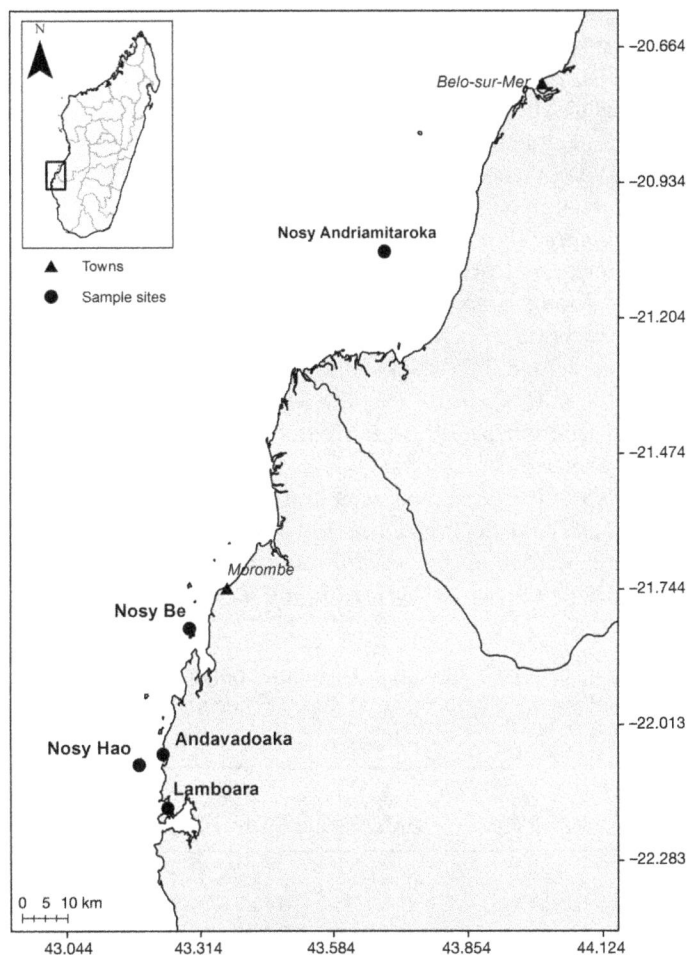

Fig. 1. The study area in south-western Madagascar

tage, Thermo Scientific) coupled to an elemental analyser (Flash EA1112, Thermo Scientific). Reference gases were calibrated against International Reference Materials (IAEA-N1, IAEA-N2 and IAEA-N3 for nitrogen; NBS-21, USGS-24 and IAEA-C6 for carbon). Results are expressed in the δ notation relative to PeeDee Belemnite and atmospheric N_2 for $\delta^{13}C$ and $\delta^{15}N$, respectively, according to the equation: $\delta X = [(R_{sample}/R_{standard}) - 1] \times 10^3$, where X is ^{13}C or ^{15}N and R is the isotope ratio $^{13}C/^{12}C$ or $^{15}N/^{14}N$, respectively. Replicate measurements of a laboratory standard (acetanilide) indicated that analytical errors were <0.1‰ for $\delta^{13}C$ and $\delta^{15}N$. Percent C and N elemental composition of tissues were used to calculate the sample C:N ratio, and indicated satisfactory lipid removal efficiency (mean ± SD C:N = 3.12 ± 0.2).

Data analysis

Assumptions regarding normality and homogeneity of variance were not met following Shapiro-Wilks and F tests. Non-parametric Kruskal-Wallis (H) tests were consequently used to examine the difference in stable isotopes values ($\delta^{15}N$ and $\delta^{13}C$) among species. Wilcoxon (W) signed rank tests were performed to assess differences in $\delta^{13}C$ and $\delta^{15}N$ values between sexes and ANCOVA was used to test the influence of size and species on $\delta^{13}C$ and $\delta^{15}N$ values of the 2 most commonly sampled species (*S. lewini* and *L. macrorhinus*). The ANCOVA is a general linear model with a continuous outcome variable ($\delta^{15}N$ and $\delta^{13}C$ values) and 2 or more predictor variables where at least one is continuous (size; FL) and at least one is categorical (species). In order to compare isotopic niches and infer habitat ($\delta^{13}C$) and resource ($\delta^{15}N$) separation among the 4 species, we used the recently developed SIBER metric (Stable Isotope Bayesian Ellipses using R; Jackson et al. 2011). Corrected standard ellipses (SEAc) are calculated from the variance and covariance of the data matrix and represent core

niche or dietary isotopic space while accounting for small sample sizes per species and variable sample sizes among species. The OVERLAP command within SIBER was used to calculate the percentage of core niche overlap among species $\delta^{13}C$ and $\delta^{15}N$ ellipse space (Jackson et al. 2011). Data were analysed using R v. 2.12.0 (R Development Core Team 2010).

RESULTS

Over the 12 mo study period, 84 *Loxodon macrorhinus*, 40 *Sphyrna lewini*, 20 *Rhynchobatus djiddensis* and 7 *Carcharhinus falciformis* samples were obtained. The size range of individuals sampled was highly variable among species, with juveniles making up the majority of individuals, especially for *S. lewini* and *R. djiddensis* (Table 2, Fig. 2). The $\delta^{13}C$ value was lowest for *S. lewini* (mean ± SD; −15.9 ± 1.19‰) and highest for *C. falciformis* (−14.4 ± 2‰), while the mean $\delta^{15}N$ was lowest for *L. macrorhinus* (11.7 ± 0.59‰) and the highest for *C. falciformis* (12.9 ± 1.36‰; Table 2). Stable isotope values found in this study were graphically compared to other teleost fishes with known trophic level and habitat (see Daly et al. 2013), including offshore pelagic, coastal pelagic and coastal demersal species (Fig. 3). Based on these data, *C. falciformis* and *S. lewini* do not have truly oceanic isotopic values. Conversely, it seems that *L. macrorhinus* and *R. djiddensis* are coastal consumers foraging in benthic and inshore habitats (Fig. 3).

Mean isotopic values varied among the 4 species for both $\delta^{13}C$ (H = 10.8; df = 3; p = 0.01) and $\delta^{15}N$ (H = 10.8; df = 3; p = 0.01; Fig. 3). At the individual level, $\delta^{13}C$ and $\delta^{15}N$ data were tightly clustered for *L. macrorhinus* and *R. djiddensis* but were more dispersed for *S. lewini* and *C. falciformis* (Figs. 3 & 4). Isotopic ellipse size was highly variable among species, with *R. djiddensis* having the smallest isotopic niche (0.3) and *C. falciformis* the largest (5.4) (Fig. 4). Ellipse sizes for *S. lewini* and *L. macrorhinus* were

Table 2. Number of sampled sharks (n), mean and range fork length, sex ratio, percentage of mature individuals in sampling of our study based on length of maturity from Compagno (1984), and mean ± SD values for stable nitrogen and carbon isotopes and C:N ratios

Species	n	Fork length (cm) Mean	Range	Sex ratio (m:f)	% of mature ind.	$\delta^{15}N$ Mean ± SD	$\delta^{13}C$ Mean ± SD	C:N Mean ± SD
Loxodon macrorhinus	84	83.7	62–100	1:1	85	11.66 ± 0.59	−15.4 ± 1.01	3.14 ± 0.18
Carcharhinus falciformis	7	162.9	73–260	0.9:1	29	12.89 ± 1.36	−14.37 ± 1.2	3.02 ± 0.1
Rhynchobatus djiddensis	20	86	65–190	0.8:1	5	11.92 ± 0.42	−14.66 ± 0.43	3.13 ± 0.1
Sphyrna lewini	40	96.5	58–172	0.8:1	5	12.42 ± 1.05	−15.89 ± 1.19	3.1 ± 0.2

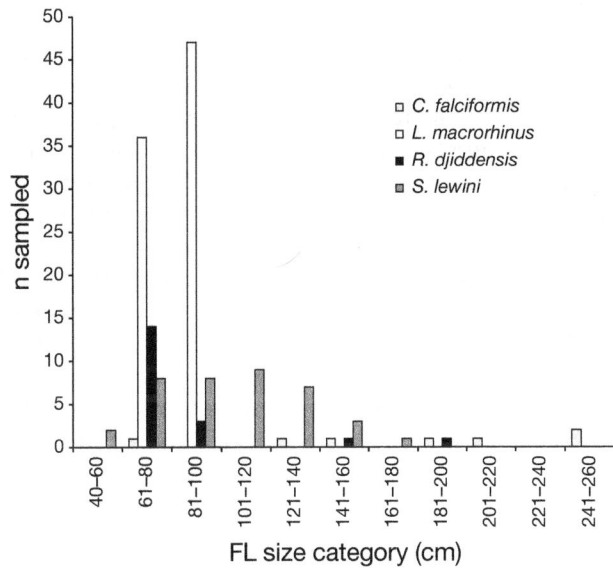

Fig. 2. Fork length (FL) distribution of *Carcharhinus falciformis*, *Loxodon macrorhinus*, *Rhynchobatus djiddensis* and *Sphyrna lewini* sampled off SW Madagascar from 2009 to 2010

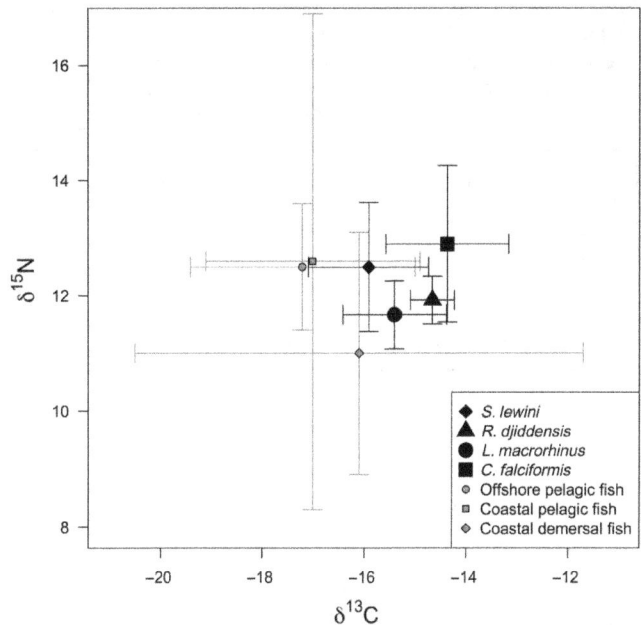

Fig. 3. Stable isotope values (δ^{15}N and δ^{13}C, ‰) in elasmobranch muscle (mean ± SD) from Madagascar (this study) compared to teleost fishes from offshore pelagic, coastal pelagic and coastal demersal habitats (from Daly et al. 2013)

4.1 and 1.9, respectively. Isotope niche overlap, based on the measure of ellipse overlap was high for *R. djiddensis* compared with *C. falciformis* (93.2%) and *L. macrorhinus* (73.2%), and to a lesser extent *S. lewini* (13.3%, Table 3). Higher isotopic niche overlap was also found between *L. macrorhinus* and both *S. lewini* (48.3%) and *C. falciformis* (46.2%), but most other comparisons yielded low niche overlap values (Fig. 4; Table 3).

For *S. lewini* and *L. macrorhinus*, ANCOVA confirmed that the 2 species were significantly distinct in their δ^{15}N values, and that there was an effect of individual size ($F_{3,116}$ = 30.75, p < 0.0001, Fig. 5). Smaller individuals of both species had similar δ^{15}N values, but there was an increase in δ^{15}N with size in *S. lewini* (R^2 = 0.49, p < 0.0001) that was not observed in the smaller-bodied *L. macrorhinus*. In contrast, ANCOVA found no influence of species and size on δ^{13}C values ($F_{3,116}$ = 2.05, p = 0.11; Fig. 5). Intraspecific variation in δ^{15}N and δ^{13}C values was high for both *S. lewini* and *L. macrorhinus* (Table 2; Fig. 5), especially for δ^{13}C, highlighting potentially inter-individual differences in diets and/or foraging habitats. There was no difference between male and female δ^{13}C values for either species (*S. lewini*: W = 161.5; p = 0.60 and *L. macrorhinus*: W = 871.5; p = 0.77) or for δ^{15}N values of *S.*

lewini (W = 150; p = 0.39). For *L. macrorhinus*, males had significantly higher δ^{15}N values (mean = 11.8 ± 0.54‰) than females (mean = 11.5 ± 0.62‰; W = 675; p = 0.04).

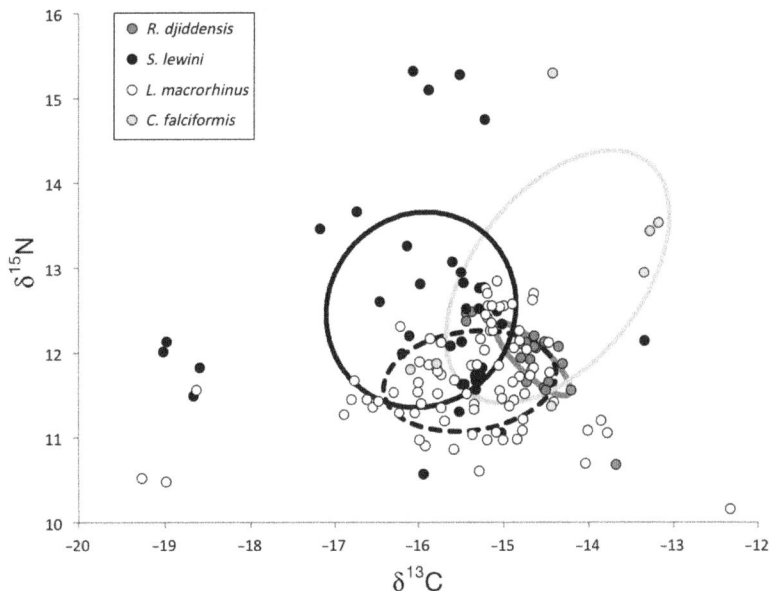

Fig. 4. Individual stable isotope values (δ^{15}N and δ^{13}C, ‰) and corrected standard ellipses (SEAc) for *Carcharhinus falciformis* (light grey continuous line), *Loxodon macrorhinus* (black dashed line), *Rhynchobatus djiddensis* (dark grey continuous line) and *Sphyrna lewini* (black continuous line)

Table 3. Percentage of isotopic niche overlap (δ^{15}N vs δ^{13}C) based on corrected standard ellipses (SEAc) for the 4 species investigated. Two-way overlap values are presented for species comparisons, i.e. a value for *R. djiddensis* overlap with *C. falciformis* (93.2%) and *C. falciformis* overlap with R. djiddensis (5.7%)

	Rhynchobatus djiddensis	*Loxodon macrorhinus*	*Sphyrna lewini*	*Carcharhinus falciformis*
R. djiddensis	–	12.8	1.1	5.7
L. macrorhinus	73.2	–	22.4	16.1
S. lewini	13.3	48.3	–	19.3
C. falciformis	93.2	46.2	26.6	–

DISCUSSION

These data provide the first investigation of trophic interactions, isotopic niches, ontogenetic and gender variation in the foraging ecology of 4 heavily exploited elasmobranch species from a data-poor region, the Mozambique Channel off south-western Madagascar. *Sphyrna lewini*, *Loxodon macrorhinus*, *Rhynchobatus djiddensis* and *Carcharhinus falciformis* are among the most commonly caught species by artisanal fishers over continental shelf waters in this region (~75% of species landed in the sampled region; F. Humber unpubl. data), and likely constitute a large proportion of the elasmobranch biomass. These species also account for a high percentage of the catch of artisanal fisheries off northern Madagascar, where concern over exploitation rates has recently been documented (Robinson & Sauer 2013). Sampling covered a range of sizes (FL range 62 to

260 cm) for *L. macrorhinus* and *C. falciformis*, but consisted of mostly juvenile *S. lewini* and *R. djiddensis*, likely reflecting artisanal shark fisheries overlapping with nursery areas. It is possible, however, that long-term exploitation of elasmobranchs in this region, primarily for the fin trade, has altered the size structure of populations.

In order to correctly interpret stable isotope values in the tissues of juvenile sharks, especially those with long turnover rates (e.g. muscle), it is critical to understand the dynamics of maternal provisioning (McMeans et al. 2010). Previous studies of mother–offspring differences of stable isotopes ratios in placentatrophic sharks have shown that embryos are generally enriched in δ^{15}N but fractionation of δ^{13}C is variable among species (McMeans et al. 2009, Vaudo et al. 2010). Based on data from the literature (size at birth and growth parameters; Compagno 1984, 1986), all animals from our study were likely more than 1 yr old. Therefore, maternal influences should have very limited impacts on isotopic values of the individuals in this study.

Stable isotope data indicated that, although the 4 species differed in their relative trophic position (indicated by mean δ^{15}N values), δ^{13}C values suggested considerable overlap in the food webs where species were foraging. There were, however, differences in how trophic interactions varied with size for *S. lewini* and *L. macrorhinus* and between sexes of *L. macrorhinus*. The 4 focal species are morphologi-

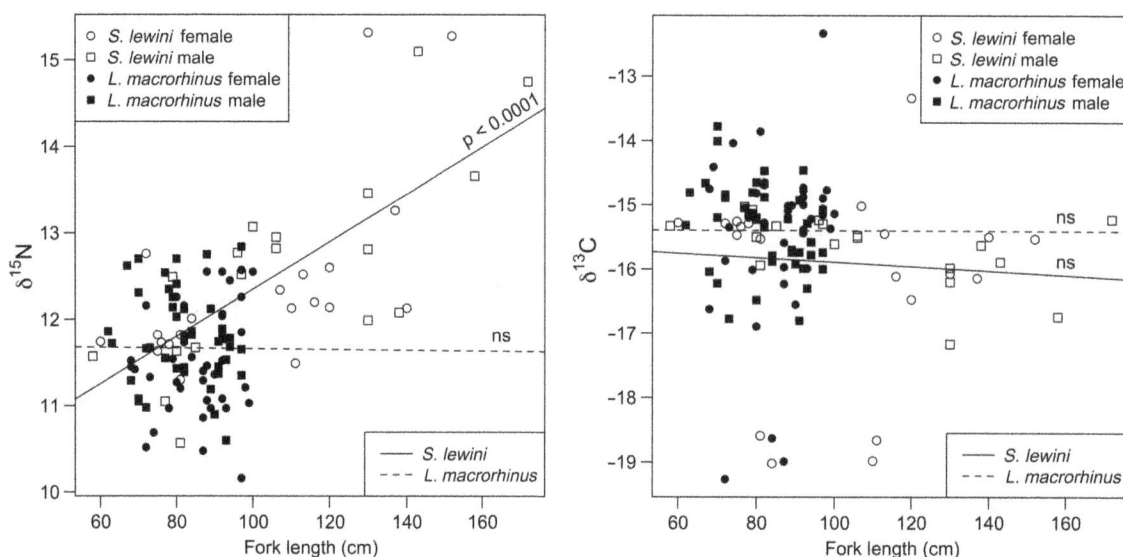

Fig. 5. Relationships between δ^{15}N and δ^{13}C and fork length in *Sphyrna lewini* and *Loxodon macrorhinus* males and females. Significance for r^2 values: $p < 0.0001$ and not significant, ns

cally and ecologically distinct in terms of body size and feeding apparatus, diet consumed, and habitat use patterns (see Table 1). *C. falciformis* and *S. lewini* are wide-ranging and occur in oceanic, epipelagic and continental shelf habitats. They feed mostly on epipelagic and pelagic fish, cephalopods and crustaceans (Compagno 1984, Bonfil 2008). Juvenile *S. lewini* reside in coastal nursery grounds (Simpfendorfer & Milward 1993) that probably extend over the continental shelf off south-western Madagascar, as observed off south-eastern Africa (Diemer et al. 2011). *R. djiddensis* also occurs over the continental shelf, but, in contrast to *S. lewini* and *C. falciformis*, is adapted to shallow coastal waters (i.e. closer to more ^{13}C-enriched benthic sources) and is thought to have relatively restricted home range and to feed mostly on molluscs and crustaceans (Darracott 1977, Compagno 1986). *L. macrorhinus* is a poorly known continental shelf species that lives in intertidal areas up to 80 m depth and which feeds on teleosts, coastal cephalopods and crustaceans (Compagno 1984). Despite these marked differences, there was considerable overlap in δ^{13}C values among the 4 species. The larger variation in δ^{13}C values observed in *L. macrorhinus* and *S. lewini* compared to *R. djiddensis* likely indicate the diversity of habitats and feeding areas encountered with larger home ranges, especially for the more pelagic *S. lewini*. The inability of δ^{13}C values to discriminate known interspecific variation in seasonal fine-scale vertical (pelagic vs. benthic) and horizontal (oceanic vs. coastal) distributions likely relates to the slow turnover rate of muscle tissue (>250 d; Kim et al. 2012). Such limitations of carbon isotopes in resolving fine-scale foraging patterns of seabirds and marine mammals have been reported in various ecological contexts (Cherel et al. 2008, Méndez-Fernandez et al. 2012). For example, despite obvious differences in habitat preferences of small cetaceans off the Iberian Peninsula (e.g. depth distributions; Pierce et al. 2010), similar δ^{13}C values were recorded among species (Méndez-Fernandez et al. 2012).

Unfortunately, isotopic data on prey from specific habitats and food webs could not be collected during our study. Such data would provide even greater insights into the diets, foraging habitats and home ranges of the species studied here (e.g. Daly et al. 2013, Kiszka et al. 2014). However, data from neighbouring areas (e.g. coastal waters of Mozambique; Daly et al. 2013) can provide insights into habitat use by our focal species off SW Madagascar. Based on these regional data, none of the most pelagic species, especially *C. falciformis* and *S. lewini* had truly oceanic isotopic values. Indeed, both species from south-western Madagascar had higher δ^{13}C values than those measured in pelagic teleosts (Daly et al. 2013). This pattern is similar to published data on carbon and nitrogen isotope values from the same region for *C. falciformis*. Both δ^{15}N and δ^{13}C were significantly higher for *C. falciformis* off south-western Madagascar than was found further north in the Mozambique Channel (Rabehagasoa et al. 2012; see Table 2). This could be due to spatial variation in diets of *C. falciformis* in this region. However, the individuals we sampled were significantly larger than those analysed by Rabehagasoa et al. (2012). Therefore, ontogenetic shifts in diets, which have been documented for this species (Compagno 1984, Cabrera-Chávez-Costa et al. 2010, Rabehagasoa et al. 2012), may also explain the observed differences with *C. falciformis* feeding on higher trophic level prey at larger sizes. Moreover, higher δ^{13}C values observed in this species in our study area may either reflect an increasing relative importance of coastal prey with age, or a larger prevalence of coastal prey in the diet of *C. falciformis* (where the continental shelf is broader). In contrast, based on a comparison of data from a variety of pelagic and coastal teleost fish species from the coastal waters of Mozambique, it seems that *L. macrorhinus* and *R. djiddensis* are truly coastal consumers foraging in inshore and benthic habitats (Daly et al. 2013). Interestingly, δ^{13}C values measured in *S. lewini* and *R. djiddensis* from south-western Madagascar were similar to those from north-western India (Borrell et al. 2011) and KwaZulu-Natal, South Africa (for *S. lewini* only; Hussey et al. 2011). However, δ^{15}N values of both *S. lewini* and *R. djiddensis* were significantly lower off south-western Madagascar (Table 2), compared to those measured in India for the same species (Borrell et al. 2011). This regional variation highlights the need for region-specific isotopic data to evaluate the trophic interactions, including for wide-ranging consumers such as large elasmobranchs.

Differences in trophic interactions among these 4 highly exploited elasmobranchs suggest these species are not trophically redundant. This is consistent with previous studies from other geographic locations, where elasmobranchs were segregated based on their relative trophic position or mean δ^{15}N values (Cortés 1999, Borrell et al. 2011, Heithaus et al. 2013, Hussey et al. 2014). Such isotopic segregation within a community has been documented in a number of coastal, oceanic and reef-associated elasmobranch species (e.g. Kinney et al. 2011, Speed et al. 2011, Rabehagasoa et al. 2012), but not in assemblages where species show morphological, taxonomic and habitat similarities over small spatial scales (Vaudo & Heithaus 2011). The highest degree of overlap in $\delta^{15}N$ values was found between individuals of *R. djiddensis* and *L. macrorhinus,* species that both occur in shallow waters of the continental shelf. Overlap in diet would therefore be expected, although fine-scale niche partitioning (concealed by isotope analyses) may occur. Detailed stomach content data collected over a seasonal cycle would be required to determine this. *S. lewini* and *L. macrorhinus* also exhibited a moderately high degree of niche overlap, but the former had a significantly higher mean $\delta^{15}N$ value (related to an ontogenetic diet shift with size). Juvenile *S. lewini* and *L. macrorhinus* of all size classes had similar $\delta^{15}N$ values, also related to their occurrence in coastal waters and the potential for high niche overlap. It is important to note that establishing isotopic baselines and system end points for carbon and nitrogen stable isotopes in this region would be required to further elucidate the trophic ecology and roles of these species. The isotopic niche sizes varied among species, with $\delta^{15}N$ and $\delta^{13}C$ values tightly clustered for *L. macrorhinus* and *R. djiddensis* but more dispersed for *S. lewini* and *C. falciformis*. The large variability in *S. lewini* and *L. macrorhinus* isotopic values may indicate either a high level of generalist feeding behaviour or, more likely, because of the long turnover times of muscle tissue, consistent differences in average trophic interactions of individuals within generalist populations. Such behaviour has been found in juvenile bull sharks *C. leucas* in a nursery area of Florida (Matich et al. 2011) and has been suggested for a number of other elasmobranch taxa such as batoids (Vaudo & Heithaus 2011). In contrast, *L. macrorhinus* and *R. djiddensis* may be more specialized at the population level or individuals may all have similarly generalized diets over the time period that muscle isotopic values are integrated. The small sample size for *C. falciformis* may have influenced the standard ellipse (SEAc) size through artificially inflating the size of the niche area (Jackson et al. 2011). Given the observed variability in the data, however, this bias is likely minimal. Stable isotope analyses of multiple tissues with different turnover rates would address questions related to individual foraging specialisation (Matich et al. 2011) and elucidate more intricate seasonal differences (Kinney et al. 2011).

Ontogenetic and sex differences in diets and habitat preferences are related to age- and sex-specific energy requirements, vulnerability to predators, and social considerations, and have been documented for a diversity of taxa (e.g. Beier 1987, Breed et al. 2006), including elasmobranchs (Lowe et al. 1996, Estrada et al. 2006, Hussey et al. 2011, Rabehagasoa et al. 2012). Such ontogenetic niche shifts can impact population dynamics, community structure, and ecosystem function (Hammerschlag-Peyer et al. 2011, Hussey et al. 2011). Relative trophic level, inferred by $\delta^{15}N$ values, significantly increased with body length in *S. lewini*, suggesting a dietary change with age in this species that likely reflects moving from more coastal to pelagic habitats and foraging on higher trophic level prey (Compagno 1984, Borrell et al. 2011, Hussey et al. 2011). Spatial segregation has been documented for a diversity of elasmobranch species (Compagno 1984), including through the use of stable isotopes (e.g. Hussey et al. 2011). In our study, differences in $\delta^{15}N$ values between male and female *L. macrorhinus* indicate that males feed at a higher relative trophic position compared to females. Whether this reflects foraging on different prey taxa or sexual segregation remains to be elucidated. We did not detect sex differences in isotopic values of *S. lewini*, which contrasts with results from individuals captured off the coast of KwaZulu-Natal, South Africa (Hussey et al. 2011). Off South Africa, $\delta^{13}C$ values of males and females between 120 and 160 cm (precaudal length) supported sexual segregation, with females most likely spending more time in oceanic waters (Hussey et al. 2011).

CONCLUSIONS

In summary, these data suggest that complex trophic structuring occurs in this highly exploited elasmobranch assemblage. Differences in trophic interactions appear to be driven by a combination of interacting factors including habitat use, home-range size, diets, and variation in all of these factors across size classes, sexes and individual behaviours. Consequently, species-specific population declines as a result of continued unregulated exploitation have the potential to lead to intricate species-specific cascades within the coastal waters of south-western Madagascar. More detailed sampling of the food web, however, is required to examine isotopic variation in prey items consumed by the 4 species and to determine if benthic and pelagic and coastal/offshore ecosystems can be readily distinguished. Given the IUCN Red list categories for 2 of the species ('Vulnerable', *R. djiddensis*; 'Endangered', *S. lewini*), local management is imperative to regulate regional fisheries. More data is also required on the biological parameters of these populations and other large predators in the system to incorporate in food web models to examine long-term effects of removing trophically distinct species within artisanal fisheries.

Acknowledgements. Our particular thanks are addressed to sample and data collectors in south-western Madagascar, working on behalf of Blue Ventures. We are grateful to G. Guillou and P. Richard (University of La Rochelle, UMR LIENSs), for running the stable isotope analyses. This work has been supported financially by LIENSs and the CPER 13 (Contrat de Projet Etat-Région).

LITERATURE CITED

Baum J, Clarke S, Domingo A, Ducrocq M and others (2007) *Sphyrna lewini*. In: IUCN 2012. IUCN Red List of Threatened Species. Version 2012.2. www.iucnredlist.org

Beier P (1987) Sex differences in quality of white-tailed deer diets. J Mammal 68:323–329

Bonfil R (2008) The biology and ecology of the silky shark, *Carcharhinus falciformis*. In: Camhi MD, Pikitch EK, Babcock EA (eds) Sharks of the open ocean. Blackwell, Oxford, p 114–127

Borrell A, Cardona L, Kumarran RP, Aguilar A (2011) Trophic ecology of elasmobranchs caught off Gujarat, India, as inferred from stable isotopes. ICES J Mar Sci 68: 547–554

Breed GA, Bowen WD, McMillan JI, Leonard ML (2006) Sexual segregation of seasonal foraging habitat in a non-migratory marine mammal. Proc R Soc Lond Ser B Biol Sci 273:2319–2326

Burkholder DA, Heithaus MR, Fourqurean JW, Wirsing A, Dill LM (2013) Patterns of top-down control in a seagrass ecosystem: Could a roving apex predator induce a behaviour-mediated trophic cascade? J Anim Ecol 82: 1192–1202

Cabrera-Chávez-Costa AA, Galván-Magaña F, Escobar-Sánchez OE (2010) Food habits of the silky shark *Carcharhinus falciformis* (Müller & Henle, 1839) off the western coast of Baja California Sur, Mexico. J Appl Ichthyol 26:499–503

Carlisle AB, Kim SL, Semmens BX, Madigan DJ and others (2012) Using stable isotope analysis to understand the migration and trophic ecology of northeastern Pacific white sharks (*Carcharodon carcharias*). PLoS ONE 7: e30492

Caut S, Angulo E, Courchamp F (2009) Variation in discrimination factors (Δ^{15}N and Δ^{13}C): the effect of diet isotopic values and applications for diet reconstruction. J Appl Ecol 46:443–453

Cherel Y, Le Corre M, Jaquemet S, Ménard F, Richard P, Weimerskirch H (2008) Resource partitioning within a tropical seabird community: new information from stable isotopes. Mar Ecol Prog Ser 366:281–291

Compagno LJV (1984) FAO species catalogue, Vol. 4. Sharks of the world. An annotated and illustrated catalogue of shark specimen known to date. Part 2. Carcharhiniformes. FAO Fish Synop 125:251–655

Compagno LJV (1986) Rhinobatidae. In: Smith MM, Heemstra PC (eds) Smiths' sea fishes. Springer-Verlag, Berlin, p 128–131

Cortés E (1999) Standardized diet compositions and trophic levels of sharks. ICES J Mar Sci 56:707–717

Daly R, Froneman PW, Smale MJ (2013) Comparative feeding ecology of bull sharks (*Carcharhinus leucas*) in the coastal waters of the southwest Indian Ocean inferred from stable isotope analysis. PLoS ONE 8:e78229

Darracott A (1977) Availability, morphometrics, feeding and breeding activity in a multi-species, demersal fish stock of the Western Indian Ocean. J Fish Biol 10:1–16

Diemer KM, Mann BQ, Hussey NE (2011) Distribution and movement of scalloped hammerhead *Sphyrna lewini* and smooth hammerhead *Sphyrna zygaena* sharks along the east coast of southern Africa. Afr J Mar Sci 33: 229–238

Dulvy NK, Baum JK, Clarke S, Compagno LJV and others (2008) You can swim but you can't hide: the global status of and conservation of pelagic sharks and rays. Aquat Conserv 18:459–482

Estes JA, Terborgh J, Brashares JS, Power ME and others (2011) Trophic downgrading of planet earth. Science 333:301–306

Estrada JA, Rice AN, Natanson NJ, Skomal GB (2006) Use of isotopic analysis from vertebrae in reconstructing ontogenetic feeding ecology in white sharks. Ecology 87: 829–834

Ferretti F, Worm B, Britten GL, Heithaus MR, Lotze HK (2010) Patterns and ecosystem consequences of shark declines in the ocean. Ecol Lett 13:1055–1071

Hammerschlag-Peyer CM, Yeager LA, Araújo MS, Layman CA (2011) A hypothesis-testing framework for studies investigating ontogenetic niche shifts using stable isotope ratios. PLoS ONE 6:e27104

Heithaus MR, Frid A, Wirsing AJ, Worm B (2008) Predicting ecological consequences of marine top predator declines. Trends Ecol Evol 23:202–210

Heithaus MR, Frid A, Vaudo JJ, Worm B, Wirsing AJ (2010) Unravelling the ecological importance of elasmobranchs. In: Carrier JC, Musick JA, Heithaus MR (eds) Sharks and their relatives. II. Biodiversity, adaptive physiology and conservation. CRC Press, Boca Raton, FL, p 611–636

Heithaus MR, Vaudo J, Kreicker S, Layman CA and others (2013) Apparent resource partitioning and trophic structure of large-bodied predators in a relatively pristine seagrass ecosystem. Mar Ecol Prog Ser 481:225–237

Hobson KA (1999) Tracing origins and migration of wildlife using stable isotopes: a review. Oecologia 120:314–326

Hussey NE, Dudley SFJ, McCarthy ID, Cliff G, Fisk AT (2011) Stable isotope profiles of large marine predators: viable indicators of trophic position, diet and movement in sharks? Can J Fish Aquat Sci 68:2029–2045

Hussey NE, Olin JA, Kinney MJ, McMeans BC, Fisk AT (2012a) Lipid extraction effects on stable isotope values (δ^{13}C and δ^{15}N) of elasmobranch muscle tissue. J Exp Mar Biol Ecol 434-435:7–15

Hussey NE, MacNeil MA, Olin JA, McMeans BC, Kinney MJ, Chapman DD, Fisk AT (2012b) Stable isotopes and elasmobranchs: tissue types, methods, applications and assumptions. J Fish Biol 80:1449–1484

Hussey NE, MacNeil MA, McMeans BC, Olin JA and others (2014) Rescaling the trophic structure of marine food webs. Ecol Lett 17:239–250

Jackson AL, Inger R, Parnell AC, Bearhop S (2011) Comparing isotopic niche widths among and within communities: SIBER – Stable Isotope Bayesian Ellipses in R. J Anim Ecol 80:595–602

▶ Kim SL, del Rio CM, Casper D, Koch PL (2012) Isotopic incorporation rates for shark tissues from a long-term captive feeding study. J Exp Biol 215:2495–2500

▶ Kinney MJ, Hussey NE, Fisk AT, Tobin AJ, Simpfendorfer CA (2011) Communal or competitive? Stable isotope analysis provides evidence of resource partitioning within a communal shark nursery. Mar Ecol Prog Ser 439:263–276

▶ Kiszka J, Méndez-Fernandez P, Heithaus MR, Ridoux V (2014) The foraging ecology of coastal bottlenose dolphins based on stable isotope mixing models and behavioural sampling. Mar Biol 161:953–961

▶ Lowe C, Wetherbee BM, Crow GL, Tester AL (1996) Ontogenetic dietary shifts and feeding behavior of tiger sharks, *Galeocerdo cuvier*, in Hawaiian waters. Environ Biol Fishes 47:203–211

▶ Matich P, Heithaus MR, Layman CA (2011) Contrasting patterns of individual foraging specialization and trophic coupling in two marine apex predators. J Anim Ecol 80: 294–305

▶ McMeans BC, Olin JA, Benz GW (2009) Stable-isotope comparisons between embryos and mothers of a placenta-trophic shark species. J Fish Biol 75:2464–2474

▶ McMeans BC, Svavarsson J, Dennard S, Fisk A (2010) Diet and resource use among Greenland sharks (*Somniosus microcephalus*) and teleosts sampled in Icelandic waters using δ^{13}C, δ^{15}N, and mercury. Can J Fish Aquat Sci 67: 1428–1438

▶ McVean AR, Walker SCJ, Fanning E (2006) The traditional shark fisheries of southwest Madagascar: a study in the Toliara region. Fish Res 82:280–289

▶ Méndez-Fernandez P, Bustamante P, Bode A, Chouvelon T and others (2012) Foraging ecology of five toothed whale species in the Northwest Iberian Peninsula, inferred using δ^{13}C and δ^{15}N isotopic signatures. J Exp Mar Biol Ecol 413:150–158

▶ Myers RA, Baum JK, Shepherd TD, Powers SP, Peterson PH (2007) Cascading effects of the loss of apex predatory sharks for a coastal ocean. Science 315:1846–1850

Pauly D (2006) Major trends in small-scale fisheries, with emphasis on developing countries, and some implications for the social sciences. Marit Stud 4:7–22

▶ Pierce GJ, Caldas M, Cedeira J, Santos MB and others (2010) Trends in cetacean sightings along the Galician coast, north-west Spain, 2003–2007, and inferences about cetacean habitat preferences. J Mar Biol Assoc UK 90: 1547–1560

▶ Pinnegar JK, Engelhard GH (2008) The shifting baseline phenomenon: a global perspective. Rev Fish Biol Fish 18: 1–16

R Development Core Team (2010) R: A language and environment for statistical computing. R Foundation for Statistical Computing, Vienna

▶ Rabehagasoa N, Lorrain A, Bach P, Potier M, Jaquemet S, Richard P, Ménard F (2012) Isotopic niches of the blue shark *Prionace glauca* and the silky shark *Carcharhinus falciformis* in the southwestern Indian Ocean. End Spec Res 17:83–92

▶ Robinson J, Sauer WHH (2013) A first description of the artisanal shark fishery in northern Madagascar: implications for management. Afr J Mar Sci 35:9–15

▶ Simpfendorfer CA, Milward NE (1993) Utilisation of a tropical bay as a nursery area by sharks of the families Carcharhinidae and Sphyrnidae. Environ Biol Fishes 37: 337–345

▶ Speed CW, Meekan MG, Field IC, McMahon CR, Abrantes K, Bradshaw CJA (2011) Trophic ecology of reef sharks determined using stable isotopes and telemetry. Coral Reefs 31:357–367

▶ Vaudo JJ, Heithaus MR (2011) Dietary niche overlap in a nearshore elasmobranch mesopredator community. Mar Ecol Prog Ser 425:247–260

▶ Vaudo JJ, Matich P, Heithaus MR (2010) Mother–offspring isotope fractionation in two species of placentatrophic sharks. J Fish Biol 77:1724–1727

Wetherbee B, Cortés E (2004) Food consumption and feeding habits. In: Carrier JC, Musick JA, Heithaus MR (eds) Sharks and their relatives. CRC Press, Boca Raton, FL, p 225–246

▶ Worm B, Davis B, Ketterner L, Ward-Paige CA and others (2013) Global catches, exploitation rates, and rebuilding options for sharks. Mar Policy 40:194–204

9

Latitudinal variation in local productivity influences body condition of South American sea lion pups

Maritza Sepúlveda[1,2,*], Danai Olea[1], Pablo Carrasco[3,4],
Macarena Santos-Carvallo[1,2], Jorge Castillo[5], Renato A. Quiñones[3,4]

[1]Centro de Investigación y Gestión de los Recursos Naturales (CIGREN), Instituto de Biología, Facultad de Ciencias, Universidad de Valparaíso, Gran Bretaña 1111, Playa Ancha, Valparaíso 2360102, Chile
[2]Centro de Investigación Eutropia, Ahumada 131 Oficina 912, Santiago 8320238, Chile
[3]Programa de Investigación Marina de Excelencia (PIMEX), Facultad de Ciencias Naturales y Oceanográficas, Casilla 160-C, Universidad de Concepción, Concepción 4070205, Chile
[4]Interdisciplinary Center for Aquaculture Research (INCAR-FONDAP), Universidad de Concepción, O'Higgins 1695, Concepción 4070007, Chile
[5]Instituto de Fomento Pesquero, Casilla 8v, Valparaíso, Almte. Manuel Blanco Encalada 839, Valparaíso 2361827, Chile

ABSTRACT: In otariids, the condition and growth of offspring are linked to prey availability because females show a strong dependence on local food availability. Thus, it is expected that body condition of sea lion pups will vary spatially and/or temporally as a response to variations in the abundance of prey species on which females feed. The objective of this study was to analyze the geographic and temporal variation in the body condition of South American sea lion *Otaria flavescens* pups in Chile, and to relate it to spatio-temporal variations in prey availability. We captured 340 live pups in 2 distant colonies, Punta Patache/Punta Negra and Cobquecura, along the Chilean coast during consecutive breeding seasons. A morphometric index of pup body condition was estimated by comparing pups in all years using least-squares linear regression of the $\log_{10}$-transformed measurements of standard length vs. body mass. We analyzed the relationship between this index and estimates of fish biomass (as a proxy of prey availability) at each locality. We found that body condition was significantly different between years and between colonies, suggesting that animals of the central-south area were in better condition than those in the north. A positive relationship between body condition and fish biomass was found, suggesting that differences in body condition may be explained by spatial and temporal differences in prey availability.

KEY WORDS: Body condition index · Body mass · Environmental effects · Morphometrics · *Otaria flavescens* · Southern sea lion

INTRODUCTION

In marine environments, productivity varies both in spatial and temporal scales, influencing marine predator distribution, diet, foraging, diving behavior, and body condition (Lea et al. 2006, Womble et al. 2009). In particular, body condition is one of the main parameters affected by local productivity because it is a relative measure of the nutritional state or the level of energy reserves of an organism, commonly interpreted as a measure of general well-being of an individual or population (Labocha et al. 2014).

Local productivity is essentially important for colonial species, such as seabirds and otariids (fur seals and sea lions). These species are central place foragers and income breeders that combine foraging

*Corresponding author: maritza.sepulveda@uv.cl

activity at sea with the raising and provisioning of their young ashore (Lea et al. 2006, Womble et al. 2009). Otariids in particular are characterized by lengthy parental care and nursing periods. As such, the mother must return frequently ashore to provision her pup and is therefore limited to trips lasting from hours to a few days, and to distances from tens to a few hundreds of kilometers from the rookery (Costa & Shaffer 2012). Consequently, for a female, food availability in the area around the rookery is critical for the condition and survival of her offspring, thus creating a direct link between local prey availability and fitness (Womble et al. 2009).

The South American sea lion *Otaria flavescens* (Shaw, 1800), like other otariid species, shows a strong dependence upon local prey availability. Although mainly composed of fish and squid, its diet can be quite variable, as sea lions adapt easily to locally abundant prey (Cappozzo & Perrin 2009). After the birth of her pup, the mother alternates between periods on land nursing the pup and foraging trips in the ocean (Muñoz et al. 2011) and has a lengthy nursing period lasting from 7 mo up to 1 or even 2 yr. Foraging trips usually last 1 or 2 d; the female remains near the rookery and usually does not venture farther than 50 to 100 km (Rodríguez et al. 2013). Considering that growth and maintenance of the pups depends almost exclusively on the ability of their mothers to obtain enough food (Lea et al. 2006), it is expected that the body condition of South American sea lion pups will vary spatially and/or temporally as a response to the distribution and abundance of the prey on which females feed during the nursing period.

The Humboldt Current System (HCS) extends latitudinally from central-south Chile (~42° S) to northern Peru (~4 or 5° S) (Quiñones et al. 2010). The HCS off Chile is one of the most productive ecosystems in the world, with primary production rates of up to 25.8 g C m^{-2} d^{-1} (Daneri et al. 2000). This system, characterized by a south–north flow of surface water of subantarctic origin and by intense upwelling of nutrient-rich/low-oxygen equatorial subsurface waters, is a highly dynamic coastal environment whose primary productivity sustains an extremely high biomass of fish, which in turn supports one of the most productive fishing activities in the world (Daneri et al. 2000). However, although the coastal waters off Chile are in general very productive, significant spatial variations in primary productivity exist along the coast, evident in measurements of chlorophyll *a* (chl *a*) concentration, phytoplankton abundance, analyses of mesozooplankton samples,

and estimates of fish biomass (Daneri et al. 2000). From 2009 to 2012, fish biomass along the central-south coast of Chile averaged 1.4 million t yr^{-1}, and was never less than 850 000 t (www.fip.cl/proyectos. aspx). By contrast, fish biomass in the north averaged 235 000 t yr^{-1} in the same years, and never exceeded 380 000 t (www.fip.cl/proyectos.aspx). These estimates indicate that the productivity in central-south Chile is notably higher than in the north (Quiñones et al. 2010).

Considering the marked differences in primary productivity between the north and central-south coast of Chile, it is expected that body condition of sea lions pups from central-south Chile should be better than that of pups from the less productive coast in the north. The aim of this study was to analyze spatial and temporal variation in the body condition of the South American sea lion pups in Chile by using a morphometric index, and to relate this variation to spatial and temporal changes in prey availability.

MATERIALS AND METHODS

Study area

This study was carried out in 2 areas of the Chilean coast, separated by a distance of about 1730 km. One of the areas is located in the north of Chile and is formed by 2 rocks, Punta Patache (20° 48′ S, 70° 12′ W) and Punta Negra (20° 50′ S, 70° 10′ W), both sustaining breeding colonies of the South American sea lion. Hereafter, this area will be referred as PP/PN. The rocks are both located about 50 to 100 m off the coast of mainland Chile (Fig. 1). Colony sizes at PP/PN were estimated by Bartheld et al. (2008) to be 950 and 1245 animals, respectively. The second study area (Cobquecura; 36° 07′ S, 72° 48′ W) is a rocky island located about 80 to 100 m off the central-south coast of Chile. This represents the most important breeding colony along the coast of central Chile, with approximately 2900 animals (Sepúlveda et al. 2011) (Fig. 1). The pupping season for the South American sea lion in the Chilean coast occurs during January and February (austral summer months). According to Acevedo et al. (2003) and Quiñones et al. (2011), the birth peak is similar in the 2 localities, extending from late January to early February. A recent study based on mtDNA and microsatellite analyses showed that no spatial genetic structure occurs in *Otaria flavescens* along the Chilean coast from 18 to 42° S (Weinberger 2013), suggesting a low

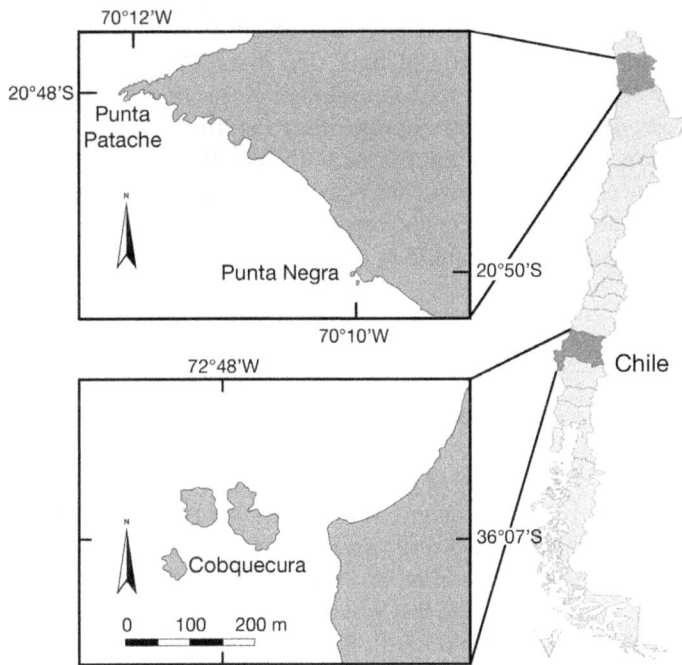

Fig. 1. Breeding colony study sites for South American sea lions in the north zone (Punta Patache/Punta Negra, PP/PN) and central-south zone (Cobquecura) of Chile

genetic differentiation and a high gene flow between sea lions from the 2 studied areas.

During the breeding season, high waves frequently sweep away healthy pups from these colonies and transport them, alive, to adjacent continental beaches. While some of these pups are capable of returning to the colony by themselves, others die on the beach from hunger and dehydration. We based our results on live pups that were captured immediately after stranding on the beach. This method has the advantage of not causing disturbances to the colony that could disrupt nursing behavior and/or reduce pup survival by trampling or abandonment of the pup (Green et al. 2010). However, it should be noted that our sampling method could be biased because not all the pups have the same probability of being swept away from the colony. This could be the case if the distribution of older and more experienced females versus younger females in the colony is not homogeneous. However, we assume that all pups have a similar probability of being stranded on the beach based on 3 criteria: (1) a similar bias operates in both colonies; (2) high waves frequently hit in different zones of the colonies; and (3) pups from different parts of the colonies are swept away from the colony to the beach. The latter has been observed during storms.

Capture and measurements of sea lion pups

In the PP/PN area, a total of 202 live pups (86 females and 116 males) in 2011 (n = 48; 17 females and 31 males), 2012 (n = 98; 44 females and 54 males) and 2013 (n = 56; 25 females and 31 males) were captured. During all years, 2 observers remained in front of the colony from 09:00 to 18:00 h for 1 mo (from the third week of January to the third week of February). Visits to the colony were done every other day, regardless of meteorological conditions. In the Cobquecura area, a total of 138 pups (56 females and 82 males) in 2009 (n = 11; 4 females and 7 males), 2010 (n = 36; 18 females and 18 males), 2011 (n = 72; 29 females and 43 males) and 2012 (n = 19; 5 females and 14 males) were captured. In this colony, 2 or 3 observers were placed in front of the colony from 09:00 to 18:00 h for 2 mo (1 January to 28 February). Visits to this colony were done on a daily basis. No other breeding colonies are close to the studied colonies, so it was assumed that all pups captured came from the studied rookeries.

To evaluate body condition, standard length (L, straight-line nose to tail, dorsal view) to the nearest 0.01 m, and body mass (M) to the nearest 0.1 kg was recorded for each pup. Pups were marked with water-resistant paint on one front flipper to avoid measuring the same individual again and set free. Body mass is proportional to body length (Bradshaw et al. 2000). In order to control for any possible effect of body size on individual variation in mass, a condition index was calculated by regressing $\log_{10}M$ against $\log_{10}L$ (Bradshaw et al. 2000). Male and female pups were analyzed separately. Applying the regression equation to $\log_{10}L$ gives $\log_{10}$ predicted mass (M_p):

$$\log_{10} M_p = a + b \times \log_{10}L \qquad (1)$$

where a is a constant and b is the regression coefficient. The ratio of observed mass (M_o) to M_p gives an index of relative condition (CI_m):

$$CI_m = M_o/M_p \qquad (2)$$

Because we sampled the pups over a 1 or 2 mo period during each year, it is possible that pup condition was subject to bias resulting from asynchrony in the date of capture (Bradshaw et al. 2000). To examine the amount of variation in pup condition that could be explained by sampling date, we used a least-squares regression of CI_m for pups from each sex–locality combination against the day of the year each pup was measured. Generalized linear models (GLMs) were used to examine the effect of colony and year of pup capture on the variation of CI_m. For a better comparison, only pups captured in 2011 and 2012 in the 2 colonies (n = 146 from PP/PN and n = 91 from Cobquecura) were used in this analysis. All statistical analyses were run in Statistica 7.0 software (StatSoft) with a significance level of $\alpha = 0.05$.

Estimation of prey availability

We also examined the relationship between fish biomass (as a proxy of prey availability) within each area and pup CI_m using a linear regression model. Data were log_{10} transformed for analysis. We analyzed the biomass of 2 pelagic clupeid species: anchovy *Engraulis ringens* and common sardine *Strangomera bentincki*, which are considered to be the most economically important resources for both industrial and small-scale fisheries in north and central-south Chile (Gutiérrez-Estrada et al. 2007, Castillo-Jordán et al. 2010), and commonly reported as one of the most important prey species in the diet of the South American sea lion in Chile, both in the north (>90% of diet composition; Sielfeld 1999) and in the central-south (30 to 60% of diet composition; Muñoz et al. 2013). Density and fish biomass data were obtained from acoustic surveys carried out onboard the RV 'Abate Molina', a stern trawler 43.2 m long and with 1400 HP. Two study areas were used to examine fish biomass, a northern area located between Arica (18° 22' S) and Punta Buitre (24° 40' S) extending from 1 nautical mile (nmi) to 20 nmi along the coast (Fig. 2a), and a central-south area located between 33° 40' S and 40° 00' S (Fig. 2b). The surveys in the north were carried out during December from 2010 to 2013, and during January from 2010 to 2013 in the central-south. In both zones, systematic acoustic sampling was performed by diurnal transects located perpendicular to the coast (east–west), separated by 10 nmi (Fig. 2a,b). Coastal areas (<5 nmi) were surveyed in a special way, with north–south sailings repeated on at least 2 occasions. A scientific echosounder SIMRAD EK 60 with split beam transducer 38 kHz calibrated according to standard methods was used. Mid-water trawl surveys were conducted to confirm species identification.

Using the acoustic information, we produced Hovmöller diagrams to illustrate the changes in fish density (t nmi^{-2}) according to latitudes (by degree) and time (by year). Fish density was interpolated by triangulation to produce a colored grid of nodes representing the relative abundance of fish using Surfer software (Golden Software). We assumed that spatial and temporal variations in prey biomass are associated with differences in prey availability to lactating females.

RESULTS

Body condition and capture date

The distribution of sampled pups was spread over the capture periods in both colonies. In PP/PN, most of the captures were registered during the first 2 wk of February, whereas in Cobquecura they were recorded during the last week of January and the first week of February (Fig. 3). Body length of both males and females from PP/PN was significantly longer than those from Cobquecura (Males: $F_{1,196} = 23.46$, p < 0.0001; Females: $F_{1,140} = 9.09$, p = 0.003) (Table 1). On the contrary, male and female pups from Cobquecura showed a significantly greater weight than those from PP/PN (Males: $F_{1,196} = 76.02$, p < 0.0001; Females: $F_{1,140} = 57.03$, p < 0.0001) (Table 1).

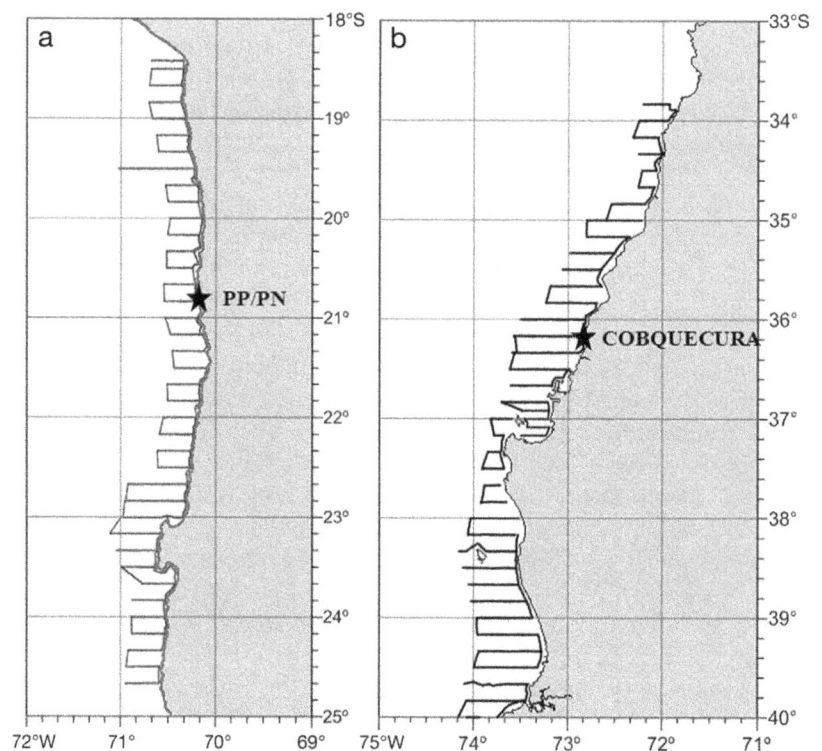

Fig. 2. Study zone and acoustic surveys in (a) north and (b) central-south zones of Chile during austral summer months (December and January)

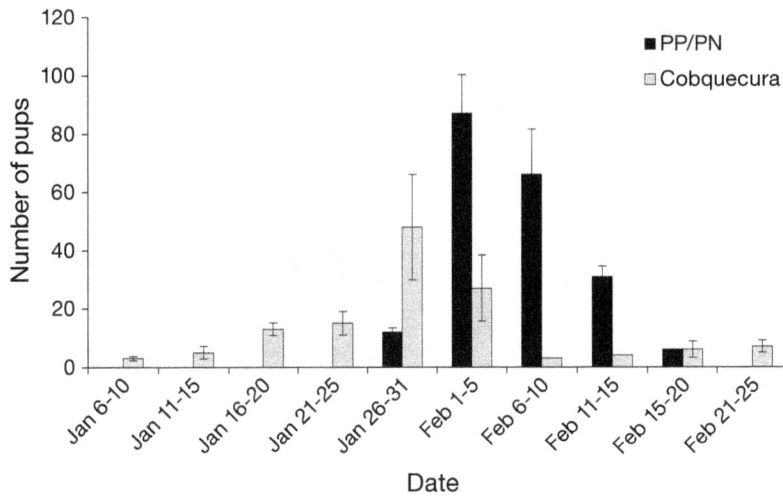

Fig. 3. Distribution of South American sea lions pups captured in the PP/PN and Cobquecura colonies. Data are combined all for years for each colony (2011 to 2013 in PP/PN and 2009 to 2012 in Cobquecura)

Using the values of the morphological measures on male and female pups for the 2 study areas, we estimated the CI_m for each sex and zone (Table 1). Three of the 4 regressions of CI_m on capture date showed no relationship; for the males and females from PP/PN ($r^2 = 0.0001$, p = 0.791; $r^2 = 0.030$, p = 0.185, respectively) and for the females from Cobquecura ($r^2 = 0.001$, p = 0.854). The only exception was found in males from Cobquecura ($r^2 = 0.056$, p = 0.032). Although significant, this result explained a very low proportion of the variance (3 %), suggesting that date of capture did not produce a relevant effect on CI_m and thus was not considered.

Body condition and prey availability

When comparing CI_m between localities, both male and female pups from Cobquecura showed a significantly greater CI_m compared to those from PP/PN (Males: $F_{1,138} = 189.15$, p < 0.0001; Females: $F_{1,91} =$

52.468, p < 0.0001) (Table 1). Female pups from both colonies were in significantly better condition in 2011 than in 2012 ($F_{1,91} = 5.11$, p = 0.026). However, no differences between years were found in males ($F_{1,138} = 2.35$, p = 0.128), and there was no interaction between year and locality for males and females ($F_{1,138} = 1.93$, p = 0.167; $F_{1,91} = 0.54$, p = 0.466, respectively).

Relatively higher acoustic densities and biomass of anchovy and sardine were recorded in 2011 for both north and central-south zones (Fig. 4, Table 2). Although food availability for both areas was higher in 2011 compared to other years, a more remarkable higher acoustic density of anchovy and sardine was found in the central-south area, especially for sardine as shown in Fig. 4. These spatial and temporal variations in prey biomass overlap with hotspots of foraging behavior by South American sea lions from the studied colonies. The foraging behavior of satellite-tracked South American sea lions from PP/PN was associated with shallow-coastal locations, mainly concentrated from 20° 43' to 21° 54' S (Hevia 2013). Foraging hotspots of sea lions from Cobquecura were also in shallow-coastal waters from 35° to 37° S (Hückstädt et al. 2014).

A significant positive relationship between $\log_{10}(CI_m)$ and $\log_{10}$(fish biomass), both for male ($r^2 = 0.447$, p < 0.00001) and female ($r^2 = 0.452$, p < 0.00001) pups was found (Table 2). When analyzed separately, each locality maintained this tendency, both for males ($r^2 = 0.030$, p = 0.036) and females ($r^2 = 0.103$, p = 0.002) in PP/PN, and for males alone in Cobquecura ($r^2 = 0.092$, p = 0.003). Only females from Cobquecura did not show a relationship between CI_m and fish biomass ($r^2 = 0.018$, p = 0.330) (Table 2).

Table 1. Means, ranges and sample size of morphological variables recorded in males and females South American sea lion pups in the localities of PP/PN (north zone) and Cobquecura (central-south zone). CI_m = body condition index

| | Males | | | | Females | | | |
| | PP/PN | | Cobquecura | | PP/PN | | Cobquecura | |
	Mean (range)	n	Mean (range)	n	Mean (range)	n	Mean (range)	n
Standard length (cm)	83.6 (72–101)	116	79.6 (71–89)	82	80.1 (63–100)	86	77.5 (69–89)	56
Body mass (kg)	11.0 (7.0–18)	116	13.4 (10.6–17.8)	82	10.1 (7.0–18.6)	86	12.3 (9.8–15)	56
CI_m	0.92 (0.6–1.3)	116	1.16 (0.9–1.4)	82	0.92 (0.6–1.4)	86	1.17 (0.9–1.4)	56

Acoustic density (t nmi^{-2})

⬤ 1 – 75   ⬤ 76 – 150   ⬤ 151 – 300   ⬤ 300 – 600   ⬤ > 600

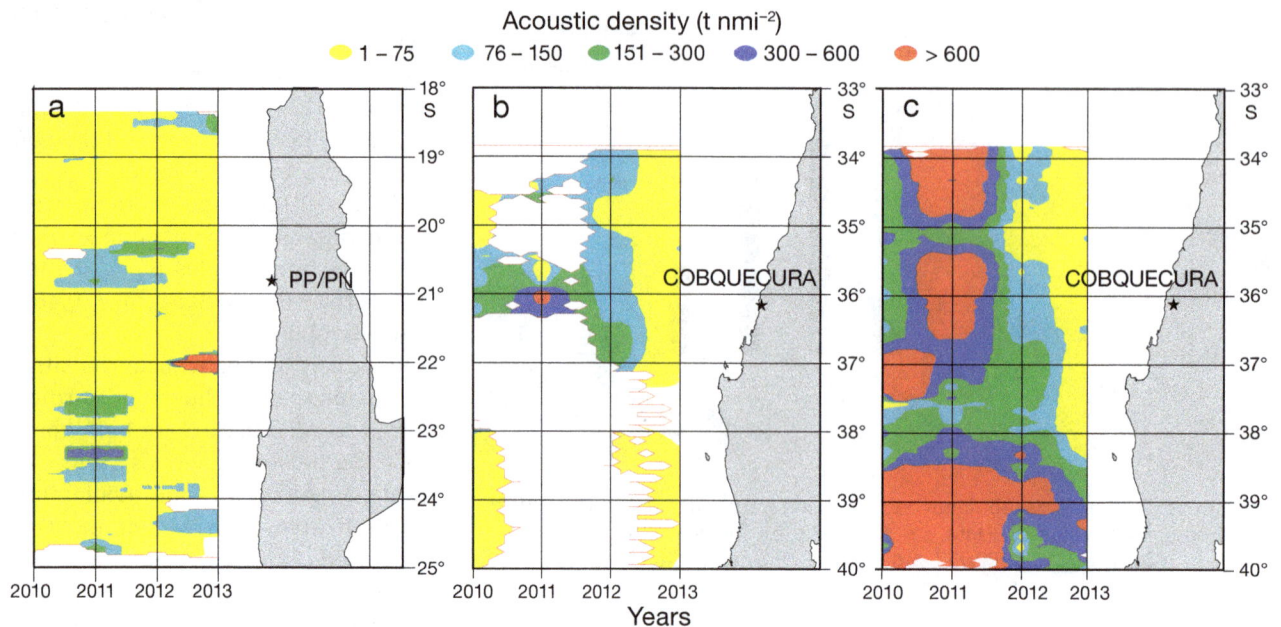

Fig. 4. Hovmöller diagrams of sardine and anchovy density by latitude during summer surveys from 2010 to 2013: (a) sardine in the north zone, (b) anchovy in central-south zone, (c) sardine in the central-south zone

Table 2. Body condition index (CI$_m$) for male and female South American sea lion pups in the localities of PP/PN (north zone) and Cobquecura (central-south zone), and estimates of fish biomass per zone from 2009 to 2013

| Year | Mean (range) CI$_m$ | | | | Fish biomass (t × 10^3) | |
| | PP/PN | | Cobquecura | | North | Central-south |
	Male	Female	Male	Female		
2009			1.07 (0.9–1.2)	1.14 (1.0–1.2)		2032.5
2010			1.11 (1.0–1.4)	1.20 (1.0–1.4)		2632.3
2011	0.90 (0.7–1.1)	0.96 (0.8–1.2)	1.20 (1.0–1.4)	1.16 (1.0–1.3)	253.9	3620.4
2012	0.90 (0.7–1.1)	0.87 (0.6–1.2)	1.14 (1.0–1.4)	1.09 (0.9–1.2)	162.3	3943.9
2013	0.98 (0.6–1.3)	0.98 (0.6–1.4)			299.5	

DISCUSSION

Our results showed that body condition of the *Otaria flavescens* pups varied over time (variation between years), and between areas (north vs. central-south Chile). Also, a close relationship was found between prey availability and body condition of the pups. As previously indicated by other authors for different species of pinnipeds (e.g. Bradshaw et al. 2000), our results show that in spite of its simplicity, the relative pup condition index used here is useful for comparing body condition of pups among colonies (Bradshaw et al. 2000). However, because we sampled pups outside the colonies, our results should be taken with caution.

One major finding of this study is the spatial variation in the body condition of pups. The body length of the pups measured in the north zone (PP/PN) was longer than that recorded in the central-south zone (Cobquecura), and higher than values reported by Cappozzo et al. (1991) (range: 79 to 82 cm) and Vaz-Ferreira (1982) (mean length: 81.7 and 75.0 cm for males and females, respectively) for *O. flavescens* pups on the Atlantic coast. In spite of this, body weight, and consequently body condition estimated by the morphometric index, suggest that body condition of the pups in the north was significantly lower than that of pups in the central-south zone. Because animals in better condition are predicted to maintain a more competent immune system and have higher potential survival (Boltnev et al. 1998, Green et al. 2010), differences in pup body condition between PP/PN and Cobquecura may have important consequences for both overall health state and first year survival (Boltnev et al. 1998).

The direct benefit of higher survival during the lactation period may have positive effects that persist during the first several years of life (Baker & Fowler 1992), thus probably affecting population trends of sea lions. The abundance of sea lions (including pups) in the Cobquecura colony has remained relatively constant for the last 12 yr (Quiñones et al. 2011), while the abundance of sea lions (especially pups) in the PP/PN colonies (particularly in Punta Negra) has decreased in the last 15 yr (Sielfeld & Guzmán 2002). These differences in population trends between the areas cannot be explained by diet composition, because sea lions from the 2 colonies are both important consumers of common sardine and anchovy (Sielfeld 1999, Muñoz et al. 2013), showing a clear pattern of epipelagic and coastal foraging (Hevia 2013, Hückstädt et al. 2014). In addition, no effect of prey energetic content is expected on sea lion populations because these 2 prey species are of a similar size and have a similar diet composition, suggesting a similar utilization of resources and energetic content (Llanos et al. 1996). Further research should be focused on evaluating whether a relationship exists between body condition and abundance in these populations, or if other factors, such as the interactions with artisanal fisheries and/or latitudinal movements of individuals to other places may explain the decrease in the sea lion abundance in the PP/PN colonies.

Our results suggest that spatial variation in the pups' body condition may be associated with differences in biological productivity around the breeding colonies. Boyd (1998) indicated that the productivity of an area is of great relevance for pinnipeds, since these species have adapted to exploit locally abundant resources. In the South American sea lion, females tend to remain in the same place and show a high degree of philopatry (Feijoo et al. 2011). Females nurse their young on land for 8 to 10 mo (Cappozzo & Perrin 2009), and as evidenced from satellite data, they forage close to the breeding sites during their foraging trips (Rodríguez et al. 2013), thus local productivity is an important factor for them. Considering this species is a central forager that shows a strong dependence upon local prey availability, it can be expected that the greater productivity of the central-south zone compared to that of the north zone would allow female O. flavescens to obtain more food in less time, thus decreasing the time they spend away from their offspring and increasing the lactating period (Muñoz et al. 2011). The difference

in productivity may explain, at least partly, the lower body condition of the pups in the north zone compared to those of the central-south zone. According to Gerrodette & DeMaster (1990) the knowledge of trends in both abundance and condition indices could be used to deduct changes in carrying capacity. The lower body condition of pups together with a decline in the abundance trend of sea lions in the PP/PN colonies may suggest an overall lower carrying capacity for O. flavescens at those localities due to lower productivity.

It is important to consider that pup body condition not only may vary in relation to food availability, but also due to other factors, such as size, age, and experience level of the mothers, and also parturition date (Boltnev & York 2001). For instance, Boyd & McCann (1989) and Boltnev & York (2001) found a positive relationship between maternal size and pup size in some pinniped species. Similarly, a positive relationship between maternal age and pup size has been shown (e.g. Bowen et al. 1994). Because we did not consider maternal effects in this study, we cannot rule out if these factors may explain, at least in part, the spatial variation in the pups' body condition.

The variability of physical climate, such as El Niño or La Niña events (Barber & Chavez 1983) may also explain the spatial and temporal differences in body condition of the pups. These perturbations produce changes in primary productivity which lead to abrupt changes in the distribution and abundance of prey species (Barber & Chavez 1983, Daneri et al. 2000). Temporal fluctuations in prey abundance may affect the foraging behavior of lactating females and eventually the body condition of their pups in different years (Soto et al. 2006). The higher values of female pup body condition in 2011 may be explained by a cold La Niña phenomenon affecting the Chilean coast between July 2010 and March 2011, which was associated with greater productivity in the coastal area. According to hydroacoustic surveys, the fish biomass in central-south Chile was at its highest in 2011, suggesting greater food availability for sea lions in that year, and probably reflecting better body condition of pups born in 2011. This fact highlights the importance of multi-year sampling in order to detect the relationship between productivity variation and the potential responses of top predators such as the South American sea lion.

In conclusion, spatial and temporal variations in prey availability appear to strongly affect the body condition of South American sea lion pups. These results corroborate the findings in other pinniped species (e.g. Bradshaw et al. 2000). Future studies should focus on the relationship of these spatial and temporal differences to sea lion population survival and reproductive aspects, and also evaluate if these differences represent a different strategy to confronting environment fluctuations (such as El Niño) that strongly affect the sea lion populations of north Chile. Additionally, due to the fact that other potential factors such as maternal influences seem to also affect the body condition of pups, their incorporation in further studies is crucial.

Acknowledgements. We thank D. Alegría, O. Muñoz, E. Pedreros, J. Barraza, C. Arancibia and F. Velásquez for their help in the field, and L. Eaton and P. Zarate for language corrections. Fish biomass estimates were obtained from Instituto de Fomento Pesquero (IFOP) and funded by the Fondo de Investigación Pesquera (FIP). Financial support for this research was obtained from Compañía Minera Doña Inés de Collahuasi (CMDIC) through the project GMS 2010/22, and from the Programa de Investigación Marina de Excelencia (PIMEX-Nueva Aldea) of the Faculty of Natural and Oceanographic Sciences of the University of Concepción, funded by Celulosa Arauco and Constitución S.A. Animal captures followed the requirements of Resoluciones Exentas No. 2396 (year 2009), 2794 (2010), 786 (2011), 1479 (2011) and 172 (2012) of the Undersecretariat of Fisheries, Ministry of Economy, Chile.

LITERATURE CITED

▶ Acevedo J, Aguayo-Lobo A, Sielfeld W (2003) Eventos reproductivos del león marino común, *Otaria flavescens* (Shaw, 1800), en el norte de Chile (Pacífico suroriental). Rev Biol Mar Oceanogr 38:69–75 (in Spanish with English abstract)

▶ Baker JD, Fowler CW (1992) Pup weight and survival of northern fur seals, *Callorhinus ursinus*. J Zool (Lond) 227:231–238

▶ Barber RT, Chavez FP (1983) Biological consequences of El Niño. Science 222:1203–1210

Bartheld JL, Pavés H, Manque C, Vera C, Miranda D (2008) Cuantificación poblacional de lobos marinos en el litoral de la I a IV Región. Informe FIP 2006-50. www.fip.cl/resultadosProyectos.aspx?sub=OQ&an=MjAwNg&rec=&tit=

▶ Boltnev AI, York AE (2001) Maternal investment in northern fur seals (*Callorhinus ursinus*): interrelationships among mothers' age, size, parturition date, offspring size and sex ratios. J Zool (Lond) 254:219–228

▶ Boltnev AI, York AE, Antonelis G (1998) Northern fur seal young: interrelationships among birth size, growth, and survival. Can J Zool 76:843–854

▶ Bowen WD, Oftedal OT, Boness DJ, Iverson SJ (1994) The effect of maternal age and other factors on birth mass in the harbour seal. Can J Zool 72:8–14

▶ Boyd IL (1998) Time and energy constraints in pinniped lactation. Am Nat 152:717–728

▶ Boyd IL, McCann TS (1989) Pre-natal investment in reproduction by female Antarctic fur seals. Behav Ecol Sociobiol 24:377–385

▶ Bradshaw CJA, Davis LS, Lalas C, Harcourt RG (2000) Geographic and temporal variation in the condition of pups of the New Zealand fur seal (*Arctocephalus forsteri*): evidence for density dependence and differences in the marine environment. J Zool (Lond) 252:41–51

Cappozzo HL, Perrin WF (2009) South American sea lion *Otaria flavescens*. In: Perrin W, Würsig B, Thewissen J (eds) Encyclopedia of marine mammals. Academic Press/Elsevier, Amsterdam, p 1076–1079

▶ Cappozzo HL, Campagna C, Monserrat J (1991) Sexual dimorphism in newborn southern sea lions. Mar Mamm Sci 7:385–394

▶ Castillo-Jordán C, Cubillos LA, Navarro E (2010) Inter-cohort growth rate changes of common sardine (*Strangomera bentincki*) and their relationship with environmental conditions off central southern Chile. Fish Res 105: 228–236

Costa DP, Shaffer SA (2012) Seabirds and marine mammals. In: Sibly RM, Brown JH, Kodric-Brown A (eds) Metabolic ecology: a scaling approach. John Wiley & Sons, Oxford, p 225–233

▶ Daneri G, Dellarosa V, Quiñones R, Jacob B, Montero P, Ulloa O (2000) Primary production and community respiration in the Humboldt Current System off Chile and associated oceanic areas. Mar Ecol Prog Ser 197:41–49

▶ Feijoo M, Lessa EP, Loizaga de Castro R, Crespo EA (2011) Mitochondrial and microsatellite assessment of population structure of South American sea lion (*Otaria flavescens*) in the southwestern Atlantic ocean. Mar Biol 158:1857–1867

▶ Gerrodette T, DeMaster DP (1990) Quantitative determination of optimum sustainable yield. Mar Mamm Sci 6:1–16

▶ Green DS, Young JK, Hernández-Camacho CJ, Gerber LR (2010) Developing a non-invasive indicator of pinniped health: neonate behavior and growth in California sea lions (*Zalophus californianus*). Cienc Mar 36:311–321

▶ Gutiérrez-Estrada JC, Silva C, Yáñez E, Rodríguez N, Pulido-Calvo I (2007) Monthly catch forecasting of anchovy *Engraulis ringens* in the north area of Chile: non-linear univariate approach. Fish Res 86:188–200

Hevia K (2013) Areas de alimentación del lobo marino común *Otaria flavescens* en la zona norte de Chile. Thesis, Universidad de Valparaíso

▶ Hückstädt LA, Quiñones RA, Sepúlveda M, Costa DP (2014) Movement and diving patterns of juvenile male South American sea lions off the coast of central Chile. Mar Mamm Sci 30:1175–1183

▶ Labocha MK, Schutz H, Hayes JP (2014) Which body condition index is best? Oikos 123:111–119

▶ Lea MA, Guinet C, Cherel Y, Duhamel G, Dubroca L, Pruvost P, Hindell M (2006) Impacts of climatic anomalies on provisioning strategies of a Southern Ocean predator. Mar Ecol Prog Ser 310:77–94

Llanos A, Herrera G, Bernal P (1996) Análisis del tamaño de las presas en la dieta de las larvas de cuatro clupeiformes en un área costera de Chile central. Sci Mar 60:435–442

Muñoz L, Pavez G, Inostroza P, Sepúlveda M (2011) Foraging trips of female South American sea lions (*Otaria flavescens*) from Isla Chañaral, Chile. Lat Am J Aquat Mamm (LAJAM) 9:140–144

▶ Muñoz L, Pavez G, Quiñones RA, Oliva D, Santos M, Sepúlveda M (2013) Diet plasticity of the South American sea lion in Chile: stable isotope evidence. Rev Biol Mar Oceanogr 48:613–622

Quiñones RA, Gutiérrez MH, Daneri G, Gutiérrez DA, González HE, Chávez F (2010) Pelagic carbon fluxes in the Humboldt Current System. In: Liu K, Atkinson L,

Quiñones R, Talaue-McManus L (eds) Carbon and nutrient fluxes in global continental margins: a global synthesis. IGBP Series Book, Springer-Verlag, New York, NY, p 44–64

Quiñones RA, Sepúlveda M, Carrasco P, Pérez-Álvarez MJ, Moraga R, Hückstädt L, Pedreros E (2011) Ecología y biología del lobo marino común, Otaria flavescens, en el Santuario Islote Lobería de Cobquecura. Informe Mayo 2008—Diciembre 2011. Programa de Investigación Marina de Excelencia (PIMEX), Facultad de Ciencias Naturales y Oceanográficas, Universidad de Concepción

▶ Rodríguez DH, Dassis M, Ponce de León A, Barreiro C, Farenga M, Bastida RO, Davis RW (2013) Foraging strategies of southern sea lion females in the La Plata river estuary (Argentina–Uruguay). Deep-Sea Res II 88-89:120–130

▶ Sepúlveda M, Oliva D, Urra A, Pérez-Álvarez MJ and others (2011) Distribution and abundance of the South American sea lion Otaria flavescens (Carnivora: Otariidae) along the central coast off Chile. Rev Chil Hist Nat 84: 97–106

Sielfeld W (1999) Estado del conocimiento sobre conservación y preservación de Otaria flavescens (Shaw, 1800) y Arctocephalus australis (Zimmermann, 1783) en las costas de Chile. Estud Oceanol 18:81–96

Sielfeld W, Guzmán A (2002) Effect of El Niño 1997/98 on a population of the Southern sea lion (Otaria flavescens Shaw) from Punta Patache/Punta Negra (Iquique, Chile). Invest Mar 30:157–160

▶ Soto KH, Trites AW, Arias-Schreiber M (2006) Changes in diet and maternal attendance of South American sea lions indicate changes in the marine environment and prey abundance. Mar Ecol Prog Ser 312:277–290

Vaz-Ferreira R (1982) Otaria flavescens (Shaw), South American sea lion. In: Gordon J (ed) Mammals of the seas: small cetaceans, seals, sirenians and otters. FAO Fisheries Series Vol 4, FAO, Rome, p 488–495

Weinberger C (2013) El lobo marino común, Otaria flavescens, en Chile: distribución espacial, historia demográfica y estructuración génica. PhD thesis, Pontificia Universidad Católica de Chile, Santiago

▶ Womble JN, Sigler MF, Willson MF (2009) Linking seasonal distribution patterns with prey availability in a central-place forager, the Steller sea lion. J Biogeogr 36:439–451

10

Effect of attractant stimuli, starvation period and food availability on digestive enzymes in the redclaw crayfish *Cherax quadricarinatus* (Parastacidae)

Hernán J. Sacristán[1,2], Héctor Nolasco-Soria[3], Laura S. López Greco[1,2,*]

[1]Biology of Reproduction and Growth in Crustaceans, Department of Biodiversity and Experimental Biology, FCE y N, University of Buenos Aires, Cdad. Univ. C1428EGA, Buenos Aires, Argentina

[2]Instituto de Biodiversidad y Biología Experimental y Aplicada (IBBEA), CONICET-UBA, Argentina

[3]Centro de Investigaciones Biológicas del Noroeste, S. C. La Paz, Baja California Sur 23090, México

ABSTRACT: Chemical stimuli in crayfish have been extensively studied, especially in the context of social interactions, but also to a lesser extent in relation to food recognition and the physiological response of digestive enzymes. This is particularly important in commercial species in order to optimize the food supplied. The first objective of this study was to determine whether incorporation of squid meal (SM) in food (base feed, BF) acts as an additional attractant for *Cherax quadricarinatus* and, if so, the concentration required for optimal effectiveness. Incorporation of SM was evaluated through individual and group behavioral tests. The second objective was to analyze the effect of food availability on behavior and level of digestive enzyme activity after short-term (48 h) and long-term (16 d) starvation periods. To assess the effect of either starvation period, 3 different treatments were conducted: no feed (control), available BF, and BF present but not available. Individual and group behavior showed no differences among treatments with different percentages of SM inclusion in BF. The time spent in chambers with different percentages of SM was similar in all treatments. Levels of amylase activity and soluble protein, as a function of food availability after a short- or long-term starvation period, were not altered. Digestive enzyme activity was not affected after 2 d of starvation in response to the treatment. However, change was observed in enzymatic profiles after juveniles were deprived of food for 16 d. The main responses were given by lipase, protease and trypsin activity. Based on previous studies and the present results, we propose a hypothesis for a possible regulation of the digestive and intracellular lipase activities depending on food availability.

KEY WORDS: Chemical stimuli · Crustaceans · Digestive enzyme · Food searching behavior · Food attractants · Starvation

INTRODUCTION

Redclaw crayfish *Cherax quadricarinatus* is a freshwater decapod crustacean, native to northern Queensland (Australia) and southeast Papua New Guinea. This species possesses many desirable biological characteristics for successful aquaculture, such as ease of reproduction, tolerance of crowding, relatively rapid growth rate and flexible eating habits (Gillespie 1990, Merrick & Lambert 1991, Gu et al. 1994). In natural ecosystems, crayfish have polytrophic feeding habits and have been described as predators, omnivores and/or detritivores, consuming a variety of macrophytes, benthic invertebrates, algae and detritus (Saoud et al. 2013). The flexible feeding habits of crayfish suggest that they might re-

*Corresponding author: laura@bg.fcen.uba.ar

spond to a very broad spectrum of chemicals (Tierney & Atema 1988). Indeed, aquatic organisms use chemical signals as sources of information for a number of ecological decisions such as food localization (Moore & Grills 1999), mate searching (Ameyaw-Akumfi & Hazlett 1975, Tierney & Dunham 1982, Dunham & Oh 1992), predator detection (Hazlett 1989), shelter choice (Tamburri et al. 1996) and advertisement of social status (Karavanich & Atema 1998, Zulandt Schneider et al. 1999, Kozlowski et al. 2003).

Crustaceans exhibit relatively slow and intermittent feeding activity and this has an impact on food acquisition and processing. These behavioral characteristics affect the physical properties of feed pellets, such as water stability (hydrostability), and as a consequence, water quality (Saoud et al. 2012). Inasmuch as food is a significant expense in aquaculture production systems, the need to maximize food consumption and reduce wasted food is fundamental for economic success (Lee & Meyers 1996).

Considering the importance of chemical signals during the development of crustaceans, it might be assumed that the incorporation of attractants to food would allow individuals to find potential food in a shorter period of time, increasing the possibility of ingestion (Mendoza et al. 1997). It has been demonstrated that squid meal acts as a stimulant, increasing food consumption in *Homarus gammarus* (Mackie & Shelton 1972), *Penaeus stylirostris* and *P. setiferus* (Fenucci et al. 1980), *P. monodon* (Smith et al. 2005), and *Litopenaeus vannamei* (Nunes et al. 2006). Similarly, shrimp protein hydrolysates stimulate feed consumption in *C. quadricarinatus* (Arredondo-Figueroa et al. 2013). There are few studies regarding the use of chemoattractant substances incorporated into the diets of cultured freshwater decapod crustaceans (Arredondo-Figueroa et al. 2013) and their effect on feeding responses (Tierney & Atema 1988, Lee & Meyers 1996, Kreider & Watts 1998).

Under natural conditions where crayfish may feed ad libitum on foods appearing in various forms and compositions, differences in digestive processes are likely to occur (Kurmaly et al. 1990). Crustaceans alternate between periods of feeding and non-feeding during their development as a result of sequential molting (Vega-Villasante et al. 1999). Molting involves several stages with different feeding behaviors, including the cessation of external food intake from late premolt through early postmolt; therefore, energy needs can be met with different available external food sources or lipid reserves. Digestive enzymes are used as a physiological response to fasting (Cuzon et al. 1980, Jones & Obst 2000, Muhlia-

Almazán & García-Carreño 2002, Rivera-Pérez & García-Carreño 2011, Calvo et al. 2013). Artificially-induced fasting and starvation may allow elucidation of the metabolic routes used in hierarchical order during molting and may initiate alternative biochemical and physiological adaptation mechanisms (Barclay et al. 1983, Comoglio et al. 2008). The midgut gland of crustaceans is the main organ for synthesis and secretion of digestive enzymes (including proteinase, lipase and carbohydrase), absorption and storage of nutrients (lipids and glycogen), which can be mobilized during the non-feeding periods (Icely & Nott 1992, Ong & Johnston 2006). The level of the digestive enzymes in decapod crustaceans does not remain constant during the molt cycles (van Wormhoudt 1974) as a result of both internal and external factors such as starvation and the availability, quantity and quality of food. In *C. quadricarinatus*, Loya-Javellana et al. (1995) demonstrated that crayfish are potentially capable of regulating their digestive processes according to food availability.

In the present study, we focused on factors affecting feeding in *C. quadricarinatus*. Our main hypothesis was that chemical signals from food affect digestive enzyme activity, and this response is modulated by food availability and starvation periods. Our first objective was to determine whether squid additives make food more attractive to crayfish and, if so, what concentration of additives elicits maximum food searching behavior. The second objective was to analyze the effect of food availability on digestive enzyme activity after short- and long-term starvation periods. This information may be useful to understand food searching behavior, and to determine the modulating effect of food presence on digestive physiology in order to design new diets and maximize food handling for the species.

MATERIALS AND METHODS

Live specimens

Juvenile redclaw crayfish were hatched from a reproductive female stock supplied by Centro Nacional de Desarrollo Acuícola (CENADAC), Corrientes, Argentina (27° 22' 42.09" S, 58° 40' 52.41" W). Each ovigerous female (mean wet body weight ± SD = 59.8 ± 3.2 g) with 100 to 150 eggs was maintained in an individual glass aquarium (length × width × height = 60 × 40 × 30 cm) containing 30 l of dechlorinated tap water, under continuous aeration (5 mg O_2 l^{-1}). The temperature was maintained at 27 ± 1°C by

Altman water heaters (100 W, precision 1°C), and the photoperiod was 14 h light:10 h dark cycle. Each aquarium was provided with a PVC tube (10 cm diameter, 25 cm long) as a shelter (Jones 1995). Females were fed daily ad libitum with *Elodea* sp. and commercial TetraColor granules (TETRA®, min. crude protein 47.5%, min. crude fat 6.5%, max. crude fiber 2.0%, max. moisture 6.0%, min. phosphorus 1.5%, and min. ascorbic acid 100 mg kg^{-1}) according to Bugnot & López Greco (2009) and Sánchez De Bock & López Greco (2010). At stage 3, juveniles became independent (Levi et al. 1999) and were separated from 6 mothers, then pooled and maintained under the same conditions mentioned above (based on previous studies; Vazquez et al. 2008, Stumpf et al. 2010, Tropea et al. 2010, Calvo et al. 2013) until they reached the desired weight.

For all experiments, we used a base food (BF, Table 1), specially formulated for *C. quadricarinatus* by Gutiérrez & Rodríguez (2010). Crude protein, total lipids, ash, and moisture contents of diets were determined at National Institute for Fisheries Research and Development (INIDEP), Mar del Plata, Argentina according to AOAC (1990); the proximal composition of the BF was 37.98 ± 0.94% crude protein, 6.05 ± 0.08% lipid, 16.05 ± 0.11% ash and 4.03 ± 0.03% moisture.

Effect of squid attractant on juvenile ability to detect food

For the behavioral experiment, a 30 × 40 × 20 cm glass aquarium without water flow was designed (Fig. 1A) based on Jaime-Ceballos et al. (2007). The

Table 1. Formulation of the reference diet for *Cherax quadricarinatus* prepared as in Gutiérrez & Rodríguez (2010). Mineral premix (mg kg^{-1}): ZnSO$_4$, 50; MgSO$_4$, 35; MnSO$_4$, 15; CoSO$_4$, 2.5; CuSO$_4$, 3; KI, 3. Vitamin premix (mg kg^{-1}, unless otherwise noted): A (retinol), 3000 UI kg^{-1}; D, 600 UI kg^{-1}; E (alpha tocopheryl acetate), 60; K, 5; C (ascorbic acid), 150; B1 (thiamin), 10; B (riboflavin), 10; Vitamin B6 (piridoxin), 7; B12, 0.02; biotin, 0.4; pantothenic acid, 35; folic acid, 6; niacin, 80; choline, 500; inositol, 100

Ingredients	Percentage
Fish meal	28
Soybean meal	39
Pre-jellified starch	19
Soybean oil	6
Bentonite	6
Mineral premix	1
Vitamin premix	1

aquarium was divided into 3 similarly-sized, parallel chambers: the middle chamber was used for acclimation, and the right and left compartments were used as 'attractant chambers'. The aquarium was placed inside a white box to minimize disturbance to crayfish behavior. Food containers (4.5 × 4.5 × 6 cm, Fig. 1B) consisted of an acrylic box surrounded by nylon mesh (1 mm mesh pore). There was a net tube (1.5 × 4.5 cm, diam. × length) inside the container to prevent small particles of food from falling out when the acrylic structure was moved by the animals.

The ingredient tested as a food attractant was squid meal (SM, *Illex argentinus*), and its inclusion in BF was analyzed. The protein concentrate extraction of SM was performed by the Soxhlet method, with isopropyl alcohol as a solvent. The protein residue was then oven-dried at 80°C for 24 h according to

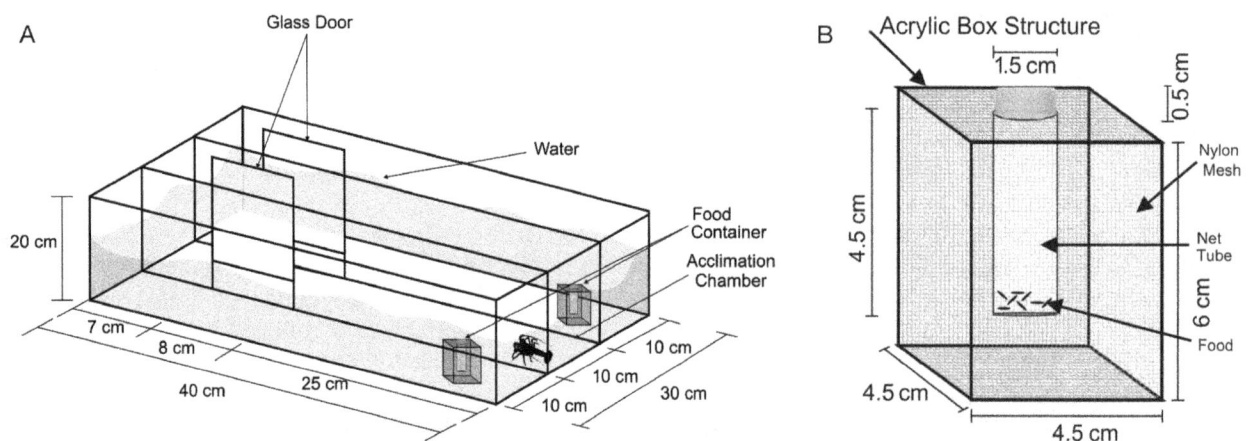

Fig. 1. Aquarium and container design for tests of food attraction with the redclaw crayfish *Cherax quadricarinatus*. (A) glass aquarium device showing the position of the glass doors, acclimation chamber, water level and food container positions; (B) food container device

Díaz et al. (1999). The SM integration in BF was performed according to Díaz et al. (1999) and the chemoattractant concentrations in the BF of the different treatments were: 0 (control), 0.25, 1, 2.5 and 10% (w/w); TetraColor granules were used as a reference positive control. The paired comparisons (Treatments) were: (1) 0% (feed control): no food versus BF; (2) 0.25%: BF + 0.25% SM versus BF; (3) 1%: BF + 1% SM versus BF; (4) 2.5%: BF + 2.5% SM versus BF; (5) 10%: BF + 10% SM versus BF; and (6) reference positive control: TetraColor granules versus no food.

The ability to detect food was evaluated under 2 experimental conditions: individually (April 2012) and in groups (April 2013). Twenty individual juvenile behaviors were observed in the glass aquarium per treatment, except for the reference positive control (N = 10) (weight: 1.35–3.25 g; N = 110) and group behavior was observed with 4 juveniles (weight: 1.21 to 3.75 g) per experiment with 5 replicates for each treatment (N = 60). The group behavior experiment was only performed for Treatments (1), (3) and (5) due to the results of individual behavior experiments.

Test specimens were acclimated to BF for 1 wk prior to the assays, and behavioral experiments were always performed between 09:00 and 13:00 h in the presence of artificial light, in order to avoid any effects of circadian rhythms (Sacristán et al. 2013). All crayfish were starved for 48 h prior to behavioral evaluation, and all were at intermolt, since it has been suggested that the level of responsiveness varies from stage to stage of the molt cycle (Harpaz et al. 1987). Only test specimens with complete sensory appendages (i.e. antennae and antennules) were selected.

At the beginning of each assay, juveniles were maintained in the acclimation chamber for 10 min as in Nunes et al. (2006). After each trial, water was discarded completely, the aquarium was washed with tap water and refilled with new filtered water. Water quality parameters were measured in order to avoid water quality effects on responses by test specimens to the chemoattractant (Lee & Meyers 1996). These parameters, i.e. dissolved oxygen (6 ± 1 mg l^{-1}), pH (7.7 ± 0.5), hardness (80 ± 10 mg l^{-1} as $CaCO_3$ equivalents), and temperature (27 ± 1°C) were within the ranges recommended for aquaculture (Jones 1997, Boyd & Tucker 1998).

Behavioral response to the presence of the attractant was recorded visually by 1 observer positioned in front of the glass aquarium. The location of SM (i.e. left or right chamber) was chosen randomly for each behavior session. After acclimation, the glass doors of

the chamber were opened and the following variables were evaluated: (1) first choice (SM or no SM) of the juveniles, and (2) residence time in each chamber for 10 min (a period established in a preliminary bioassay). The food amount (BF, SM+BF or TetraColor) offered in each trial was 5% of the mean body weight of all crayfish. The percentage of positive choice was calculated as: positive choice (%) = (total number of positive choices / total number of comparisons) × 100, as in Nunes et al. (2006). The % residence time was calculated as: residence time (%) = (total time of positive residence / total assay time) × 100.

Effect of food availability on digestive enzyme activity

To evaluate the effect of food availability on digestive enzymes, 2 experiments were performed according to length of starvation period (short or long). In both experiments, treatments were: (1) no BF (control), (2) available BF (ABF), (3) BF present but not available (NABF). For each treatment, an 18 × 35 × 19 cm plastic aquarium was used; food was unprotected in the ABF treatment but was protected by a food container in the NABF treatment. Either the food or the food container was placed in the middle of the aquarium. In the ABF and NABF treatments, the amount of food offered was 5% of the juvenile's weight.

Expt 1: short-term starvation period

For this experiment, 144 intermolt phase crayfish were selected (weight: 1.14 – 3.99 g). Juveniles were starved in a common aquarium (53 × 40 × 12 cm) at a constant temperature (27 ± 1°C) for 48 h. Each treatment consisted of 4 replicates (N = 48). Before the beginning of the experiment, 12 starved crayfish were randomly placed and acclimated for 1 h in each aquarium. For each treatment (control, ABF and NABF), 8 crayfish were anesthetized in cold water after 0, 5, 10, 30, 60 and 120 min, and the midgut gland was dissected.

Expt 2: long-term starvation period

A total of 72 intermolt phase crayfish (weight: 1.75 – 5.17 g) were selected and starved for 16 d in individual plastic containers (500 cm^3) filled with 350 ml of dechlorinated water under continuous aeration. These containers were placed in 53 × 40 ×

12 cm aquaria with water maintained at 27 ± 1°C. Starvation days were established in preliminary studies. During this period, the plastic containers were cleaned and water was renewed 3 times a week (during experiments no molting organisms were observed). Thereafter, the same procedure as in Expt 1 was performed, but the analysis times were 0, 30 and 120 min; at each time 8 crayfish were anesthetized in cold water and the midgut gland was dissected.

Enzymatic preparation and activity assays

At the end of short- and long-term starvation experiments, the midgut glands were dissected, weighed (±0.1 mg) and immediately frozen at –80°C. Each midgut gland was homogenized in Tris-HCl buffer (50 mM, pH 7.5, 1:4 w/v) in an ice-water bath, with a Potter homogenizer. After centrifugation at 10 000 × g for 30 min at 4°C (Fernández Gimenez et al. 2009), the lipid layer fraction was removed and the supernatant was stored at –80°C until used as an enzyme extract for the enzymatic analysis. The absorbance of enzymatic assays was read on a JASCO CRT-400 spectrophotometer.

The amount of total soluble protein was evaluated with the Coomassie blue dye method according to Bradford (1976) using serum bovine albumin as the standard. Total proteinase activity was assayed using 1% azocasein as the substrate in 50 mM Tris-HCl, pH 7.5 (García-Carreño 1992). One proteinase unit was defined as the amount of enzyme required to increase 0.01 optical density (OD) units min^{-1} at 440 nm (López-López et al. 2005). Lipase activity of each enzyme extract was determined according to Versaw et al. (1989). The assay mixture consisted of 100 µl of sodium taurocholate 100 mM, 1900 µl of buffer Tris-HCl (50 mM, pH 7.5) and 20 µl of enzyme extract. After pre-incubation (25°C for 5 min), 20 µl of β-naphthyl caprylate substrate (Goldbio N-100) 200 mM in dimethyl sulfoxide (DMSO) was added. The mixture was incubated at 25°C for 30 min. Then 20 µl Fast Blue BB (100 mM in DMSO) was added. The reaction was stopped with 200 µl of trichloroacetic acid (TCA, 0.72 N), and clarified with 2.76 ml of ethyl acetate:ethanol (1:1 v/v). Absorbance was recorded at 540 nm. One lipase unit was defined as the amount of enzyme required to increase 0.01 OD units min^{-1} at 540 nm (López-López et al. 2005).

Amylase activity of each extract was determined according to Vega-Villasante et al. (1993). The assay mixture consisted of 500 µl Tris-HCl (50 mM, pH 7.5) and 5 µl enzyme extract; 500 µl starch solution (1% in Tris-HCl, 50 mM, pH 7.5) was added to start the reaction. The mixture was incubated at room temperature for 10 min. Amylase activity was determined by measuring the production of reducing sugars resulting from starch hydrolysis as follows: immediately after incubation, 200 µl of sodium carbonate (2 N) and 1.5 ml DNS reagent were added to the reaction mixture and the mixture was boiled for 15 min in a water bath. The volume was adjusted to 10 ml with distilled water, and the colored solution was read at 550 nm. Reference tubes were prepared similarly, but crude extract was added after the DNS reagent. One amylase unit was defined as the amount of enzyme required to cause an increase of 0.01 OD units min^{-1} at 550 nm (López-López et al. 2005).

Trypsin activity was assayed according to Erlanger et al. (1961). The substrate solution was prepared using 100 mM benzoyl Arg-p-nitroanilide (BAPNA) dissolved in 1 ml of DMSO and brought to a volume of 100 ml with Tris-HCl 50 mM, pH 8.2 containing 10 mM CaCl$_2$. Activity was measured by mixing 80 µl enzyme extract and 1.25 ml of substrate solution, and then the mixture was incubated for 20 min at 37°C. Subsequently, 0.25 ml of acetic acid was added, and the hydrolysis of BAPNA was determined by measurement of free p-nitroaniline at 410 nm. The trypsin activity was measured at 0, 30, and 120 min for Expts 1 and 2.

Statistical analysis

The positive choice and residence time data derived from paired comparisons of feeding behaviors were tested using the chi-squared test of independence (Zar 1999) and 1-way ANOVA (Sokal & Rohlf 1995) respectively. Digestive enzyme data from the short- and long-term starvation experiments were analyzed using generalized linear mixed models (GLMMs) with the statistical program R and the GLMMs package (Zuur et al. 2009), including treatments (control, ABF and NABF) and time as fixed factors. The significance level was set at α = 0.05.

RESULTS

Effect of chemoattractant on juvenile response

The results of individual and group crayfish behaviors are shown in Table 2. For individual crayfish response, no significant differences were found among

Table 2. Number of comparisons, positive choices and residence times of juvenile *Cherax quadricarinatus* individual and group behavior. BF: base food, SM: squid meal. Treatments: 0% (control): no food versus BF; 0.25%: BF + 0.25% SM versus BF; 1%: BF + 1% SM versus BF; 2.5%: BF + 2.5% SM versus BF; 10%: BF + 10% SM versus BF; positive control: TETRA® TetraColor granules versus no food

Treatment (% of SM in BF)	Individual behavior			Group behavior		
	No. of comparisons	Positive choices (%)	Residence time (%)	No. of comparisons	Positive choices (%)	Residence time (%)
0	20	55	41.28 ± 4.18	5	50	38.02 ± 10.96
0.25	20	50	52.94 ± 4.24	–	–	–
1	20	40	47.99 ± 6.33	5	40	40.23 ± 5.35
2.5	20	50	42.36 ± 3.62	–	–	–
10	20	40	38.90 ± 5.61	5	70	37.70 ± 3.30
Positive control	10	20	34.36 ± 8.39	–	–	–

treatments with different percentages of SM included in the BF. Residence times in the chambers with different percentages of SM were similar in all treatments (p = 0.22). Group behavior showed that the percentage of positive choice was the same for all treatments (p > 0.05). Additionally, the crayfish did not preferentially stay in the chamber with the attractant (p = 0.91).

Expt 1: short-term starvation period

The results of specific enzyme activity for amylase, lipase, protease, trypsin and soluble protein level in the short-term starvation experiment are presented in Fig. 2. The digestive enzyme profiles and soluble protein from midgut gland extracts showed a similar pattern among treatments. Specifically, crayfish from the NABF treatment had significantly lower levels (p < 0.05) of amylase activity at 5 and 120 min (5.24 and 5.14 U mg protein^{-1} respectively) than the control and ABF (Fig. 2A). No significant difference was found between ABF and the control group (p > 0.05). Lipase activity of crayfish was not significantly affected (p = 0.19) by the treatments over the 120 min period of the experiment (Fig. 2B). Protease activity in the midgut gland of the juveniles in the NABF treatment was significantly lower (1.02 U mg protein^{-1}) than those in the control and ABF treatments at 5 min (p < 0.05) (Fig. 2C). Moreover, the crayfish in the ABF group differed significantly from the control (p < 0.05) only at 30 min.

Trypsin activity showed significant differences (p < 0.05) among control, ABF and NABF at 120 min (Fig. 2D); furthermore, the soluble protein level of the crayfish was not significantly affected (p = 0.47) by the treatments over 120 min (Fig. 2E).

Expt 2: long-term starvation period

The effect of food availability on the digestive enzyme activity of crayfish after long-term starvation is shown in Fig. 3. There was no significant difference in amylase activity (p = 0.37) among treatments over the 120 min observation period (Fig. 3A). Lipase activity of ABF exhibited a significantly lower activity (p < 0.05) than NABF at 30 and 120 min (61.52 and 48.31 U mg protein^{-1} respectively) (Fig. 3B). There were significant differences in protease activity (p < 0.05) among NABF, control and ABF at the initial time (Fig. 3C). At 30 min, the protease activity in NABF (1.15 U mg protein^{-1} min^{-1}) and ABF (1.18 U mg protein^{-1}) was significantly (p < 0.05) lower than the control group. The ABF decreased significantly (p < 0.05) to 0.70 U mg protein^{-1} at 120 min. Trypsin activity for ABF showed significant differences (p < 0.05) compared to the control and NABF treatments at the end of the experiment (Fig. 3D), but the levels of soluble protein concentration in the crayfish starved for 16 d did not show significant differences (p = 0.13) among treatments during the time assayed (Fig. 3E). Similar results were found when total enzyme activity (U mg midgut gland^{-1}) was calculated (data not shown).

DISCUSSION

Although crayfish have polytrophic feeding habits (Saoud & Ghanawi 2013), this study showed that squid protein extract in the tested concentration range did not increase the attractiveness of feed to *Cherax quadricarinatus*. This result disagrees with other studies on *Pleoticus muelleri*, *Homarus gammarus*, *Litopenaeus vannamei*, *Penaeus monodon*, *P. setiferus* and *P. stylirostris* (Mackie & Shelton 1972,

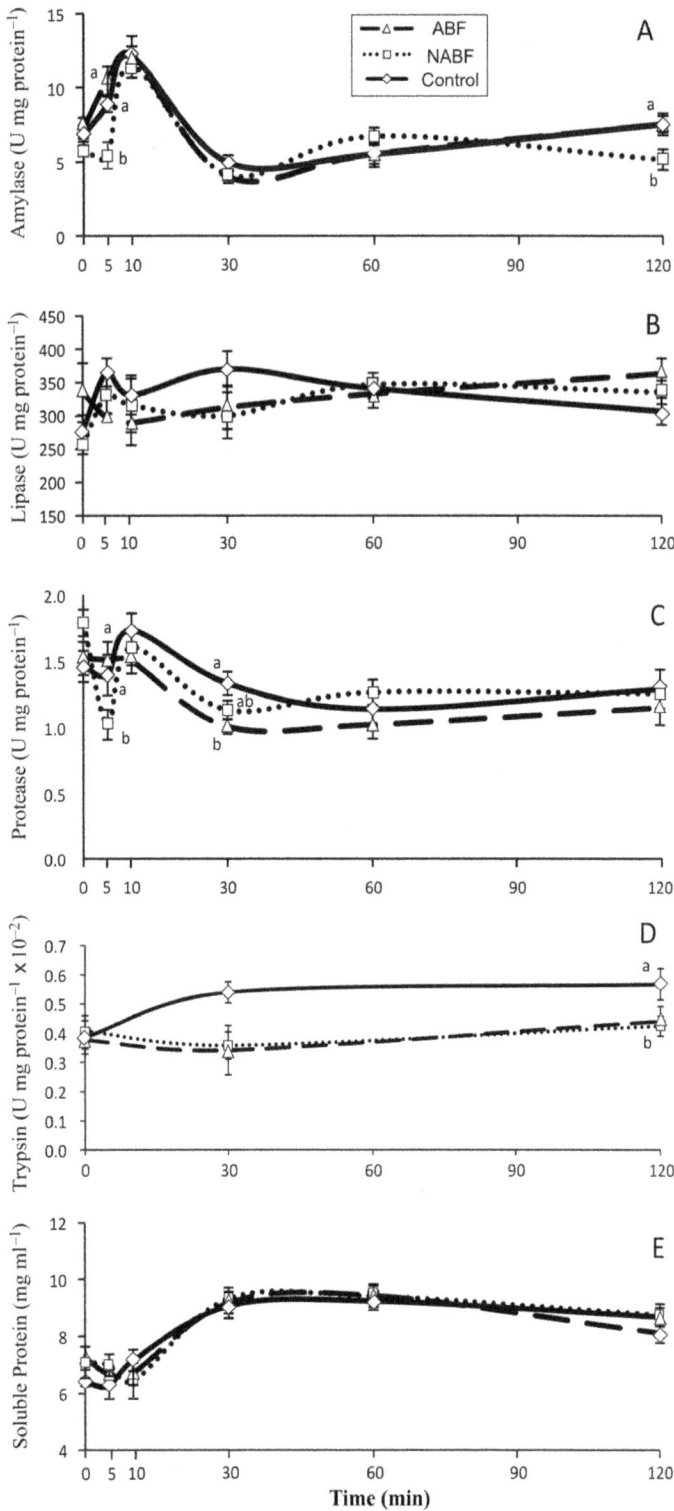

Fig. 2. Specific enzyme activities in the midgut gland of *Cherax quadricarinatus* juveniles after the short-term (48 h) starvation experiment. (A) Amylase, (B) lipase, (C) protease, (D) trypsin, (E) soluble protein. Values are shown as mean ± SE. Different letters indicate significant differences (p < 0.05). ABF: available base food; NABF: no available base food; control: no base food

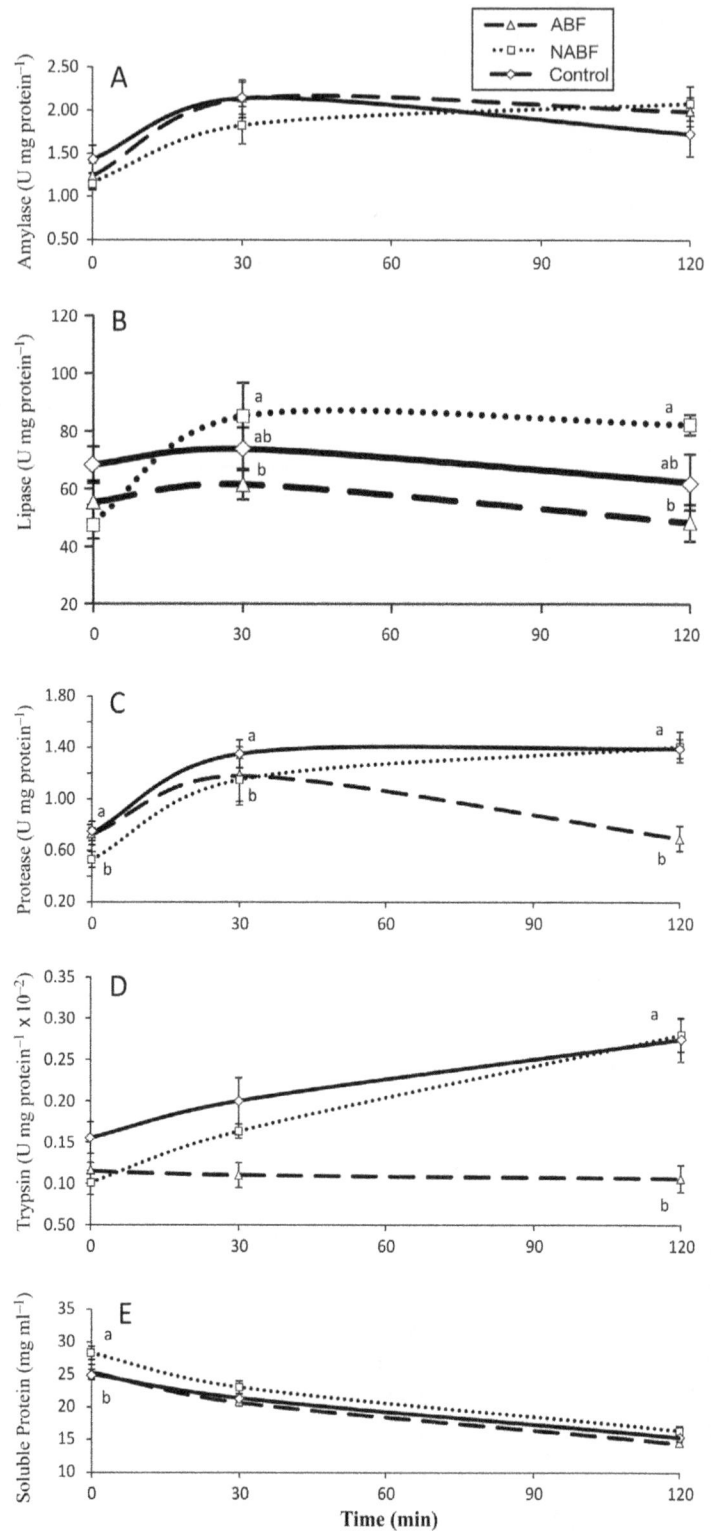

Fig. 3. Specific enzyme activities in the midgut gland of *Cherax quadricarinatus* juveniles after the long-term (16 d) starvation experiment. (A) Amylase, (B) lipase, (C) protease, (D) trypsin, (E) soluble protein. Values are shown as mean ± SE. Different letters indicate significant differences (p < 0.05). ABF: available base food; NABF: no available base food; control: no base food

Fenucci et al. 1980, Díaz et al. 1999, Smith et al. 2005, Nunes et al. 2006) even though the SM and the method for incorporating it into the experimental feed were the same across all studies. However, these other studies were performed on marine crustaceans (lobsters and shrimp), whereas this study is the first to test the effectiveness of SM as an attractant for a freshwater decapod.

The experiment on food detection shows that crayfish wander around the aquarium regardless of where food is localized. Subsequently, when the animal is sufficiently close to or in contact with potential food, their chemoreceptors play a fundamental role in acceptance or rejection of potential food (Heinen 1980). Thus, given the feeding habits of *C. quadricarinatus* and our results, we propose the hypothesis that *C. quadricarinatus* mainly finds food due to the time they invest in environmental wandering. Given that our food searching behavior experiments were carried out in stagnant water, and that odor plumes are strongly affected by flow dynamics for detection of chemical signals (Weissburg 2011), crayfish food detection should also be studied using flowing water. Additionally, food searching behavior could be analyzed in a Y-maze apparatus to enhance the sensitivity of the experimental setup. Mackie (1973) determined that the squid-soluble extract is rich in proline, glycine, alanine and arginine. According to Heinen (1980), glycine and arginine act as attractants in decapod crustaceans. Tierney & Atema (1988) found that cellobiose and sucrose are important feeding stimulants for *Orconectes rusticus*; the amino acids glycine and glutamate elicited feeding movements in *O. rusticus* and *O. virilis* crayfish. Amino acids are abundant in the tissues of marine organisms and they probably guide predators and scavengers to food. Amino acids may likewise signal food availability to crayfish (Tierney & Atema 1988). Further research is needed to determine if the use of complex ingredients such as krill meal, or simple components like amino acids can act as signals for food detection in *C. quadricarinatus*.

Under natural conditions where crayfish may feed ad libitum on food of various forms and compositions, differences in digestive processes are likely to occur (Kurmaly et al. 1990). In the study on foregut evacuation, return of appetite and gastric fluid in *C. quadricarinatus*, Loya-Javellana et al. (1995) demonstrated that ingested food was evacuated linearly with time in crayfish fed daily, and in a somewhat curvilinear trend in those fed every 2 d. This would indicate that crayfish are potentially capable of regulating their digestive processes according to food availability (Loya-Javellana et al. 1995).

Our results on starvation period and food availability demonstrated that crayfish starved for 48 h had a higher digestive enzyme specific activity than crayfish starved for 16 d. Relative to the highest activity level of amylase, lipase, protease and trypsin recorded during the analysis period of both starvation treatments, we found 4.95, 3.70, 0.29 and 1.12 times, respectively, more activity in crayfish starved in the short-term experiment compared to crayfish starved for the long-term. However, in terms of total activity, the differences were smaller, with similar total protease and trypsin activity, and only 60% higher lipase and amylase activity. Therefore, we can conclude that digestive enzyme activity is not affected after 2 d of starvation and in response to treatment. However, different enzymatic profiles were observed in *C. quadricarinatus* juveniles deprived of food for 16 d. The main responses occurred in lipase, protease and trypsin activity, which were higher in control and NABF groups; however, this may have been due to the protein provided by the food (in the case of specific activity), or due to the additional weight provided by the ingested food to the tissue (midgut gland), in the case of total activity; in both cases decreasing the enzyme activity.

The levels of amylase activity and soluble protein as a function of food availability after short- or long-term starvation were not altered. Calvo et al. (2013), analyzing *C. quadricarinatus* juveniles (1 g), observed that starvation did not have an effect on amylase activity, but an accentuated tendency to decrease after 50 d of starvation and to increase after 40 d of re-feeding. Our results are opposed to those of Clifford & Brick (1983), who found that fasting *Macrobrachium rosenbergii* use the energy from the oxidation of carbohydrates.

Our research demonstrates that *C. quadricarinatus* juveniles respond differently to food availability after a long-term starvation period (16 d). These results agree with Calvo et al. (2013), who observed low levels of lipase activity after a 50 d starvation period, suggesting that lipase is not synthesized when food is not available. In the same species, Yudkovski et al. (2007) demonstrated that lipase transcripts decrease in the hepatopancreas during non-feeding stages. Studying the effect of starvation on the expression of lipase transcripts in *Litopenaeus vannamei*, Rivera-Pérez & García-Carreño (2011) showed that 2 types of lipase exist: a digestive lipase and an intracellular lipase (lysosomal). The digestive lipase is only found in the digestive gland and is negatively regulated during fasting by the absence of food, whereas the intracellular lipase is expressed in various tissues (digestive gland, uropods, pleopods, digestive tube, gills, hemocytes, muscle and gonad), and is positively regulated during starvation, suggesting that it is responsible for lipid mobilization from lipid depots (energy reserves).

Based on these previous studies and our present results, we propose a possible regulation pathway for the digestive lipase activity, which is summarized in Fig. 4. We hypothesize that the detection of food promotes de novo synthesis of digestive lipase (ABF and NABF treatments). Subsequently, manipulation, ingestion, stomach food content and nutritive molecules in the digestive gland stimulate digestive lipase secretion into the digestive gland ducts and stomach, interacting with food and carrying out degradation in both sections of the digestive tract (ABF). Therefore, the digestive lipase present in the stomach, intestine and digestive gland is acting on the food and cannot be fully quantified when the lipase activity is measured solely in the digestive gland (Fig. 4A). This would agree with the fact that digestive lipase activity does not increase when food is available (ABF). Therefore, the presence of food inside the stomach, and subsequent products of stomach digestion in the digestive gland and intestine would stimulate further degradation of the food. However, detection of food inhibits the intracellular lipase synthesis pathway, and thus stored lipids are not used as an energy source.

When there is no food in the environment for a long period of time, the intracellular lipase de novo synthesis is likely to be stimulated, and as a consequence, lipid stores mobilized. The pathway of digestive lipase synthesis is inhibited and digestive lipase remains at basal levels (Fig. 4B). In the present study, this assumption is supported by our observation of the low level of digestive lipase activity recorded in the control group after 16 d of fasting. However, when food is present but not available (only possible under experimental conditions), digestive lipase is synthesized and stored inside digestive gland cells and is not secreted. This could be due to the fact that there is no handling stimuli, ingestion, food content in the stomach, and food stomach digestion products in the digestive gland. However, because the synthesis of intracellular lipase is likely to be inhibited in the short-term, there may or may not be a mechanism to counteract this experimental effect (Fig. 4C) i.e. when *C. quadricarinatus* detects the presence of food, the intracellular lipase synthesis is not active to restrict mobilization of lipid reserves and the synthesis of extracellular digestive lipase is also stimulated, in spite of the presence or absence of food in the

stomach and digestive gland. This assumption is supported by our observation of increased activity of digestive lipase in the NABF treatment of this study.

Protease and trypsin activities reflected a similar response to food availability after 16 d of fasting. This result supports the concept that trypsin (together with chymotrypsin) is one of the main proteases of the digestive gland in decapod crustaceans, and it is believed to be responsible for 40 to 60% of total protein digestion in penaeids (Galgani et al. 1984, Tsai et al. 1986).

Our results show that the presence of available food stimulates trypsin secretion at 120 min after long-term starvation but not after short-term starvation. It is possible that during short fasts, the levels of digestive capacity do not decrease significantly because of the history of previous food. The result of the differential response of trypsin secretion under different fasting periods is related to the findings of Muhlia-Almazán & García-Carreño (2002), who reported that trypsin activity from the hepatopancreas in *L. vannamei* was diminished in response to fasting. Cuzon et al. (1980) demonstrated that trypsin activity in shrimp decreases during starvation. In turn, *C. quadricarinatus* juveniles exposed to non-accessible food after 16 d of fasting show an increase in trypsin activity relative to the ABF group, which would indicate that it might be stored inside digestive gland cells until food entry to the digestive system. This is also related to the results reported by Hernández-Cortés et al. (1999), who demonstrated the presence of trypsinogen in the digestive gland of the crayfish *Pacifastacus leniusculus*. Furthermore, Sainz et al. (2004), in their study on trypsin synthesis and storage as zymogen in fed and fasted individuals of *L. vannamei*, revealed that trypsinogen is not totally secreted from a single cell, but rather appears to be secreted partially as an effect of ingestion. It must be considered that trypsinogen can be spontaneously reactivated during the preparation of enzyme extracts and therefore be quantified as an active enzyme, such that it might not be distinguished from the enzyme that is activated and secreted for food digestion by natural causes (Sánchez-Paz et al. 2003). Therefore, more studies in *C. quadricarinatus* are needed to clarify this issue. More studies are also needed regarding possible changes in messenger expression, as well as immunohistochemical studies of digestive tract cells as a reflection of physiological changes in the digestive enzyme secretion in this species. We observed a differential response (in terms of reaction time) in lipase, protease and trypsin activity when food became available again after pro-

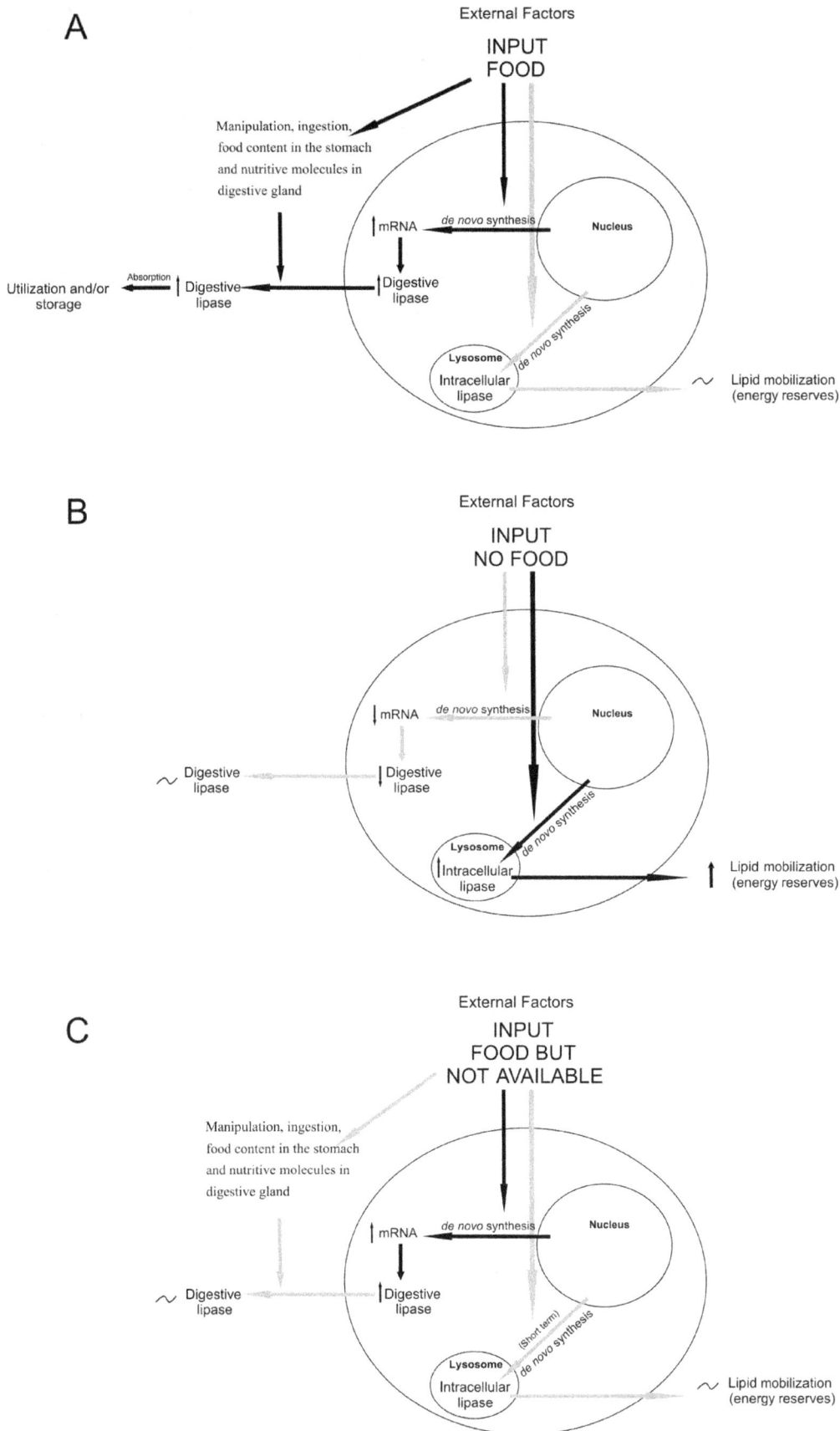

Fig. 4. Possible digestive and intracellular lipase regulation in the midgut gland of *Cherax quadricarinatus* according to Yudkovski et al. (2007), Rivera-Pérez & García-Carreño (2011) Calvo et al. (2013) and the present study. (A) Feeding, (B) long-term starvation and (C) feed stimuli without ingestion. Black and grey arrows: stimulation and inhibitory pathways, respectively

longed fasting (16 d). Lipase responded rapidly (after only 30 min) to the presence of food, whereas protease and trypsin levels responded only after 120 min. Therefore, these enzymes could be used as a tool to analyze the nutritional status of *C. quadricarinatus*.

During periods of starvation, crustaceans must use their energy reserves to meet their needs, and so enzymatic activity must be finely regulated to degrade the necessary energy reserves while preserving cell integrity as much as possible (Sánchez-Paz et al. 2006). Hence, changes in food intake during development may have important consequences for life history (Brzek & Konarzewski 2001). Although instantaneous ecological consequences of poor and sporadic nutrition are sometimes difficult to identify, the reproductive potential of any organism experiencing such conditions may be reduced (Sánchez-Paz et al. 2006).

Acknowledgements. This study was part of H.J.S.'s postgraduate scholarships (ANPCYT and CONICET) and PhD dissertation (University of Buenos Aires, Argentina). We are grateful to Dr. Raymond Bauer, University of Louisiana, Lafayette, LA, USA, and to the reviewers for their comments to improve the manuscript, to Lic. Amir Dyzenchauz for language revision, to Centro Nacional de Desarrollo Acuícola (CENADAC, Argentina) for providing the reproductive stock and to Dr. Gerado Cueto for his help with the statistical analysis. L.S.L.G. is grateful to Agencia Nacional de Promoción Científica y Tecnológica (PICT 2007, project 01187 and 2012 project 01333), CONICET (PIP 2009-2011, number 129 and PIP 2012-2014), UBACYT (projects X458 and 20020100100003), MINCYT-CONACYT (México) MX/09/07 and MINCYT-CAPES BR/11/21 for funding this research.

LITERATURE CITED

▶ Ameyaw-Akumfi CE, Hazlett BA (1975) Sex recognition in the crayfish *Procambarus clarkii*. Science 190:1225–1226

AOAC (Association of Official Analytical Chemists) (1990) Official methods of analysis of the Association of Official Analytical Chemists. AOAC, Arlington, VA

▶ Arredondo-Figueroa JL, Ponce-Palafox JT, Shirai-Matsumoto K, Pérez-Zavaleta A, Barriga-Sosa I, Luna AR (2013) Effects of including shrimp protein hydrolysate in practical diets on the growth and survival of redclaw crayfish hatchlings *Cherax quadricarinatus* (von Martens, 1868). Aquacult Res 44:966–973

▶ Barclay MC, Dall W, Smith DM (1983) Changes in lipid and protein during starvation and the moulting cycle in the tiger prawn, *Penaeus esculentus* (Haswell). J Exp Mar Biol Ecol 68:229–244

Boyd CE, Tucker CS (1998) Pond aquaculture water quality management. Kluwer Academic Publishers Group, Norwell, MA

▶ Bradford MM (1976) A rapid and sensitive method for the quantitation of microgram quantities of protein utilizing the principle of protein–dye binding. Anal Biochem 72:248–254

▶ Brzek P, Konarzewski M (2001) Effect of food shortage on the physiology and competitive abilities of sand martin (*Riparia riparia*) nestlings. J Exp Biol 204:3065–3074

▶ Bugnot A, López Greco LS (2009) Sperm production in the freshwater crayfish *Cherax quadricarinatus* (Decapoda, Parastacidae). Aquaculture 295:292–299

▶ Calvo N, Stumpf L, Sacristán HJ, López Greco LS (2013) Energetic reserves and digestive enzyme activities in juveniles of the red claw crayfish *Cherax quadricarinatus* nearby the point-of-no-return. Aquaculture 416-417:85–91

▶ Clifford HC, Brick RW (1983) Nutritional physiology of the freshwater shrimp *Macrobrachium rosenbergii* (De Man): I. Substrate metabolism in fasting juvenile shrimp. Comp Biochem Physiol A 74:561–568

▶ Comoglio L, Goldsmit J, Amin O (2008) Starvation effects on physiological parameters and biochemical composition of the hepatopancreas of the southern king crab *Lithodes santolla* (Molina, 1782). Rev Biol Mar Oceanogr 43:345–353

▶ Cuzon G, Cahu C, Aldrin JF, Messager JL, Stephan G, Mevel M (1980) Starvation effect on metabolism of *Penaeus japonicus*. Proc World Maric Soc 11:410–423

Díaz AC, Fernández Gimenez AV, Fenucci JL (1999) Evaluación del extracto proteico de calamar en la nutrición del langostino argentino *Pleoticus muelleri* Bate (Decapoda, Penaeoidea). Acuicultura en Armonía con el Ambiente, Acuicultura Venezuela 1999, 1:84–192

▶ Dunham DW, Oh JW (1992) Chemical sex discrimination in the crayfish, *Procambarus clarkii*: role of antennules. J Chem Ecol 18:2363–2372

▶ Erlanger BF, Kokowsky N, Cohen W (1961) The preparation and properties of two new chromogenic substrates of trypsin. Arch Biochem Biophys 95:271–278

▶ Fenucci JL, Zein-Eldin ZP, Lawrence AL (1980) The nutritional response of two penaeid species to various levels of squid meal in a prepared feed. Proc World Maricul Soc 11:403–409

Fernández Gimenez AV, Díaz AC, Velurtas SM, Fenucci JL (2009) Partial substitution of fishmeal by meat and bone meal, soybean meal, and squid concentrate in feeds for the prawn, *Artemesia longinaris*: effect on digestive proteinases. Isr J Aquacult Bamidgeh 61:48–56

Galgani ML, Benyamin Y, Ceccaldi HJ (1984) Ideation of digestive proteinases of *Penaeus kerathurus* (Forskal): a comparison with *Penaeus japonicus* Bate. Comp Biochem Physiol Part B 78:355–361

García-Carreño FL (1992) The digestive proteases of langostilla (*Pleuroncodes planipes*, Deapoda): their partial characterization, and the effect of feed on their composition. Comp Biochem Physiol B 103:575–578

Gillespie J (1990) Redclaw: a hot new prospect. Aust Fish 49:2–3

▶ Gu H, Mather PB, Capra MF (1994) The relative growth of chelipeds and abdomen and muscle production in male and female redclaw crayfish, *Cherax quadricarinatus* von Martens. Aquaculture 123:249–257

Gutiérrez ML, Rodríguez EM (2010) Effect of protein source on growth of early juvenile redclaw crayfish *Cherax quadricarinatus* (Decapoda, Parastacidae). Freshw Crayfish 17:23–29

▶ Harpaz S, Kahan D, Galun R, Moore I (1987) Responses of freshwater prawn, *Macrobranchium rosenbergii*, to chemical attractants. J Chem Ecol 13:1957–1965

▶ Hazlett BA (1989) Additional sources of disturbance pheromone affecting the crayfish *Orconectes virilis*. J Chem Ecol 15:381–385

▶ Heinen JM (1980) Chemoreception in decapod Crustacea and chemical feeding stimulants as potential feed additives. Proc World Maric Soc 11:317–334

Hernández-Cortés P, Cerenius L, García-Carreño FL, Soderhall K (1999) Trypsin from *Pacifastacus leniusculus* hepatopancreas: purification and cDNA cloning of the synthesized zymogen. Biol Chem 380:499–501

Icely JD, Nott JA (1992) Digestion and absorption: digestive system and associated organs. In: Harrison FW, Humes AG (eds) Microscopic anatomy of invertebrates, Vol 10: decapod Crustacea. Wiley-Liss, New York, NY, p 147–201

Jaime-Ceballos B, Civera Cerecedo R, Villarreal H, Galindo López J, Pérez-Jar L (2007) Use of *Spirulina platensis* meal as feed attractant in diets for shrimp *Litopenaeus schmitti*. Hidrobiológica 17:113–117

▶ Jones CM (1995) Production of juvenile redclaw crayfish, *Cherax quadricarinatus* (von Martens) (Decapoda, Parastacidae) II. Juvenile nutrition and habitat. Aquaculture 138:239–245

Jones CM (1997) The biology and aquaculture potential of the tropical freshwater crayfish, *Cherax quadricarinatus*. Department of Primary Industries, Report No. QI90028, Brisbane

▶ Jones PL, Obst JH (2000) Effects of starvation and subsequent refeeding on the size and nutrient content of the hepatopancreas of *Cherax destructor* (Decapoda: Parastacidae). J Crustac Biol 20:431–441

▶ Karavanich C, Atema J (1998) Individual recognition and memory in lobster dominance. Anim Behav 56:1553–1560

▶ Kozlowski O, Voigt R, Moore P (2003) Changes in odour intermittency influence the success and search behaviour during orientation in the crayfish (*Orconectes rusticus*). Mar Freshw Behav Physiol 36:97–110

▶ Kreider JL, Watts SA (1998) Behavioral (feeding) responses of the crayfish, *Procambarus clarkii*, to natural dietary items and common components of formulated crustacean feeds. J Chem Ecol 24:91–111

▶ Kurmaly K, Jones DA, Yule AB (1990) Acceptability and digestion of diets fed to larval stages of *Homarus gammarus* and the rule of dietary conditioning behavior. Mar Biol 106:181–190

▶ Lee PG, Meyers SP (1996) Chemoattraction and feeding stimulation in crustaceans. Aquacult Nutr 2:157–164

▶ Levi T, Barki A, Hulata G, Karplus I (1999) Mother–offspring relationships in the redclaw crayfish *Cherax quadricarinatus*. J Crustac Biol 19:477–484

▶ López-López S, Nolasco H, Villarreal-Colmenares H, Civera Cerecedo R (2005) Digestive enzyme response to supplemental ingredients in practical diets for juvenile freshwater crayfish *Cherax quadricarinatus*. Aquacult Nutr 11:79–85

▶ Loya-Javellana GL, Fielder DR, Thorne MJ (1995) Foregut evacuation, return of appetite and gastric fluid secretion in the tropical freshwater crayfish, *Cherax quadricarinatus*. Aquaculture 134:295–306

▶ Mackie AM (1973) The chemical basis of food detection in the lobster *Homarus gammarus*. Mar Biol 21:103–108

Mackie AM, Shelton RG (1972) A whole animal bioassay for the determination of the food attractants of the lobster *Homarus gammarus*. Mar Biol 14:217–221

▶ Mendoza R, Montemayor J, Verde J (1997) Biogenic amines and pheromones as feed attractants for the freshwater

prawn *Macrobrachium rosenbergii*. Aquacult Nutr 3: 167–173

Merrick JR, Lambert CN (1991) The yabby, marron and red claw: production and marketing. Macarthur Press, Parramatta

▶ Moore PA, Grills JL (1999) Chemical orientation to food by the crayfish *Orconectes rusticus*: influence of hydrodynamics. Anim Behav 58:953–963

▶ Muhlia-Almazán A, García-Carreño FL (2002) Influence of molting and starvation on the synthesis of proteolytic enzymes in the midgut gland of the white shrimp *Penaeus vannamei*. Comp Biochem Physiol B 133:383–394

▶ Nunes AJP, Sá MVC, Andriola-Neto FF, Lemos D (2006) Behavioral response to selected feed attractants and stimulants in Pacific white shrimp, *Litopenaeus vannamei*. Aquaculture 260:244–254

▶ Ong BL, Johnston D (2006) Influence of feeding on hepatopancreas structure and digestive enzyme activities in *Penaeus monodon*. J Shellfish Res 25:113–121

▶ Rivera-Pérez C, García-Carreño F (2011) Effect of fasting on digestive gland lipase transcripts expression in *Penaeus vannamei*. Mar Genomics 4:273–278

Sacristán HJ, Franco Tadic LM, López Greco LS (2013) Influencia de la alimentación sobre el ritmo circadiano de las enzimas digestivas en juveniles de engorde de la langosta de agua dulce *Cherax quadricarinatus* (Parastacidae). Lat Am J Aquat Res, 41:753–761

▶ Sainz JC, García-Carreño F, Sierra-Beltrán S, Hernández-Cortés P (2004) Trypsin synthesis and storage as zymogen in the midgut gland of the shrimp *Litopenaeus vannamei*. J Crustac Biol 24:266–273

▶ Sánchez de Bock M, López Greco LS (2010) Sex reversal and growth performance in juvenile females of the freshwater crayfish *Cherax quadricarinatus* (Parastacidae): effect of increasing temperature and androgenic gland extract in the diet. Aquacult Int 18:231–243

Sánchez-Paz A, García-Carreño F, Mulhia-Almazán A, Hernández-Saavedra N, Yepiz-Plascencia G (2003) Differential expression of trypsin mRNA in the white shrimp (*Penaeus vannamei*) midgut gland under starvation conditions. J Exp Mar Biol Ecol 292:1–17

Sánchez-Paz A, García-Carreño F, Muhlia-Almazán A, Peregrino-Uriarte AB, Hernández-López J, Yepiz-Plascencia G (2006) Usage of energy reserves in crustaceans during starvation: status and future directions. Insect Biochem Mol Biol 36:241–249

▶ Saoud IP, Ghanawi J (2013) A review of the culture and diseases of redclaw crayfish *Cherax quadricarinatus* (von Martens 1868). J World Aquacult Soc 44:1–29

▶ Saoud IP, Garza De Yta A, Ghanawi J (2012) A review of nutritional biology and dietary requirements of redclaw crayfish *Cherax quadricarinatus* (von Martens 1868). Aquacult Nutr 18:349–368

▶ Saoud IP, Ghanaw IJ, Thompson KR, Webster CD (2013) A review of the culture and diseases of redclaw crayfish *Cherax quadricarinatus* (von Martens 1868). J World Aquacult Soc 44:1–29

▶ Smith DM, Tabrett SJ, Barclay MC, Irvin SJ (2005) The efficacy of ingredients included in shrimp feeds to stimulate intake. Aquacult Nutr 11:263–271

Sokal RR, Rohlf FJ (1995) Biometry: the principles and practice of statistics in biological research, 3rd edn. WH Freeman, New York, NY

▶ Stumpf L, Calvo NS, Pietrokovsky S, López Greco LS (2010) Nutritional vulnerability and compensatory growth in early juveniles of the 'redclaw' crayfish *Cherax quadricarinatus*. Aquaculture 304:34–41

▶ Tamburri MN, Finelli CM, Wethey DS, Zimmer-Faust RK (1996) Chemical induction of larval settlement behaviour in flow. Biol Bull 191:367–373

▶ Tierney AJ, Atema J (1988) Behavioral responses of crayfish (*Orconectes virilis* and *Orconectes rusticus*) to chemical feeding stimulants. J Chem Ecol 14:123–133

▶ Tierney AJ, Dunham DW (1982) Chemical communication in the reproductive isolation of the crayfishes *Orconectes propinquus* and *Orconectes virilis* (Decapoda, Cambaridae). J Crustac Biol 2:544–548

▶ Tropea C, Piazza Y, López Greco LS (2010) Effect of long-term exposure to high temperature on survival, growth and reproductive parameters of the 'redclaw' crayfish *Cherax quadricarinatus*. Aquaculture 302:49–56

Tsai IH, Chuang KL, Chuang JL (1986) Chymotrypsin in digestive tract of crustacean decapods (Shrimps). Comp Biochem Physiol Part B 85:235–239

▶ van Wormhoudt A (1974) Variations of the level of the digestive enzymes during the intermolt cycle of *Palaemon sereratus*: influence of the season and effect of the eyestalk ablation. Comp Biochem Physiol A 49:707–715

▶ Vazquez FJ, Tropea C, López Greco LS (2008) Development of the female reproductive system in the freshwater crayfish *Cherax quadricarinatus* (Decapoda, Parastacidae). Invertebr Biol 127:433–443

▶ Vega-Villasante F, Nolasco-Soria H, Civera R (1993) The digestive enzymes of the Pacific brown shrimp *Penaeus californiensis*.: I — Properties of amylase activity in the digestive tract. Comp Biochem Physiol B 106:547–550

Vega-Villasante F, Fernández I, Oliva M, Preciado M, Nolasco-Soria H (1999) The activity of digestive enzymes during the molting stages of the arched swimming *Callinectes arcuatus* Ordway, 1863 (Crustacea: Decapoda: Portunidae). Bull Mar Sci 65:1–9

▶ Versaw KW, Cupper LS, Winters DD, Williams EL (1989) An improved colorimetric assay for bacterial lipase in nonfat dry milk. J Food Sci 54:1557–1568

Weissburg M (2011) Waterborne chemical communication: stimulus dispersal dynamics and orientation strategies in crustaceans. In: Breithaupt T, Thiel M (eds) Chemical communication in crustaceans. Springer-Verlag, New York, NY, p 63–83

▶ Yudkovski Y, Shechter A, Chalifa-Caspi V, Auslander M and others (2007) Hepatopancreatic multitranscript expression patterns in the crayfish *Cherax quadricarinatus* during the moult cycle. Insect Mol Biol 16:661–674

Zar JH (1999) Biostatistical analysis, 4th edn. Prentice-Hall, Upper Saddle River, NJ

▶ Zulandt Schneider RA, Schneider RWS, Moore PA (1999) Recognition of dominance status by chemoreception in the crayfish, *Procambarus clarkii*. J Chem Ecol 25:781–794

Zuur AF, Ieno EN, Walker NJ, Saveliev AA, Smith GM (2009) Mixed effects models and extensions in ecology with R. In: Gail M, Krickeberg K, Samet JM, Tsiatis A, Wong W (eds) Statistics for biology and health. Springer-Verlag, New York, NY, p 323–339

11

Behavioural and reflex responses of mottled mojarra *Eucinostomus lefroyi* (Gerreidae) to cold shock exposure

Emma Samson, Jacob W. Brownscombe*, Steven J. Cooke

Fish Ecology and Conservation Physiology Laboratory, Ottawa-Carleton Institute for Biology, Carleton University, 1125 Colonel By Dr., Ottawa, ON K1S 5B6, Canada

ABSTRACT: Global climate change is predicted to increase incidences of abrupt declines in oceanic temperatures due to storms or upwelling. Fish occupying shallow, near-shore marine habitats may be vulnerable to mortality or sub-lethal fitness effects due to cold shock. Mottled mojarra *Eucinostomus lefroyi* are a ubiquitous prey species associated with subtropical and tropical coastal habitats and thus serve as a model for evaluating the consequences of cold shock events. Here, we conducted one of the first studies of cold shock on a tropical–subtropical fish species. Fish acclimated to 24°C were exposed to acute temperature drops to 16, 18, or 20°C for 1 h. Ventilation rates were assessed every 15 min during exposure, and equilibrium loss every 5 min. Thirty minutes after cold shock exposure, fish were placed in a circular swim flume and chased by hand until exhaustion to measure chase time and distance chased as proxies for swimming capabilities. Fish exposed to 18 and 20°C had significantly higher ventilation rates than those exposed to 16°C or controls held at 24°C. Exposure to 16 and 18°C caused reflex impairment (e.g. no response to tail grabbing), while exposure to 20°C caused no impairment. After 30 min of recovery at ambient temperature, no reflex impairments were detected in any of the treatments, and the swimming capabilities of fish exposed to cold shock were similar to, or better than control fish. Our findings suggest that abrupt changes in water temperature >4°C below ambient can cause behavioural impairments that may lead to mortality in this species.

KEY WORDS: Cold shock · Thermal biology · Subtropics · Mottled mojarra · Reflex Action Mortality Predictors · RAMP

INTRODUCTION

Temperature has been described as the 'master' abiotic factor for fish (Brett 1971) given its manifold effects on behaviour, biochemical processes, locomotion, bioenergetics, growth, life history, and cardiorespiratory physiology. In subtropical and tropical habitats, fish are adapted to a more narrow temperature range in comparison to their temperate counterparts (Larsen et al. 2011), which suggests that tropical fish species may be more sensitive to large fluctuations in ambient water temperature. Further, climate change is predicted to increase the intensity of extreme weather conditions such as tropical cy-

clones and heavy precipitation events (Michener et al. 1997), which will likely cause sudden declines in ambient water temperature, particularly in shallow waters (IPCC 2007). Moreover, coastal upwelling may become more intense, transporting cooler, deep waters to coastal areas (Bakun 1990). In the context of climate change, most efforts have been devoted to considering the biological consequences of warming waters (Roessig et al. 2004), but sudden decreases in water temperature are also relevant. For example, in Butternut Cay, Florida, USA, in January 2010 water temperatures declined 11°C, from 19 to 8°C, resulting in a large fish kill for many near-shore dwelling species (FWCC 2010, A. Adams pers. comm.).

*Corresponding author: jakebrownscombe@gmail.com

There are numerous studies examining the effects of an increase in temperature on tropical marine fish (e.g. Munday et al. 2008, Nilsson et al. 2009, Donelson et al. 2011), yet virtually no research on the effects of decreases in water temperature, including the phenomenon known as cold shock (Lamadrid-Rose & Boehlert 1988, Donaldson et al. 2008, Meyer-Rochow 2013). Studies focusing on temperate freshwater species, such as rainbow trout *Oncorhynchus mykiss*, have found that they can adapt physiologically to ecologically relevant acute or chronic cold water exposure (Shiels et al. 2003, Haverinen & Vornanen 2007). Yet for tropical marine fish, even basic information on thermal tolerances (especially lower) is rare (Fangue & Bennett 2003, Eme & Bennett 2009, Murchie et al. 2011). There is a need for more information on the effects of thermal change, including cold shock, on tropical marine fish (Wilson et al. 2010).

Cold shock in fish occurs when the temperature suddenly drops below a level the fish is acclimated to (Donaldson et al. 2008, Meyer-Rochow 2013). Previous research, largely on freshwater fish, has shown that cold shock may cause both behavioural and physiological consequences, in addition to fatalities (Fry 1947, reviewed in Donaldson et al. 2008). Inhibition of the spinal reflexes often occurs (Roots & Prosser 1962), which reduces the individuals' swimming abilities, eventually leading to a loss of equilibrium (Lamadrid-Rose & Boehlert 1988). The cause of this inhibition appears to be a result of effects of cold temperatures on the hypothalamic–pituitary–interrenal axis and the brain (Donaldson et al. 2008). An experiment on the tropical species *Mugil cephalus* revealed that a cold shock 15°C below ambient resulted in many individuals losing equilibrium, and 75% mortality (Lamadrid-Rose & Boehlert 1988). In addition to spinal reflexes, other conditioned reflexes appear to become impaired. Such inhibition, however, is reversible if the fish survive the temperature change (Prosser & Farhi 1965, Montgomery & Macdonald 1990) and return to ambient water. The loss of equilibrium and reflex impairment is a particularly important factor for prey species. If the tolerance of prey species to cold shock is lower than that of their predators, it puts the prey species at risk due to reduced predator avoidance capabilities (Lamadrid-Rose & Boehlert 1988).

Mottled mojarra *Eucinostomus lefroyi* inhabit shallow subtropical and tropical marine waters on the east coast of North America, where they (along with congenerics) are a major component of fish communities in estuaries and coastal habitats (Kerschner et al. 1985). They are a prey species for predators such as lemon sharks *Negaprion brevirostris*, for which mojarra species were found to compromise over 50% of the consumed prey biomass in a study carried out in 2003 at Bimini, The Bahamas (Newman et al. 2012). Since temperature fluctuations in shallow waters may be more extreme than in deeper waters (IPCC 2007), mottled mojarra may experience more frequent and more extreme temperature decreases in the future. Therefore, it is important to understand the effects of such decreases on the behaviour and physiological responses of the species, and whether their tolerance is within the threshold of the predicted decreases. Given their large distribution, the fact that mottled mojarra or other similar small-bodied fishes are so prominent in coastal tropical marine systems, and the apparent important role of such fish in marine food webs, research on the effects of cold shock on such fishes is warranted (Kerschner et al. 1985).

The objective of this study was to examine the sensitivity of mottled mojarra to cold shock exposure, by (1) determining the critical thermal (CT) minimum temperature for the species and (2) measuring the effects of a range of ecologically relevant, abrupt temperature declines on the behaviour, ventilation rate and reflex responses. We aimed to determine cold shock tolerance thresholds and relevant behavioural responses of an ecologically important nearshore marine fish, which may serve as a model for understanding the potential effects of this stressor on similar species.

MATERIALS AND METHODS

Fish collection

A total of 64 mottled mojarra (89 ± 7 mm total length, TL; range 75 to 100 mm) were collected by seine net from shallow sandy habitats in Kemps Creek (24° 48′ 54.29″ N, 76° 18′ 03.09″ W) and Page Creek (24° 49′ 4.7″ N, 76° 18′ 51.6″ W), and transported to a holding tank at the Cape Eleuthera Institute in Eleuthera, The Bahamas on 15 January 2013. Water temperature at time of capture ranged from 23 to 25°C. Fish were held for 24 h prior to examination in circular tanks (3.7 m diameter × 1.25 m height, 13 180 l) with 1 m water depths, supplied with constant flow of fresh seawater. The seawater intake was from 1.5 m of water such that the water temperature was stable (~24°C during the study period) despite more extreme diel thermal variation in the nearshore tidal creeks from which the fish were captured.

CT minimum

Mottled mojarra (n = 14) were exposed to a linear temperature decrease (~0.2°C min⁻¹) in individual aerated 1 l containers (containing 750 ml of seawater) from ambient temperature (24°C) until equilibrium was lost, following the methods of Murchie et al. (2011). Two separate trials were conducted with 7 fish in each. Fish were considered to have lost equilibrium when they were unable to maintain upright posture for a minimum of 5 s, at which point time to loss was recorded. Determination of CT minimum was needed to inform the design of subsequent experiments. That is, our interest was in sub-lethal outcomes, so we needed to identify the level of lower thermal tolerance of this species.

Cold shock

Cold shock trials were conducted from 09:00 to 18:00 h from 16 to 29 January 2013. Each trial consisted of 8 fish, all kept in water of the same temperature, with 2 trials for each treatment (16 fish in total). Trials alternated randomly amongst treatments, which consisted of 16, 18, and 20°C temperatures (all below ambient temperature), and controls that remained at ambient temperature (24°C).

To begin trials, individual opaque plastic containers (1 l volume) were filled with 750 ml of seawater at ambient temperature. A small air stone was placed in each container to provide oxygen and ensure that the water was well mixed. Containers were then partially submerged in an ice bath (100 l cooler), or in the case of the controls, ambient seawater. For cold shock treatments, the temperature in the containers was lowered to the desired temperature by managing the amount of ice in the cooler. Once treatment temperatures were attained, mottled mojarra were transferred by dip net from the holding tank into individual containers. Trials were then initiated and lasted for 1 h. Temperature was recorded every 15 min during trials using a YSI meter (EcoSense pH10 A, Xylem).

During cold shock trials, ventilation rates were recorded by the same observer every 15 min, by visually counting operculum beats for 30 s. Reflex Action Mortality Predictors (RAMP; Davis 2010) were also assessed at the beginning and end of each trial, or at the time of equilibrium loss. A total of 5 RAMP indices were used: tail grab, body flex, head complex, vestibular–ocular response (VOR), and equilibrium. These metrics were chosen because they are strong indicators of vitality in other fish species in the context of fisheries-related stress (Raby et al. 2012, Brownscombe et al. 2013, 2014, Cooke et al. 2014), as well as cold shock (Szekeres et al. 2014). Tail grab was assessed by grabbing the fish's tail by hand, and was considered impaired if the fish did not immediately attempt to swim away. Body flex was assessed by briefly holding the fish out of the water; a lack of attempt to struggle free indicated impairment. While held out of the water, head complex was considered impaired if regular opercular beats were not observed. VOR was assessed by rolling the fish from side to side to determine whether the eyes were able to track the handler. Equilibrium was assessed by flipping the fish upside down in the water, and was considered impaired if the fish failed to recover within 3 s. Higher RAMP scores indicated greater impairment.

Once trials were completed, mottled mojarra were transferred into recovery containers (1 l volume, containing 750 ml of water) at ambient temperature for 30 min. After this recovery period, fish were transferred individually to 5 l circular containers with 8 cm of water depth, which were divided into 4 equal quadrants by lines drawn on the bottom of the container. Fish were chased by hand until exhaustion by one researcher, while another researcher recorded the number of lines mottled mojarra crossed prior to exhaustion. Fish were considered exhausted when the chaser could touch the fish's tail 3 times without an attempt to escape. RAMP was assessed both immediately before and after chase trials; ventilation rates were measured immediately after. Once completed, the fish were allowed to recover for a minimum of 1 h before being released into the wild. Although this approach to evaluating swimming ability does not generate an actual swimming speed per se, it does provide a robust approach for making inter-treatment comparisons (Portz 2007).

Statistical analysis

The relationship between mottled mojarra TL and CT minimum was analysed with Pearson's correlation coefficient. Mottled mojarra ventilation rates were compared between treatments and over time with a linear mixed effect (LME) model with individual fish as a random factor. Significant terms were determined using backwards model selection with log-ratio tests. A variance structure was used to correct for variance heterogeneity (Pinheiro & Bates 2000). Tukey's HSD post hoc tests were used to determine significant differences between treatments. Mottled mojarra RAMP score frequencies were compared among treatments using a chi-squared test for independence. For the chase to exhaustion experiment, the effects of temperature on the number of lines crossed and time chased were analysed using 1-way ANOVA. Tukey's HSD tests were used to determine significant differences. Assumptions of normality and homogeneity were tested prior to analysis. Results were considered significant at p ≤ 0.05, and all analyses were conducted using R studio (R Core Team 2012).

RESULTS

CT minimum

The CT minimum (at which mottled mojarra lost equilibrium) was 12.29 ± 0.75°C (mean ± SE). There was no significant association between TL and CT minimum within the size range of 75 to 100 mm (r = 0.44, p = 0.56).

Cold shock

During cold shock exposure, mottled mojarra ventilation rates were highly variable among treatments, with fish in both the 16 and 24°C treatments generally exhibiting lower ventilation rates than those in the 18 and 20°C treatments (Fig. 1). Comparing ventilation rates among treatments and over time, there was a significant effect of time (LME; $F_{1,254}$ = 79.3, p < 0.001), and significant interaction between treatment and time ($F_{1,254}$ = 16.6, p < 0.001). At the start of the exposure period, ventilation rates were significantly higher at 18 than at 24°C (Tukey's HSD; p = 0.01). After 15 min of exposure, ventilation rates were significantly higher at 18 and 20°C than at 16°C (p < 0.01). For the remaining time periods (i.e. 30, 45 and 60 min) ventilation rates were significantly higher at 18 and 20°C than at 16 and 24°C (p < 0.05).

Prior to the cold shock exposure period, all mottled mojarra had RAMP scores of 0 (no impairment). During exposure, no fish lost equilibrium at 24 or 20°C, however, 3 (19%) and 9 (56%) fish lost equilibrium at 18 and 16°C respectively (Fig. 2). After the 60 min exposure period, there was a significant difference in the amount of reflex impairment between treatments (χ^2 = 44, df = 12, p < 0.001). While mottled mojarra in the 20 and 24°C treatments exhibited no reflex impairment, fish in the 16 and 18°C treatments con-

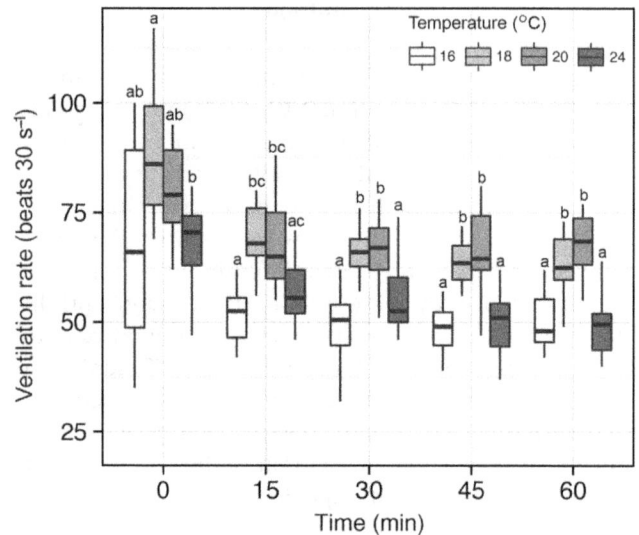

Fig. 1. Ventilation rates (beats per 30 s) of mottled mojarra across a range of temperatures for a 1 h period. Boxes: 1st and 3rd quartiles; horizontal lines: medians; whiskers: 1.5 × inner quartile range; dots: outliers. Dissimilar letters indicate statistically significant differences using Tukey's HSD post hoc test

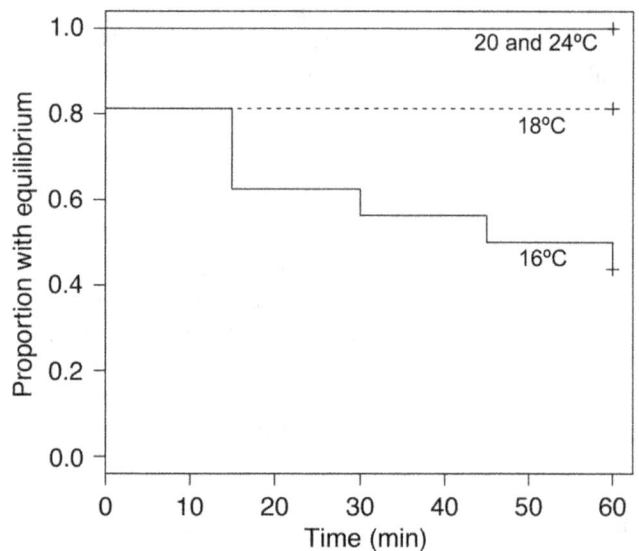

Fig. 2. Proportion of mottled mojarra that lost equilibrium during the exposure period across temperature treatments

tinued to experience reflex impairment (Fig. 3), which was more severe at 16°C (Table 1). After the cold shock exposure period, fish in all treatments had zero RAMP impairment, and those exposed to 16°C exhibited the best swimming performance, with the longest chase times and most lines crossed of any treatment (Fig. 4). There was a significant effect of

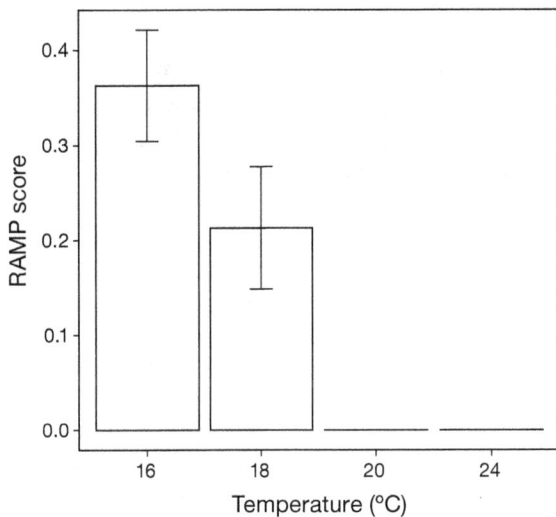

Fig. 3. Reflex Action Mortality Predictor (RAMP) score (±SE) for mottled mojarra immediately after exposure to a range of temperatures for 60 min

Table 1. Proportion of mottled mojarra that exhibited reflex impairment after 60 min of cold shock exposure at 16 or 18°C. RAMP: Reflex Action Mortality Predictors; VOR: vestibular–ocular response

RAMP	16°C	18°C
Body flex	0.81	0.38
Tail grab	0.25	0.31
VOR	0.13	0.00
Equilibrium	0.19	0.19
Head complex	0.44	0.25

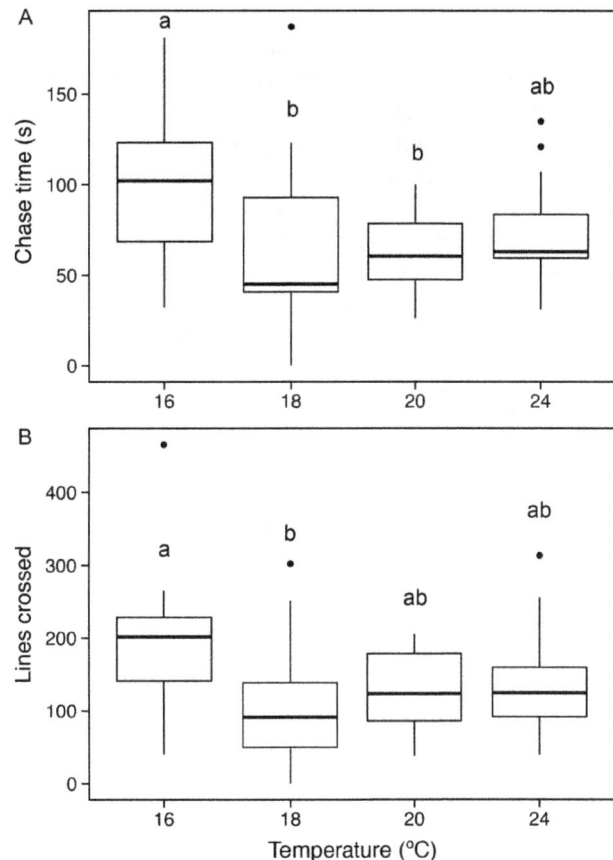

Fig. 4. (A) Time chased and (B) number of lines crossed by mottled mojarra after a 60 min exposure to various temperatures. Boxes: 1st and 3rd quartiles; horizontal lines: medians; whiskers: 1.5 × inner quartile range; dots: outliers. Dissimilar letters indicate statistically significant differences between treatments using Bonferroni-corrected pairwise t-tests

treatment on the amount of time chased ($F_{3,60}$ = 3.6, p = 0.02); mottled mojarra in the 16°C treatment swam significantly longer than those in the 18 and 20°C treatments (p < 0.05; Fig. 4). There was also a significant effect of treatment on the number of lines crossed ($F_{3,60}$ = 4.1, p = 0.01); fish in the 16°C treatment crossed significantly more lines than those in the 18°C treatment (p = 0.01). After the chase to exhaustion, all fish exhibited impaired tail grab response (integral to the chasing process), but no other RAMP impairment.

DISCUSSION

Abrupt declines in water temperature had significant impacts on the ventilation rates, reflexes, and swimming performance of mottled mojarra. Ventilation rates displayed some surprising patterns, where at relatively moderate decreases in temperature (18 and 20°C; 4 and 6°C below ambient), mottled mojarra exhibited higher ventilation rates than at ambient, while at the greatest decline in temperature (16°C; 8°C below ambient) fish had similar ventilation rates to ambient. Abrupt changes in temperature typically cause physiological disturbances that increase metabolic demands in ectotherms (Wendelaar Bonga 1997, Jentoft et al. 2005, Donaldson et al. 2008), which is likely why mottled mojarra exhibited increased ventilation rates at 18 and 20°C. However, some fish species can compensate physiologically to abrupt declines in temperature (Fry 1947, Shiels et al. 2003, Narum et al. 2013), which may explain why ventilation rates remained similar in fish exposed to 16°C to those at ambient temperature (Fig. 1). However, it is surprising that such compensation would not occur at lesser degrees of temperature change. Ventilation rates are often assumed to be reflective of cardio-respiratory demand, but there is growing evidence to suggest that when exposed to stressors, ventilation rates may not always be a strong predictor of physiological status (Barreto & Volpato 2004).

A decline in temperature of 4°C caused significantly higher ventilation rates, but mottled mojarra exhibited no equilibrium loss during exposure, nor any other reflex impairment after the 1 h exposure period. Conversely, declines of 6 and 8°C caused rapid equilibrium loss and significant reflex impairment, particularly at 8°C below ambient temperature. Equilibrium loss is a common result of cold shock exposure in fish, and is an indication of physiological stress that may lead to mortality (Griffith 1978, Hyvärinen et al. 2004, Donaldson et al. 2008, Davis 2010, Raby et al. 2012). Loss of ability to swim upright also greatly reduces the capacity of fish to move into deeper water to escape further cold exposure (in the case of storm-induced perturbation of shallow waters) or avoid predators, further increasing the probability of mortality. Interestingly, tidal creek environments in Eleuthera (where fish were collected for this study) experience daily fluctuations of up to 11°C (Murchie et al. 2011), while the lower sub-lethal thermal limit for mottled mojarra in this study was 12°C on average. However, the cold shock scenarios tested here exposed fish to more abrupt declines in temperature that may be more similar to those resulting from storms or ocean upwelling than typical diel patterns associated with tidal cycles and solar radiation.

Post-exposure swimming capabilities yielded some surprising results: 30 min after 1 h of exposure to 4 and 6°C temperature declines, mojarra exhibited similar swimming capabilities to those maintained at ambient temperature, while those exposed to 8°C declines actually swam longer and crossed more lines than those in the other treatments. Despite the fact that mottled mojarra experienced significant reflex impairment at more severe temperature declines during exposure, swimming performance quickly returned after a 30 min recovery period at ambient temperature. Exposure to cold shock initiates a physiological stress response, a component of which is increased capacity to deliver oxygen to muscles, and an increase in available energy reserves (Wendelaar Bonga 1997, Shiels et al. 2003). This may explain why, when placed back into water at ambient temperature, mottled mojarra exposed to 8°C colder water had similar, if not greater swimming capabilities compared to other treatments. This is consistent with findings in another subtropical marine species, the bonefish (*Albula vulpes*; Szekeres et al. 2014),

which collectively supports the concept that if fish are able to escape cold water during cold shock events, swimming capabilities quickly return and the chances of mortality due to predation are minimal. However, the ability of fish to escape near-shore areas affected by cold shock events (e.g. upwellings or storms) ultimately depends on the behaviour of the fish and the nature of the environment. In environments such as Kemps Creek in Eleuthera, with its expansive shallow flats, fish must travel long distances to reach deeper water.

In summary, exposing mottled mojarra to abrupt cold shock caused significant behavioural impairment; the level of impairment was generally associated with the magnitude of temperature change. Exposure to 16 and 18°C (8 and 6°C below ambient) caused significant reflex impairment, potentially reflecting a physiological stress response. Indeed, extended exposure to water >4°C below ambient may cause mortality in this species. However, if mottled mojarra are able to escape abrupt declines in temperature, swimming capabilities are quickly regained and therefore predation risk would likely be low. To date, very little research has focused on the consequences of cold shock on tropical or subtropical species. Future studies should extend our understanding of the impacts of cold shock to other species in these areas, including critical thermal thresholds and durations. Moreover, there is a need for research on the impacts of cold shock on a greater size range of individuals, especially smaller fish, which may be more sensitive to temperature variability (Donaldson et al. 2008). Such research would be timely, given that adverse weather conditions and changes in ocean currents driven by global climate change may expose near-shore dwelling tropical fish species to increasingly frequent cold shock events in the coming decades.

Acknowledgements. We thank the staff at Cape Eleuthera Institute for supporting our research. J.W.B. was supported by the Natural Sciences and Engineering Research Council of Canada (NSERC). S.J.C. was supported by the Canada Research Chair's program and NSERC. E.S. was supported by the University of Manchester with travel support from NSERC.

LITERATURE CITED

▶ Bakun A (1990) Global climate change and intensification of coastal ocean upwelling. Science 247:198–201

Barreto RE, Volpato GL (2004) Caution for using ventilatory frequency as an indicator of stress in fish. Behav Process 66:43–51

Brett JR (1971) Energetic responses of salmon to temperature. A study of some thermal relations in the physiology and freshwater ecology of sockeye salmon (*Oncorhynchus nerkd*). Am Zool 11:99–113

▶ Brownscombe JW, Thiem JD, Hatry C, Cull F, Haak CR, Danylchuk AJ, Cooke AJ (2013) Recovery bags reduce post-release impairments in locomotory activity and behavior of bonefish (*Albula* spp.) following exposure to angling-related stressors. J Exp Mar Biol Ecol 440:207–215

▶ Brownscombe JW, Nowell L, Samson E, Danylchuk AJ, Cooke SJ (2014) Fishing-related stressors inhibit refuge-seeking behaviour in released sub-adult great barracuda *Sphyraena barracuda*. Trans Am Fish Soc 143:613–617

▶ Cooke SJ, Messmer V, Tobin AJ, Pratchett MS, Clark TD (2014) Refuge-seeking impairments mirror metabolic recovery following fisheries-related stressors in the Spanish flag snapper (*Lutjanus carponotatus*) on the Great Barrier Reef. Physiol Biochem Zool 87:136–147

▶ Davis MW (2010) Fish stress and mortality can be predicted using reflex impairment. Fish Fish 11:1–11

▶ Donaldson MR, Cooke SJ, Patterson DA, Macdonald JS (2008) Cold shock and fish. J Fish Biol 73:1491–1530

▶ Donelson JM, Munday PL, McCormick MARK, Nilsson GE (2011) Acclimation to predicted ocean warming through developmental plasticity in a tropical reef fish. Glob Change Biol 17:1712–1719

▶ Eme J, Bennett WA (2009) Critical thermal tolerance polygons of tropical marine fishes from Sulawesi, Indonesia. J Therm Biol 34:220–225

▶ Fangue NA, Bennett WA (2003) Thermal tolerance responses of laboratory acclimated and seasonally acclimatized Atlantic stingray, *Dasyatis sabina*. Copeia 2003:315–325

FWCC (Fish and Wildlife Conservation Commission) (2010) Snook cold kill report. Fish and Wildlife Research Institute, St. Petersburg, FL

Fry FEJ (1947) Effects of the environment on animal activity. Publications of the Ontario Fisheries Research Laboratory No. 68, University of Toronto Studies Biological Series No. 55. University of Toronto Press, Toronto

▶ Griffith JS (1978) Effects of low temperature on the survival and behaviour of threadfin shad, *Dorosoma petenense*. Trans Am Fish Soc 107:63–70

▶ Haverinen J, Vornanen M (2007) Temperature acclimation modifies sinoatrial pacemaker mechanism of the rainbow trout heart. Am J Physiol 292:R1023–R1032

▶ Hyvärinen P, Heinimaa S, Rita H (2004) Effects of abrupt cold shock on stress responses and recovery in brown trout exhausted by swimming. J Fish Biol 64:1015–1026

IPCC (2007) Summary for policymakers. In: Solomon SD, Qin M, Manning Z, Chen M and others (eds) Climate change 2007: the physical science basis. Contribution of Working Group I to the Fourth Assessment Report of the Intergovernmental Panel on Climate Change. Cambridge University Press, Cambridge

▶ Jentoft S, Aastveit AH, Torjesen PA, Andersen Ø (2005) Effects of stress on growth, cortisol and glucose levels in non-domesticated Eurasian perch (*Perca fluviatilis*) and domesticated rainbow trout (*Oncorhynchus mykiss*). Comp Biochem Physiol Part A Mol Integr Physiol 141:353–358

▶ Kerschner BA, Peterson MS, Gilmore RG Jr (1985) Ectopic and ontogenetic trophic variation on mojarras (Pisces: Gerreidae). Estuaries 8:311–322

▶ Lamadrid-Rose Y, Boehlert GW (1988) Effects of cold shock on egg, larval, and juvenile stages of tropical fishes: potential impacts of ocean thermal energy conversion. Mar Environ Res 25:175–193

Larsen TH, Brehm G, Navarrete H, Franco P, Gomez H (2011) Range shifts and extinctions driven by climate change in the tropical Andes: synthesis and directions. In: Herzog SK, Martínez R, Jørgensen PM, Tiessen H (eds) Climate change and biodiversity in the tropical Andes. Inter-American Institute for Global Change Research (IAI) and Scientific Committee on Problems of the Environment (SCOPE), p 47–67

Meyer-Rochow VB (2013) Thermal pollution: general effects and effects on cellular membranes and organelles in particular. In: Allodi S, Nazari EM (eds) Exploring themes on aquatic toxicology. Research Signpost Publishing, Trivandrum, p 1–34

▶ Michener WK, Blood ER, Bildstein KL, Brinson MM, Gardner LR (1997) Climate change, hurricanes and tropical storms, and rising sea level in coastal wetlands. Ecol Appl 7:770–801

▶ Montgomery JC, Macdonald JA (1990) Effects of temperature on nervous system: implications for behavioral performance. Am J Physiol 259:R191–R196

▶ Munday PL, Jones GP, Pratchett MS, Williams AJ (2008) Climate change and the future for coral reef fishes. Fish Fish 9:261–285

▶ Murchie KJ, Cooke SJ, Danylchuk AJ, Danylchuk SE, Goldberg TL, Suski CD, Philipp DP (2011) Thermal biology of bonefish (*Albula vulpes*) in Bahamian coastal waters and tidal creeks: an integrated laboratory and field study. J Therm Biol 36:38–48

▶ Narum SR, Campbell NR, Meyer KA, Miller MR, Hardy RW (2013) Thermal adaptation and acclimation of ectotherms from differing aquatic climates. Mol Ecol 22:3090–3097

▶ Newman SP, Handy RD, Gruber SH (2012) Ontogenetic diet shifts and prey selection in nursery bound lemon sharks, *Negaprion brevirostris*, indicate a flexible foraging tactic. Environ Biol Fishes 95:115–126

▶ Nilsson GE, Crawley N, Lunde LG, Munday PL (2009) Elevated temperature reduces the respiratory scope of coral reef fishes. Glob Change Biol 15:1405–1412

Pinheiro JC, Bates DM (2000) Mixed-effects models in S and S-PLUS. Springer, New York, NY

Portz DE (2007) Fish-holding-associated stress in Sacramento river chinook salmon (*Oncorhynchus tshawytscha*) at South Delta fish salvage operations: effects on plasma constituents, swimming performance, and predator avoidance. PhD dissertation, University of California, San Diego, CA

▶ Prosser L, Farhi E (1965) Effects of temperature on conditioned reflexes and on nerve conduction in fish. Z Vgl Physiol 50:91–101

R Core Team (2012). R: A language and environment for statistical computing. R Foundation for Statistical Computing, Vienna

▶ Raby GD, Donladson MR, Hinch SG, Patterson DA and others (2012) Validation of reflex indicators for measuring vitality and predicting the delayed mortality of wild coho salmon bycatch released from fishing gears. J Appl Ecol 49:90–98

▶ Roessig JM, Woodley CM, Cech JJ, Hansen LJ (2004) Effects of global climate change on marine and estuarine fishes and fisheries. Rev Fish Biol Fish 14:251–275

Roots BI, Prosser CL (1962) Temperature acclimation and the nervous system in fish. J Exp Biol 39:617–629

► Shiels HA, Vornanen M, Farrell AP (2003) Acute temperature change modulates the response of ICA to adrenergic stimulation in fish cardiomyocytes. Physiol Biochem Zool 76:816–824

Szekeres P, Brownscombe JW, Cull F, Danylchuk AJ and others (2014) Physiological and behavioural conse-quences of cold shock on bonefish (*Albula vulpes*) in The Bahamas. J Exp Mar Biol Ecol 459:1–7

► Wendelaar Bonga SEW (1997) The stress response in fish. Physiol Rev 77:591–625

► Wilson SK, Adjeroud M, Bellwood DR, Berumen ML, Booth D, Bozec YM, Syms C (2010) Crucial knowledge gaps in current understanding of climate change impacts on coral reef fishes. J Exp Biol 213:894–900

12

Diel distribution and feeding habits of *Neomysis mirabilis* under seasonal sea ice in a subarctic lagoon of northern Japan

Kazutaka Takahashi[1,*], Norio Nagao[2], Satoru Taguchi[3]

[1]Department of Aquatic Bioscience, Graduate School of Agricultural and Life Sciences, The University of Tokyo 1-1-1, Yayoi, Bunkyo-ku, Tokyo 113-8657, Japan
[2]Institute of Bioscience, Universiti Putra Malaysia, 43400 Serdang, Selangor Darul Ehsan, Malaysia
[3]Faculty of Engineering, Soka University, 1-236 Tangi-cho, Hachioji, Tokyo 192-8577, Japan

ABSTRACT: We investigated the diel distribution and feeding habits of the mysid *Neomysis mirabilis* under seasonal sea ice in a subarctic lagoon of northern Japan. Although large individuals (>11 mm total length) were present in the eelgrass beds regardless of the time of day, smaller individuals only migrated to the eelgrass beds at night, possibly from shallower waters. Investigation of their stomach contents revealed that *N. mirabilis* is primarily dependent on eelgrass epiphytes as a food source, in addition to small crustaceans. Calculations based on gut pigment analysis indicated that epiphytes were sufficient for large mysids to fulfill their metabolic requirements, whereas small mysids needed to ingest additional food items, such as small crustaceans, possibly because they were less able to graze the highly adhesive prostrate epiphytes. This study suggests that grazing on epiphytes could be a third option in the feeding habits of mysids (in addition to suspension-feeding and predation), and is beneficial in maintaining their high biomass, even during the winter.

KEY WORDS: Crustaceans · Mysidacea · Overwintering · Distribution · Sea ice · Eelgrass · Epiphytes

INTRODUCTION

Mysids belonging to the genus *Neomysis* (Mysidae, Crustacea) occur abundantly in lagoons and estuaries of the subarctic north-western Pacific coast (Pecheneva et al. 2002, Yamada et al. 2007, Yusa & Goshima 2011). *Neomysis* is an omnivore, feeding on detritus, algae and animal prey (Baldo-Kost & Knight 1975, Siegfried & Kopache 1980, Fockedey & Mees 1999). The success of *Neomysis* in coastal habitats is largely attributed to its ability to adapt to variable environments, typically represented by a tolerance to low salinity and a wide range of water temperatures (Simmons & Knight 1975, Pezzack & Corey 1982, Roast et al. 1999). Mysids in coastal lagoons at high latitudes, however, experience particularly harsh en-

vironments during the winter, when, in addition to below-freezing temperatures, water surfaces are covered with seasonal sea ice, resulting in dramatically decreased primary productivity in the water column (Tada et al. 1993). Secondary production by planktonic copepods is also depressed (Fortier et al. 1995). Nevertheless, mysids are often the most dominant organisms in terms of biomass in the zooplankton community under seasonal sea ice (Fukuchi et al. 1979, Fortier & Fortier 1997). These observations indicate that mysids are important components of the ecosystem; however, the detailed ecology of mysids under sea ice has not been fully elucidated, partly due to the difficulty of quantitative sampling (see Fukuchi et al. 1979). In this study, we investigated vertical distribution and feeding habits of *Neomysis*

*Corresponding author: akazutak@mail.ecc.u-tokyo.ac.jp

mirabilis (Czerniavsky, 1882) below seasonal sea ice, in order to examine the mechanisms by which they maintain their high abundance during the productively poor winters in subarctic coastal lagoons.

MATERIALS AND METHODS

Sampling was conducted at Saroma-ko lagoon (44° 10′ N, 143° 45′ E) in eastern Hokkaido, northern Japan (Fig. 1). Saroma-ko lagoon covers an area of about 152 km², and is the southernmost locale where sea ice forms in the northern hemisphere (Taguchi & Takahashi 1993). In normal years, 30 to 40 cm thick ice covers the lagoon from December to late March, resulting in low light conditions in the water column. During this period, the production of ice algae, which is well adapted to such environments, surpasses that of phytoplankton and benthic microalgae (McMinn et al. 2005). In addition, the study site contains one of the largest eelgrass *Zostera marina* beds in the world (Katsuki et al. 2009); the eelgrass and its associated epiphytes are an important food source for aquatic animals (e.g. Aya & Kudo 2007, Yamada et al. 2010).

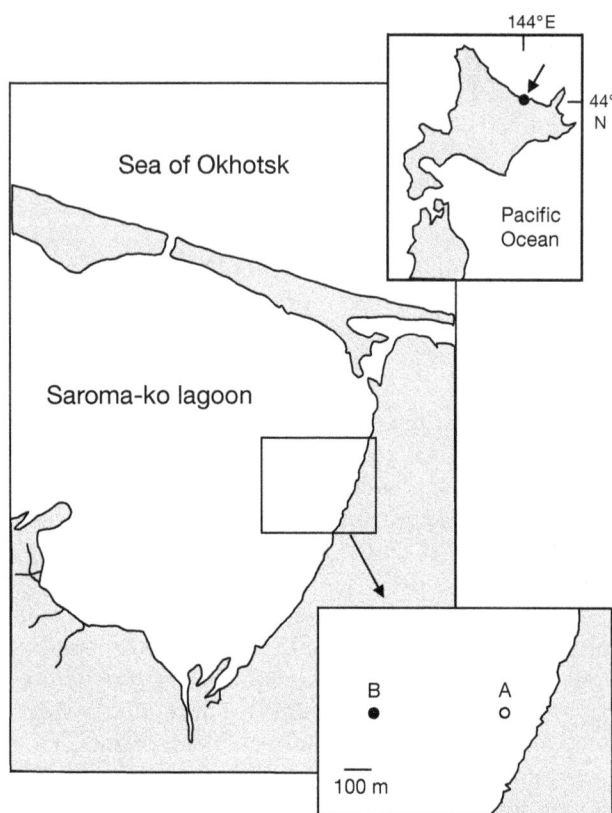

Fig. 1. Location of Saroma-ko lagoon (arrow) and sampling sites in Hokkaido, northern Japan. Stn A: near-shore eelgrass bed; Stn B: offshore muddy bottom

Vertical distribution patterns and feeding habits of *Neomysis mirabilis* were investigated at 2 sampling stations in Saroma-ko lagoon: Stn A, 100 m from shore at a depth of 2 m with a bottom covered with eelgrass beds; and Stn B, 600 m from shore at a depth of 5 m, characterized by a muddy bottom lacking vegetation (Fig. 1). Both stations were covered by 35 cm thick sea ice. Mysids were sampled from 3 layers: near the bottom, immediately under the sea ice, and in between. A conical plankton net (30 cm diameter, 330 µm mesh) was used to sample the upper 2 layers in the water column, and a sledge net (10 cm high × 30 cm wide, with 330 µm mesh) was used for near-bottom sampling. During all sampling events, the gear was attached to a rope, forming a loop between 2 holes that were 20 m apart in the ice, and towed horizontally with a constant towing speed (ca. 0.7 m s⁻¹). Spherical buoys mounted on the frame of the plankton net allowed us to sample the layer immediately under the ice. To sample the middle layer, the plankton net was towed between stainless steel eye bolts attached to a square post that was vertically set in the ice holes. The positions of the eye bolts were adjusted to the middle point of the sampling site depths (1 and 2.5 m for Stns A and B, respectively), and the looped rope was tensioned between the eye bolts when towing the plankton net, which was fastened between the looped rope, in order to sample at a constant depth in the water column. Filtered water volume was estimated as the product of the distance between holes and the area of net opening or sledge mouth. Samples were taken every 3 h for 24 h, from 07:00 h on 27 February to 07:00 h on 28 February 1998, although sampling at Stn B was not conducted after 22:00 h due to thick fog which prevented us from reaching the sampling station. All collected samples were immediately frozen in dry ice and kept at −80°C in the dark until analysis. In the laboratory, the frozen samples were thawed using chilled, filtered seawater, and the mysids were identified and counted using a dissecting microscope under dim light.

As an index of feeding activity, diel changes in gut pigment content were determined. Specimens were rinsed with filtered seawater and then dipped into 4 ml of N,N-dimethylformamide to extract the gut pigments (Suzuki & Ishimaru 1990). Between 5 and 40 extraction bottles were prepared for each sampling time, depending on the number of mysids in the samples. The extraction bottles were kept at −20°C in darkness until analysis (>24 h). Pigment concentrations were measured using a Turner Model 111 fluorometer. Gut pigment content was expressed as chlorophyll *a* (chl *a*) + phaeopigments in a chl *a*-equivalent weight per body dry weight (DW) to take into account variations in body size. Body DW was estimated after Shushkina et al. (1971) from body length, which was measured under the microscope.

Prey items of the mysids were observed using scanning electron microscopy (SEM; JSM-T100). The specimens were fixed with 4% glutaraldehyde immediately after thawing. Under a dissecting microscope, the foregut was carefully removed and divided using the dry fracturing method (Toda et al. 1989). The guts of 10 mysids from the day (13:00 h) and 10 mysids from the night (22:00 h) were examined. The mean (±SD) total lengths (TL) of the specimens from which stomach contents were examined were 12.3 ± 1.3 and 11.7 ± 1.3 mm for day and night, respectively.

In order to investigate potential food items in the environment, the taxonomic characteristics of microalgae (phytoplankton, ice algae, and epiphytes on the eelgrass) were examined at Stn A. Phytoplankton was collected from the ice–water interface (surface) and the near-bottom layer using a Niskin bottle. Ice algae were collected from an ice core that was taken in the vicinity of the sampling hole, using an ice core sampler. Leaves of *Z. marina* were obtained from the ice hole at Stn A, and the microalgal mats (epiphytes) on the leaves were collected by scraping the surface of each leaf with a knife and suspending it in filtered seawater. All samples were preserved in 4% Lugol's solution. Taxonomic identification and enumeration of algal species was conducted using light microscopy. Over 400 cells were counted in each sample to avoid the effect of sample size on the relative abundances of each species.

Based on the diel changes in mean gut pigment, the grazing rate on algal prey was estimated for different size classes using the gut fluorescence method (Mackas & Bohrer 1976). The gut evacuation rate at the *in situ* temperature (−1°C) was estimated as 0.1572 h^{-1} based on the temperature-dependent relationship in *Mysis relicta*, with a Q_{10} function of 1.93 (Chipps 1998). Although *M. relicta* is a freshwater species, its size and the temperature of its habitat are similar to that of *N. mirabilis*. Carbon grazing rates were estimated using a C:chl *a* ratio of 47.63 (De Jonge 1980). The daily

metabolic requirement of *N. mirabilis* (μg C ind.$^{-1}$ d^{-1}) at −1°C was estimated from its DW (μg ind.$^{-1}$) using an empirical allometric relationship (Ikeda 1985) for the oxygen consumption rate, RO (μl O_2 ind.$^{-1}$ h^{-1}). The estimated RO was converted to respiratory carbon equivalents, RC (μg C ind.$^{-1}$ d^{-1}) as RC = RO × RQ × 12 / 22.4 × 24, where RQ (respiratory quotient) is the molar ratio of carbon produced to oxygen utilized, 12 is the atomic weight of carbon, 22.4 is the molar volume of an ideal gas at standard temperature and pressure, and 24 is the number of hours per day. We used an RQ of 0.97, assuming a protein-based metabolism (Gnaiger 1983). The assimilation efficiency was assumed to be 80%, which is a general value for algal prey (Takahashi 2004).

RESULTS

Four species of mysids, including 3 species of *Neomysis*, were collected from under the sea ice (Table 1). *N. mirabilis* was the most dominant species, accounting for 90% of the total mysid population, and occurred almost exclusively at the near-shore station (Stn A; eelgrass bed), regardless of the time of day. Second in dominance was *N. awatschensis* (Brandt, 1851), which accounted for 8% of the total catch; its occurrence was also limited to the

Table 1. Diel and vertical variation in mean (±SD) density of mysids (ind. m^{-3}) collected at 2 sampling sites during the day and at night in Saroma-ko lagoon. The superscript lowercase letters indicate a significant difference between sampling layers (p < 0.01, 1-way ANOVA and Tukey's post-hoc test). S: surface (under sea-ice), M: middle layer, B: near bottom, n: number of sampling times, (−) no occurrence

Species	Sampling layer	Stn A (near-shore)		Stn B (off-shore)	
		Day (n = 5)	Night (n = 4)	Day (n = 4)	Night (n = 1)
Neomysis mirabilis	S	4.5 ± 4.8[a]	9.4 ± 5.9	−	−
	M	4.1 ± 2.4[a]	21.6 ± 10.8[b]	−	0.7
	B	2.7 ± 2.5[a]	8.8 ± 8.9	−	−
Neomysis awatshensis	S	1.4 ± 2.4	1.2 ± 1.4	−	−
	M	0.4 ± 0.6	0.2 ± 0.4	−	−
	B	0.3 ± 0.7	0.4 ± 0.8	−	−
Neomysis czerniawskii	S	−	−	−	1.4
	M	−	−	−	0
	B	−	−	−	1.7
Boreoacanthomysis sherenki	S	−	−	−	0.7
	M	0.1 ± 0.3	−	−	−
	B	0.7 ± 1.5	−	−	−

near-shore station day and night. In contrast, a small number of *N. czerniavskii* (Derzhavin, 1913) were only captured at night at the offshore station (Stn B, muddy bottom). Another uncommon species, *Boreoacanthomysis schrencki* (Czerniavsky, 1882), was collected from Stn A during the day and was only found at Stn B at night.

Further analyses of distribution patterns and gut pigment contents were conducted for the most abundant species, *N. mirabilis*, at Stn A. During the day, *N. mirabilis* was distributed throughout the water column, although its mean density tended to increase with increasing distance from the bottom (Table 1). At night, the density of *N. mirabilis* was 2 to 5 times

that of the day throughout the water column (Table 1). All individuals collected were immature males, and the mean TL of the mysids collected at night (10.5 mm) was significantly smaller than that of the mysids collected during the day (11.5 mm, Student's *t*-test, p < 0.05; Fig. 2A). This difference was attributed to an increase of smaller individuals in the middle and near-bottom layers at night (Fig. 2B).

The gut pigment contents of *N. mirabilis* varied widely, from 0.6 to 130.8 ng chl *a* mg DW⁻¹, and the mean value at night was significantly lower than that during the day (6.2 vs. 9.5 ng chl *a* mg DW⁻¹, respectively; Student's *t*-test, p < 0.05; Fig. 2C). A significant difference in gut pigment contents was found in

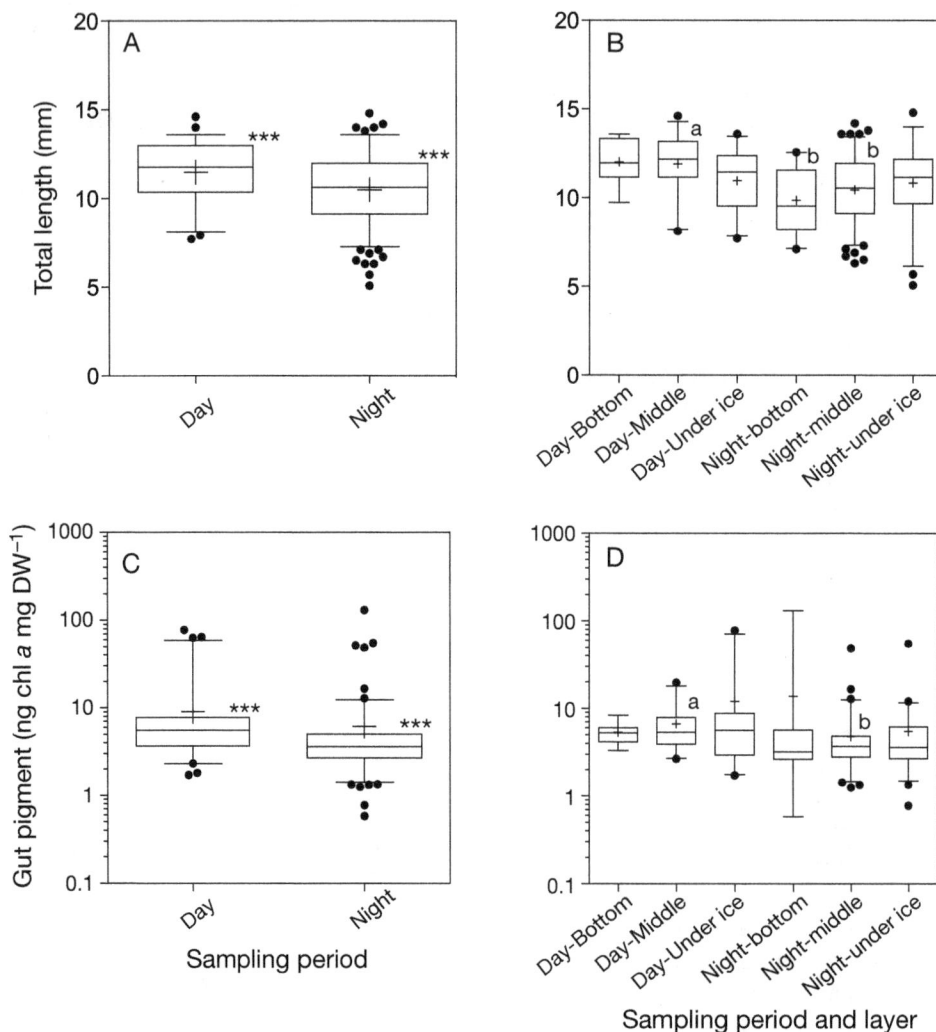

Fig. 2. Diel and spatial variations in body length and gut pigment content in a *Neomysis mirabilis* population at a near-shore eelgrass bed (Stn A) in Saroma-ko lagoon under seasonal sea ice. (A) Diel changes in mean body length; (B) spatio-temporal variation in mean body length; (C) diel changes in mean gut pigment content; (D) spatio-temporal variations in mean gut pigment content. Crosses in the box and whisker plots indicate the mean, the central line in the box is the median, and the box limits are the 25 and 75% quartiles. The whiskers cover 5 to 95% of the data, and dots indicate outliers. ***indicates a significant difference between 2 bars (p < 0.05; Student's *t*-test). Different lowercase letters indicate a significant difference between 2 bars (p < 0.05; 1-way ANOVA with Tukey's post-hoc test)

the middle layer population between day and night (1-way ANOVA with Tukey's post-hoc test, p < 0.05; Fig. 2D).

All *N. mirabilis* stomachs examined by SEM were full of prey items, regardless of sampling time (Fig. 3A). Benthic pennate diatoms *Cocconeis* spp. were the most common throughout the diel cycle (Fig. 3B). Other pennate diatoms, such as *Synedra* spp. (Fig. 3C)

and *Navicula* spp. (Fig. 3D) were found in 30 and 10% of individuals collected during the day and night, respectively. Fragments of crustaceans, including benthic harpacticoid copepods (Fig. 3E,F), were also found in some individuals, but their frequency was relatively low (<10%). A few dinoflagellates or ciliates were found in 10% of individuals collected during the day.

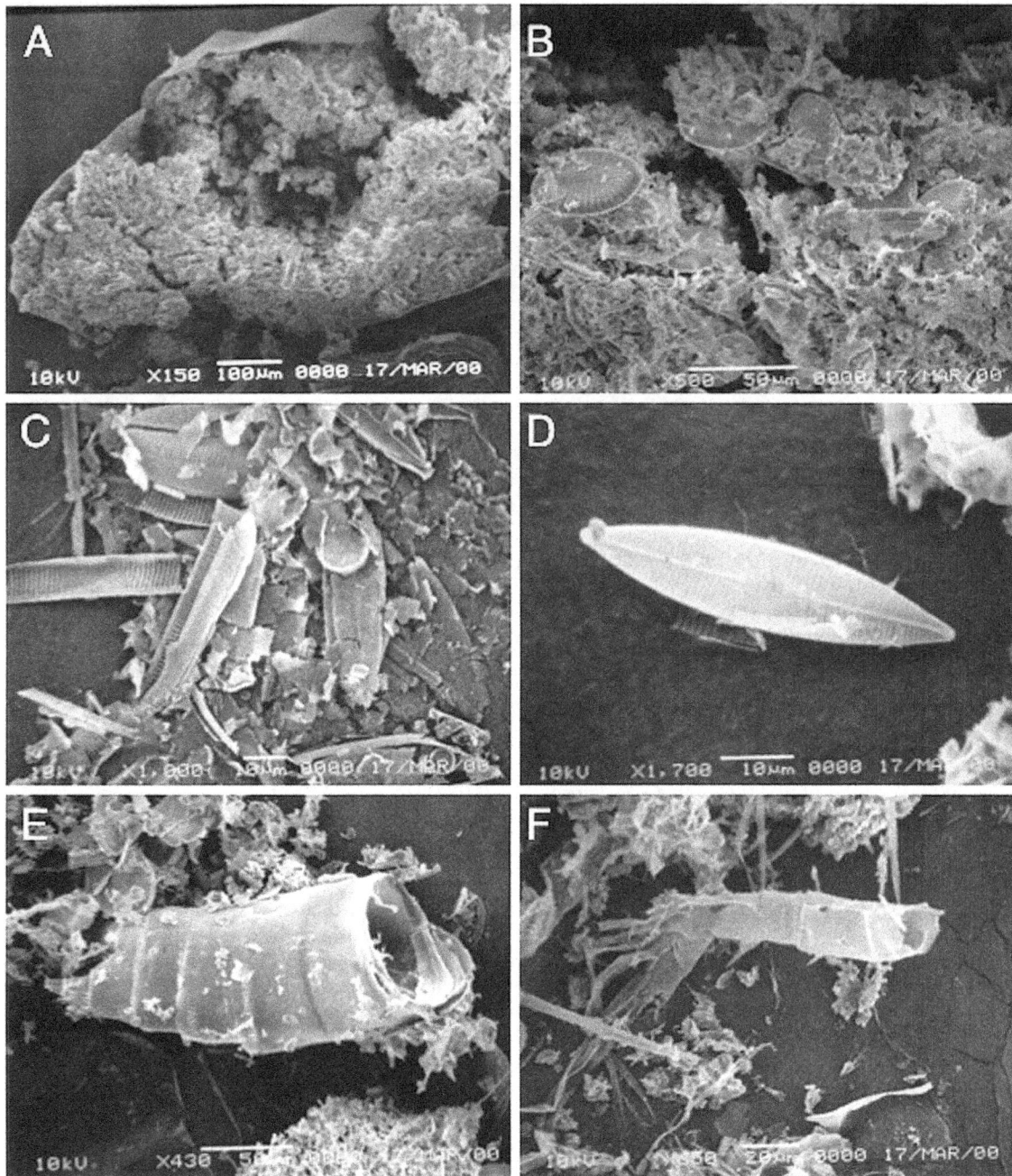

Fig. 3. Scanning electron microscope photographs of the foregut contents of *Neomysis mirabilis* collected in near-shore eelgrass beds under seasonal sea ice. (A) Whole view of stomach contents; (B) unidentified material, and the diatom *Cocconeis* spp.; (C) *Synedra* spp.; (D) *Navicula* sp.; (E) abdomen of harpacticoid copepods; (F) unidentified crustacean parts

Microscopic observations of microalgae in their various habitats under the sea ice showed that assemblages of epiphytes on the eelgrass were distinctly different from those of phytoplankton and ice algae assemblages (Fig. 4). Epiphytes on eelgrass were dominated by *Cocconeis* spp., which accounted for 85% of the total cells studied, with small numbers of *Odontera* spp., *Synedra* spp., *Navicula* spp. and *Pleurosigma* spp. In contrast, the diatoms *Fragilariopsis* spp. and *Detonula* spp. were dominant in phytoplankton and ice algae assemblages, accounting for more than 60% of the total cells investigated, whereas the occurrence of *Cocconeis* spp. in the phytoplankton and ice algae assemblages was extremely limited (<3% of the total cells studied).

Daily carbon ingestion by *N. mirabilis* in the form of algal prey was variable, depending on size, and in general their importance as an energy source increased with mysid size (Table 2). For example, for medium-sized (10.6 mm TL) and large specimens (13.1 mm), the daily ingestion of algal prey fulfilled their metabolic requirements for respiration, whereas in small specimens (7.8 mm), the daily ingestion of algae only satisfied 70% of their metabolic requirements.

DISCUSSION

Among the 4 species of mysids collected in this study, *Neomysis mirabilis* was the most dominant, accounting for 90% of the total mysid population under the sea ice. As found in this study, this species occurs abundantly in eelgrass beds along the coastal zone of the subarctic north-western Pacific (Pecheneva et al. 2002, Yamada et al. 2007, Yusa & Goshima 2011). Zelickman (1974) observed the behaviour of this species in the field, and reported that it formed ribbons or spherical swarms in the eelgrass bed, and

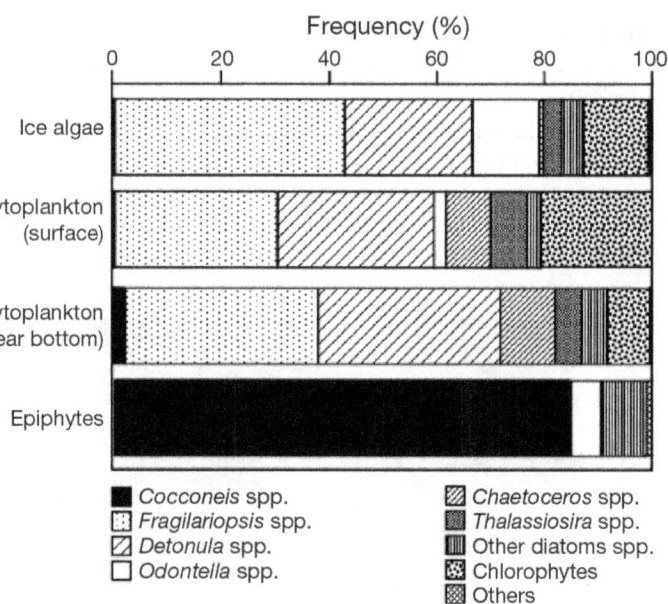

Fig. 4. Composition of algal cells from various *Neomysis mirabilis* habitats under seasonal sea ice at Stn A (near-shore eelgrass bed)

remained at the same location for several hours. The present study also found that *N. mirabilis* is located in eelgrass beds throughout the diel cycle under sea ice; however, some individuals exhibited diel migration within, or outside of, the eelgrass beds. For example, the density of *N. mirabilis* in eelgrass beds in the middle layer at night was significantly higher than that during the day (Table 1). Net avoidance by mysids may partly explain this difference; however, a large proportion of the population seemed to nocturnally migrate to the eelgrass beds, since the increase in density was mainly due to an increase in small individuals (Fig. 2A,B) which have a relatively low escape ability (Fleminger & Clutter 1965). Zelickman (1974) reported that *N. mirabilis* swarmed in the shadow of the water's edge during the day, but disappeared at night. He also observed that in the light, young *N. mirabilis* in an aquarium always stayed

Table 2. Estimated daily ingestion of algal prey based on gut pigment content in *Neomysis mirabilis* in an eelgrass bed under sea ice. Minimum carbon requirements to maintain metabolism are also noted. Daily algal ingestion was estimated with pigment destruction = 90%. Assimilated carbon was estimated with a C:chl a ratio of 47.63 and an assimilation efficiency of 80%. Metabolic requirements were estimated based on the relationship between body dry weight and oxygen consumption rate (Ikeda 1985)

Size class	Total length (mm) Mean (range)	Daily algal ingestion (ng chl *a* ind.$^{-1}$ d^{-1})	Assimilated carbon (µg C ind.$^{-1}$ d^{-1})	Metabolic requirement (µg C ind.$^{-1}$ d^{-1})	% Assimilated carbon to metabolic requirement
Small	7.8 (5.1–8.9)	128.5	4.9	6.8	73
Medium	10.6 (9.1–12.0)	360.1	13.7	13.0	106
Large	13.1 (12.2–14.8)	1263.6	48.1	20.5	235

away from larger individuals. Therefore, we suggest that small N. mirabilis migrate into the eelgrass beds at night from shallower habitats, whereas larger individuals remain in the eelgrass beds throughout the diel cycle.

The gut pigment contents of N. mirabilis were relatively constant throughout the day, but were significantly lower at night (Fig. 2C). We attributed this decrease to the occurrence of small mysids in the middle and near-bottom layers at night (Table 1, Fig. 3B,D). However, SEM observations revealed that benthic diatoms Cocconeis spp. and unidentified amorphous material were the most common prey, regardless of the time of day (Fig. 3). Microscopic analysis revealed that Cocconeis spp. only dominated in the eelgrass epiphyte assemblages (Fig. 4), as reported in previous studies (Sieburth & Thomas 1973, Nakaoka et al. 2001). Therefore, we conclude that N. mirabilis is primarily an epiphyte grazer under the sea ice during the winter. Daily ingestion, estimated by the gut pigment method, indicated that epiphytes were an adequate food source for metabolism maintenance in medium-sized and large mysids, which were located in the eelgrass beds throughout the day (Table 2). However, our calculations indicated that epiphytes alone were not sufficient to support the metabolic rates of small mysids. Cocconeis spp. have a highly adhesive, prostrate form (Moore 1975), and cannot easily be eaten by smaller mysids (ca. <10 mm TL) that do not have fully developed grazing capabilities (Table 2). Therefore, planktonic and benthic copepods are important to small mysids in fulfilling their metabolic requirements. The nocturnal increase of small mysids in the middle and near-bottom layers may be due to foraging for such prey organisms in those areas. Although ontogenetical changes in diet by mysids has been largely studied from the view point of the development of predatory ability on animal prey (Siegfried and Kopache 1980, Branstrator et al. 2000, Viherluoto et al. 2000), this study suggests that the importance of animal prey is higher in younger mysids when grazing is a primary foraging mechanism.

This study found that during the winter, N. mirabilis depends mainly on eelgrass epiphytes as a food source under the sea ice, in addition to small crustaceans. In general, many mysid species, including Neomysis, are known to alternatively adopt 2 different feeding modes: suspension feeding and predation (Mauchline 1980, Takahashi 2004). The results of this study suggest that grazing on epiphytes may be a third alternative in the feeding habits of mysids that inhabit eelgrass beds. The importance of epi-

phytes as a food source has also been reported in N. integer in brackish lakes in the United Kingdom (Irvine et al. 1993). Hasegawa et al. (2007) reported that the density of epiphytes in Akkeshi-ko estuary, south-eastern Hokkaido, reaches its annual maximum in November, just before the sea ice forms. This suggests that epiphytes are the most easily obtainable food items during the winter, when primary and secondary productivity in the water column decreases. Given their wide range of gut fluorescence values, however, dependency on the epiphytes could also vary intra-specifically, particularly in smaller mysids. The flexible feeding habits of N. mirabilis (switching feeding modes depending on food availability) are beneficial to maintaining their high biomass, even during the winter, and consequently result in mysids being important prey for commercially important fish and crustaceans, particularly during the winter and spring.

Acknowledgements. We thank H. Hattori, H. Saito and others from the Saroma Aquaculture and Research Institute for their help in the field. We are also grateful to Y. Niimura for assisting with the algal identifications. SEM observations were supported by T. Suzuki and T. Toda. The study was partly supported by a Sasakawa Scientific Research Grant from The Japan Society and a Grant-in-Aid for Scientific Research (B) from the Ministry of Education, Science, Sports and Culture to K.T. (24310007).

LITERATURE CITED

▶ Aya FA, Kudo I (2007) Isotopic determination of Japanese scallop Patinopecten (Mizuhopecten) yessonsis (Jay) tissues shows habitat-related differences in food sources. J Shellfish Res 26:295–302

Baldo-Kost ALB, Knight AW (1975) The food of Neomysis mercedis Holmes in the Sacramento-San Joaquin estuary. Calif Fish Game 61:35–46

▶ Branstrator KD, Cabana G, Mazumder A, Rasmussen JB (2000) Measuring life-history omnivory in the opossum shrimp, Mysis relicta, with stable nitrogen isotopes. Limnol Oceanogr 45:463–467

▶ Chipps SR (1998) Temperature-dependent consumption and gut-residence time in the opossum shrimp Mysis relicta. J Plankton Res 20:2401–2411

▶ De Jonge VN (1980) Fluctuations in the organic carbon to chlorophyll a ratios for estuarine benthic diatom populations. Mar Ecol Prog Ser 2:345–353

Fleminger A, Clutter RI (1965) Avoidance of towed nets by zooplankton. Limnol Oceanogr 10:94–104

▶ Fockedey N, Mees J (1999) Feeding of the hyperbenthic mysid Neomysis integer in the maximum turbidity zone of the Elbe, Westerschelde and Gironde estuaries. J Mar Syst 22:207–228

▶ Fortier M, Fortier L (1997) Transport of marine fish larvae to Saroma-ko Lagoon (Hokkaido, Japan) in relation to the availability of zooplankton prey under the winter ice cover. J Mar Syst 11:221–234

Fortier L, Fortier M, Demers S (1995) Zooplankton and larval fish community development: comparative study under first-year sea ice at low and high latitudes in the northern hemisphere. Proc NIPR Symp Polar Biol 8:11–19

Fukuchi M, Tanimura A, Hoshiai T (1979) 'NIPR-I', a new plankton sampler under sea ice. Bull Plankton Soc Japan 26:104–109

Gnaiger E (1983) Calculation of energetic and biochemical equivalents of respiratory oxygen consumption. In: Gnaiger E, Forstner H (eds) Polarographic oxygen sensors. Springer, New York, NY, p 337–345

▶ Hasegawa N, Hori M, Mukai H (2007) Seasonal shifts in seagrass bed primary producers in a cold-temperate estuary: dynamics of eelgrass Zostera marina and associated epiphytic algae. Aquat Bot 86:337–345

▶ Ikeda T (1985) Metabolic rates of epipelagic marine zooplankton as a function of body mass and temperature. Mar Biol 85:1–11

▶ Irvine K, Moss B, Bales M, Snook D (1993) The changing ecosystem of a shallow, brackish lake, Hickling Broad, Norfolk, UK. I. Trophic relationships with special reference to the role of Neomysis integer. Freshw Biol 29:119–139

▶ Katsuki K, Seto K, Nomura R, Maekawa K, Khim BK (2009) Effect of human activity on Lake Saroma (Japan) during the past 150 years: evidence by variation of diatom assemblages. Estuar Coast Shelf Sci 81:215–224

▶ Mackas DL, Bohrer R (1976) Fluorescence analysis of zooplankton gut contents and investigation of diel feeding patterns. J Exp Mar Biol Ecol 25:77–85

Mauchline J (1980) The biology of mysids and euphausiids. Adv Mar Biol 18:1–681

▶ McMinn A, Hirawake T, Hamaoka T, Hattori H, Fukuchi M (2005) Contribution of benthic microalgae to ice covered coastal ecosystems in northern Hokkaido, Japan. J Mar Biol Assoc UK 85:283–289

▶ Moore JW (1975) The role of algae diet of Asellus aquaticus L. and Gammarus pulex L. J Anim Ecol 44:719–730

▶ Nakaoka M, Toyohara T, Matsumasa M (2001) Seasonal and between-substrate variation in mobile epifaunal community in a multispecific seagrass bed of Otsuchi Bay, Japan. Mar Ecol 22:379–395

▶ Pecheneva NV, Labai VS, Kafanov AI (2002) Bottom communities of Nyivo Lagoon (northeastern Sakhalin). Russ J Mar Biol 28:225–234

▶ Pezzack DS, Corey S (1982) Effects of temperature and salinity on immature and juvenile Neomysis americana (Smith) (Crustacea; Mysidacea). Can J Zool 60:2725–2728

▶ Roast SD, Widdows J, Jones MB (1999) Respiratory responses of the estuarine mysid Neomysis integer (Peracarida: Mysidacea) in relation to a variable environment. Mar Biol 133:643–649

Shushkina EA, Kuz'micheva VU, Ostapenko LA (1971) Energy equivalent of body mass, respiration and calorific value of mysids from the Sea of Japan. Oceanology 12:880–889

▶ Sieburth J, Thomas CD (1973) Fouling on eelgrass (Zostera marina L.). J Phycol 9:46–50

▶ Siegfried CA, Kopache ME (1980) Feeding of Neomysis mercedis (Holmes). Biol Bull (Woods Hole) 159:193–205

▶ Simmons MA, Knight AW (1975) Respiratory response of Neomysis intermedia (Crustacea: Mysidacea) to changes in salinity, temperature and season. Comp Biochem Physiol A 50:181–193

▶ Suzuki R, Ishimaru T (1990) An improved method for the determination of phytoplankton chlorophyll using N,N-dimethylformamide. J Oceanogr Soc Jpn 46:190–194

Tada K, Kurata M, Nishimura Y (1993) Seasonal changes of chlorophyll a and nutrients in Lake Saroma. Bull Plankton Soc Japan 39:163–165

Taguchi S, Takahashi M (1993) Summary of the colloquium. In: Taguchi S, Takahashi M (eds) Plankton colloquium: ecosystem of Lake Saroma under sea ice covered condition in winter. Bull Plankton Soc Japan 39:152–154

Takahashi K (2004) Feeding ecology of mysids in freshwater and marine coastal habitats: a review. Bull Plankton Soc Japan 51:46–72 (in Japanese with English abstract)

▶ Toda T, Suh HL, Nemoto T (1989) Dry fracturing: a simple technique for scanning electron microscopy of small crustaceans and its application to internal observations of copepods. J Crustac Biol 9:409–413

▶ Viherluoto M, Kuosa H, Flinkman J, Viitasalo M (2000) Food utilization of pelagic mysids, Mysis mixta and M. relicta, during their growing season in the northern Baltic. Mar Biol 136:553–559

▶ Yamada K, Takahashi K, Vallet C, Taguchi S, Toda T (2007) Distribution, life history, and production of three species of Neomysis in Akkeshi-ko estuary, northern Japan. Mar Biol 150:905–917

▶ Yamada K, Hori M, Tanaka Y, Hasegawa N, Nakaoka M (2010) Contribution of different groups to the diet of major predatory fishes at a seagrass meadow in northeastern Japan. Estuar Coast Shelf Sci 86:71–82

Yusa T, Goshima S (2011) Life history of Neomysis mirabilis (Crustacea: Mysidacea) in Notoro Lagoon, Hokkaido, Japan. Crustac Res 40:81–92

▶ Zelickman EA (1974) Group orientation in Neomysis mirabilis (Mysidacea: Crustacea). Mar Biol 24:251–258

13

Histochemistry on vibratome sections of fish tissue: a comparison of fixation and embedding methods

Bianka Grunow[1,*]**, Tina Kirchhoff**[1]**, Tabea Lange**[1,2]**, Timo Moritz**[2]**, Steffen Harzsch**[1]

[1]Cytology and Evolutionary Biology, Zoological Institute and Museum, Ernst-Moritz-Arndt-University of Greifswald, Soldmannstrasse 23, 17487 Greifswald, Germany
[2]Deutsches Meeresmuseum, Katharinenberg 14–20, 18439 Stralsund, Germany

ABSTRACT: Despite improvements in imaging techniques during recent years, for many non-model systems the fixation of tissues followed by embedding and sectioning for histochemical or immunohistochemical staining remains an important technique in vertebrate histology. The present study sets out to explore the preservation of histological sections of fish tissues using different preparation techniques. The quality of transverse vibratome sections from trunk segments of the lesser-spotted dogfish *Scyliorhinus canicula*, Atlantic sturgeon *Acipenser oxyrinchus* and zebrafish *Danio rerio* were compared using different fixatives (formaldehyde, paraformaldehyde and zinc-formaldehyde) and embedding methods (gelatine, agarose and low-temperature melting agarose). Our data show that the quality of the vibratome sections for histochemical staining is strongly dependent upon fixation and embedding media. Although paraformaldehyde fixation results in a more pronounced shrinkage of the trunk segment than the other fixatives used , the quality of the sections and the histochemical staining was best with this fixative in zebrafish and dogfish. Additionally, the embedding methods have a strong influence on the quality of the sections. In the dogfish and sturgeon samples, the preferred embedding media were agarose and low-temperature melting agarose, since gelatine often caused shrinkage of the tissues. In conclusion, for histochemical examinations, the processing protocols for vibratome sectioning need to be adapted individually to each study organism.

KEY WORDS: Histochemistry · Vibratome · Bony fish · Cartilaginous fish · Fixation · Embedding media

INTRODUCTION

Methods for studying mammalian tissue using vibratome sections have been commonly used since the early 1970s (Hökfelt & Ljungdahl 1972, Lindvall et al. 1973), and from the 1980s, publications on vibratome sections using fish tissue are also available (Funch et al. 1984, Kah et al. 1986, Yulis & Lederis 1986). This technique is well-suited to prepare tissue sections, typically between 10 and 100 µm thick, that can subsequently be processed by histochemical, immunohistochemical or electron microscopic methods. This approach has been used for studying the brain (Kálmán & Ari 2002, Lieberoth et al. 2003, Lechtreck et al. 2009, Pilaz & Silver 2014, Reichmann et al. 2015), cardiovascular system (Bryson et al. 2011, Price et al. 2014), liver (Hampton et al. 1987, Satoh et al. 2005) and retina and lenses (Freel et al. 2003, Alvarez-Viejo et al. 2004, Enski et al. 2013) in fishes and other vertebrates. Whole body transverse sections are commonly used for diagnostic work to understand the full extent of the histological lesions that are caused by fish pathogens (Bruno et al. 2013). However, to our knowledge, differences among protocols of vibratome sectioning for whole body transverse sections in fish has not yet been systematically explored.

In comparison to paraffin embedding or cryo-embedding, the advantages of vibratome sectioning are manifold (Robertson 2002). For example, there is

*Corresponding author: bianka.grunow@uni-greifswald.de

no need for dehydrating and rehydrating the tissues, which decreases loss of cell constituents. There is no need for applying high temperatures or harsh chemical treatments that may lead to antigen degradation. No expensive special blades are required, so that artefacts typically caused by paraffin embedding or freezing for cryo-sections are avoided. Furthermore, although sections of paraffin- or cryo-embedded tissues provide good internal resolution, a precise orientation of the tissue can be difficult. However, there are also disadvantages of vibratome sectioning. Sections are generally thicker than those obtained with paraffin or cryo-embedding and therefore penetration of reagents such as dyes or antibodies is more time-consuming.

In fish histology, different fixatives and embedding media are established for vibratome sectioning. Most frequently, the tissue is fixed with paraformaldehyde (Kálmán & Ari 2002, Alvarez-Viejo et al. 2004, Mahler and Driever 2007, McGrail et al. 2010) or formaldehyde (McGrail et al. 2010, Nyholm et al. 2009). For embedding, the media agarose (Kálmán & Ari 2002, Mahler & Driever 2007, Nyholm et al. 2009, Jayachandran et al. 2010, McGrail et al. 2010), gelatine (Alvarez-Viejo et al. 2004, Dietrich et al. 2010) or low-temperature melting agarose (Cheung et al. 2012) are generally used.

This study sets out to analyse in detail the effects of fixative and embedding methods on the tissue preservation of trunk segments for larvae and adult specimens of cartilaginous and bony fish species. In this report, 3 fixatives, formaldehyde, paraformaldehyde and zinc-formaldehyde, and 3 standard embedding reagents, gelatine, agarose and low-temperature melting agarose are compared. Formaldehyde fixes tissue or cells by cross-linking primary amino groups in proteins and is generally used in histology for the conservation of biological material (Rolls 2012). Paraformaldehyde is a polymerization product of Formaldehyde and the most common method for fixing animal tissue in cell biology (Kuhlmann 2009). Zink-formaldehyde is gaining importance as it has been shown to be a better fixative than paraformaldehyde for vibratome sections (especially in immunochemistry) as it enables increased penetration of antibodies (Ott 2008).

The choice of the embedding medium to stabilize tissues during the sectioning process is dependent on the tissue structure. Griffioen et al. (1992) described that a gelatine-embedding protocol prevented lesion damages in brain tissue, and Fukuda et al. (2010) used gelatine-embedded brains to create 3-dimensional histological maps and reconstruction of large-sized brain tissues. Another aspect of the embedding material is its optical properties. Blackiston et al. (2010) preferred low-temperature melting agarose because of the ease of orientation of the tissue in a fully transparent block, as is also possible with agarose.

Three fish species were used in this study: (1) larvae of a cartilaginous fish, the lesser-spotted dogfish *Scylliorhinus canicula*; (2) larvae from Atlantic sturgeon *Acipenser oxyrinchus*, representing a basal actinopterygian; and (3) the zebrafish *Danio rerio* wild-type as a representative of Teleostei and as a common model in vertebrate research.

MATERIALS AND METHODS

Animals

The lesser-spotted dogfish reproduces regularly in captivity: for animals to use in this study, 6 individuals (3 males and 3 females) in the public aquaria of the Ozeaneum (Stralsund, Germany) regularly deposited eggs on artificial plants. Parents were kept at 12 to 13°C, a salinity of 30 PSU and an oxygen supply of 92.7% in a 43500 l recirculation system. The developmental stages of the embryo and larvae described by Ballard et al. (1993) were used to determine the animals' age. Eggs were kept in a separate tank of 125 l at 15°C, 30.7 PSU and 98.5% O_2. Juveniles of 320 degree-days (DD) and newly fertilized larvae of Atlantic sturgeon were obtained from the Institute for Fisheries, State Research Center Mecklenburg-Vorpommern for Agriculture and Fishery in Born/Darss, Germany. Juveniles and eggs were kept at 20°C in a 600 l continually recirculated freshwater system. Adult zebrafish were obtained prior to fixation from the Department of Anatomy and Cell Biology, Universitätsmedizin Greifswald, Greifswald, Germany. These animals were kept at 26°C in freshwater. Prior to fixation all specimens were anaesthetised until respiratory failure using benzocaine solution (ethyl p-amino-benzoate; Sigma Aldrich) dissolved in aquarium water.

Animal measurement and fixation

Table 1 provides an overview of the animals used in this study. In total, 46 dogfishes of 5 developmental stages (700 DD, 900 DD, 1200 DD, 1500 DD, and 2000 DD), 47 sturgeons of 3 different ages (80 DD, 320 DD and 620 DD) and 41 adult zebrafish were used. Specimens of all 3 species were fixed in each

Table 1. Overview of studied material. Age (degree-days, DD), length (mean ± SE) and number of animals for the different fixation methods. FOR: 4 % formaldehyde; PFA: 4 % paraformaldehyde; ZnFA: 340 mM or 800 mM zinc-formaldehyde

Species	Length	No. of animals		
Age (DD)	(mm)	Total	FOR	PFA	ZnFA
Lesser-spotted dogfish					
700	26.06 ± 1.24	8	3	3	2
900	31.89 ± 0.40	10	3	4	3
1200	42.78 ± 1.59	8	3	3	2
1500	61.96 ± 3.47	10	4	3	3
2000	89.71 ± 5.13	10	3	4	3
Atlantic sturgeon					
80	8.21 ± 0.27	15	5	5	5
320	15.80 ± 0.15	18	6	6	6
620	18.59 ± 0.46	14	4	5	5
Zebrafish					
Adult	32.39 ± 0.30	41	13	14	14

of 3 chemicals: 4 % formaldehyde in filtered aquarium water (FOR; prepared from a 37 % formaldehyde stock solution stabilized with methanol; Berlin-Brandenburger Lager- und Distributionsgesellschaft mbH); 4 % paraformaldehyde prepared following manufacturer's instructions (PFA, pH 7.8; ~1500 mOsmol kg^{-1}; Carl Roth) in 0.1 M phosphate-buffered saline (PBS); and a zinc-formaldehyde solution (ZnFA) (for details see Ott 2008). Two ZnFA solutions were prepared and applied to the samples, adjusted according to their osmolarity. Dogfishes were fixed using 800 mM ZnFA, and for the freshwater species (sturgeon and zebrafish) a 340 mM ZnFA solution was used. Before fixation, the trunk parts posterior to the pectoral fin and posterior to the dorsal fin area were cut off to ensure proper saturation of the trunk segment (Fig. 1). Size measurements of the animals and the tissue samples were taken using a digital sliding caliper (CD-20DCX absolute digimatic; Mitutoyu).

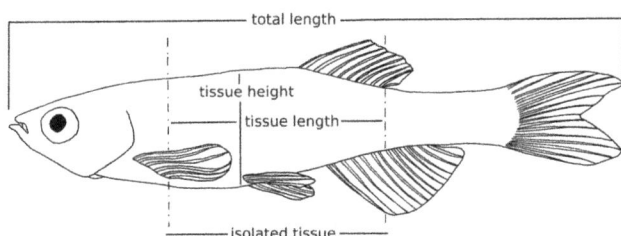

Fig. 1. Illustration of animal measurements. Total length of the animal was measured from tip of rostrum to end of tail fin. The height of the isolated trunk part was measured at the level of pelvic fin and the length was measured from one end to the other end of the trunk

These measurements included animal length (from the tip of the rostrum to the end of the tail fin), and length and height of the trunk segments (height of the tissue at the level of the pelvic fin insertion) prior and after fixation. A camera tripod was used to ensure the same angle for analyzing the size of the samples.

All samples of the trunk segments were incubated in the fixatives for 4 h at a constant temperature of 20°C followed by five 20 min washing steps in phosphate buffered saline. FOR- and PFA-fixed samples were washed in PBS whereas ZnFA-fixed samples were washed in 1 M 4-(2-hydroxyethyl)-1-piperazineethanesulfonic acid (HEPES buffer; Serva) according to the protocol of Ott (2008). All samples were kept in fresh PBS at 4°C until embedding.

Embedding and sectioning

The vibratome sections were performed from whole isolated and fixed trunk segments from dogfish (2000 DD), sturgeon (320 DD) and adult zebrafish. Specimens from each species were embedded in each of the 3 investigated embedding media prior to caudal sectioning. The embedding media were: (1) 4 % porcine skin gelatine/20 % chicken egg white albumin (Sigma) dissolved in distilled water (GEL) (Loesel et al. 2002), (2) 4 % agarose (Serva) dissolved in distilled water (AGA), and (3) 4 % low-temperature melting agarose (melting temperature ≤65°C; Sigma) dissolved in distilled water (LMA). The samples kept in gelatine blocks were post-fixed with 10 % formaldehyde (diluted from 37 % formaldehyde, Carl Roth) and 0.1 M PBS at 4°C overnight. All embedding blocks (1.5 × 1.5 × 1.0 cm) were stored at 4°C in PBS until sectioning. Transverse sections (50 μm thick) were prepared with a vibratome Hyrax V50 (Carl Zeiss MicroImaging), adjusting frequency, amplitude, and speed of the razor blade (Isana men; Rossmann) for each sample (Table 2). At least five 50 μm sections were prepared per sample for quality comparisons. Sections were cut by a single person to avoid variances in section quality due to operator changes. Analysis of the section quality was performed by 2 persons in a blinded manner. All sections were analyzed and evaluated as arithmetic mean. Sections of best quality (+++) showed no damage or deformation, whereas sections of good quality (++) had a little damage to the epidermis or the muscle tissue, and sections of poor quality (+) were highly damaged or showed deformation. Experiments of samples with poor quality sections were performed twice to exclude mistakes during processing.

Table 2. Overview of vibratome adjustment. Frequency, amplitude and speed for sectioning 50 μm thick fish trunk segments fixed in 4 % formaldehyde (FOR), 4 % paraformaldehyde (PFA), or 340 mM or 800 mM zinc-formaldehyde (ZnFA), and embedded in 4 % gelatine/20 % albumin (GEL), 4 % agarose (AGA) or 4 % low-temperature melting agarose (LMA)

| | FOR | | | PFA | | | ZnFA | | |
	GEL	AGA	LMA	GEL	AGA	LMA	GEL	AGA	LMA
Lesser-spotted dogfish									
Frequency	50	45–60	60	40–70	50	40–70	50	50	50
Amplitude	0.7	0.7–1.2	0.7	0.7–1.2	1.0	0.9	0.9	0.9	0.9–1.0
Speed	1–3	1	1	1	4–5	1	1	3–4	3–4
Atlantic sturgeon									
Frequency	50	60–65	30	50	40	30	50	30	30
Amplitude	0.5–0.7	1.2	1.0	0.9	1.1	0.3	0.9	0.9–1.0	1.0
Speed	1	1	1	10	1	1	2	1	1
Zebrafish									
Frequency	30	50	30	30	30	60	30–75	30–75	50
Amplitude	0.7	1.2	1.0–1.2	0.3	1.1	0.8	0.3–1.2	0.3–1.2	1.2
Speed	1	1	1	1	1	1	1–4	1	1

Histochemical staining of vibratome sections

To determine the optimal fixation and embedding method for fluorescence labelling, the sections which were of very good quality were stained with the actin-binding probe phalloidin conjugated to a fluorescent dye and Hoechst, a dye for staining nuclei. The sections were washed 6 times in PBS-BSA-TX (5 % BSA, 0.5 % Triton X-100, 0.05 % Na-acid; Sigma) for 90 min to remove the fixative. Specimens fixed in ZnFA were washed in HEPES buffered saline for the same time. Afterwards, sections were preincubated in PBS-BSA-TX for 2 h at room temperature. The tissues were then incubated in Phalloidin-TRITC 546 (1:1000; Sigma) and 0.5 μl ml^{-1} Hoechst H 33258 (1:2000; Sigma) diluted in PBS-BSA-TX for 15 min at room temperature. Thereafter, sections were rinsed in 6 changes of PBS-BSA-TX for 2 h and twice in PBS for 20 min to remove unbound remnants of the reagents. Finally, the sections were mounted in MOWIOL 4-88 (Carl Roth). Previously, different incubation times (5 min, 10 min, 15 min, 30 min) and concentrations (Phalloidin: 1:100, 1:500, 1:1000; 1:2000; Hoechst: 1:1000; 1:2000; 1:5000) of the fluorescence dyes were tested to obtain the best signal-to-noise ratio. Negative controls showed no specific labelling. The grading of the staining intensity was evaluated by assessing staining signal-to-noise ratio and required camera exposure time for photomicrographs. Samples graded with (+++) showed a very clear staining and had good signal-to-noise ratio (low background noise and visible structures of actin and DNA in high magnifications) as well as a short exposure time (<500 ms, to reduce bleaching). In comparison, samples with very long exposure times were graded with (+). Completely destroyed samples were graded with (–) as a further examination of the tissue was not possible.

Microscopy and image processing

Pictures were taken of the whole animal and the trunk segment isolated from head and tail prior and after fixation in order to document changes caused by the different chemicals. Images were taken using a Canon EOS 50D camera with different lenses, Canon EF-S 18-55 mm and Sigma DG Macro 105 mm or with a binocular (MZ75, Leica) equipped with a camera (DFC 425, Leica), depending on the animals' size. Embedded trunk segments were imaged with a Nikon SMZ800 Zoom stereomicroscope equipped with a digital microscope camera 'Digital Sight DS-2Mv' (Nikon). A Nikon Eclipse 90i fluorescence microscope equipped with a digital microscope camera 'Digital Sight DS-2 MBWc' (Nikon) was used for imaging sections and fluorescence samples. Auto-fluorescence tested at an excitation of 470 nm was negative. Unlabelled sections were documented with a cold light source (Schott KL200, Lighting and Imaging Schott) under polarized reflected light to reduce reflections.

Images were optimized with Adobe Photoshop CS 4 by adjusting tonal range as well as brightness and global contrast and some pictures were sharpened with the tool 'unsharp masking'. Some specimens were digitized as composite images in z-axis and x–y plane. Those images were fused using the

NIS-Elements Advanced Research Imaging software (v.3.22.15.738, Nikon). Greyscale fluorescence images were black-white inverted. Single-channel images were combined using the 'maximum intensity' or 'average intensity' tool of the freely available software ImageJ v.1.43m) to yield the merged fluorescence images. To evaluate the signal-to-noise ratio, raw images were used.

Statistical analysis

The mean ± SE was calculated from the total body length of every species for the same developmental stage (Table 1) and of the trunk segment before and after fixation for every developmental stage and fixation group as well as of the fixation group independently from the age of the specimen. Statistical analysis was performed using SPSS v.22 (IBM). Data were tested for significance via Wilcoxon rank-sum test for comparing the length and the height of tissue after fixation in the different developmental groups but also between the different fixation methods. $p < 0.05$ showed significant differences and $p < 0.001$ showed highly significant differences.

RESULTS

Influence of fixation conditions on colouration and gross morphology

Fig. 2 shows (A–C) dogfish, (D–F) sturgeon and (G–I) zebrafish trunk segments treated with FOR, PFA or ZnFA. The pigmentation of dogfish tissue did not change obviously after fixation (Fig. 2A–C``). In sturgeon tissues, fixation using FOR or ZnFA had a pronounced influence on the transparency of the fin fold, i.e. the tissue changed from transparent to white (Fig. 2D`,D``,F`,F``). In zebrafish, a change in tissue pigmentation was distinctive in all fixation groups (Fig. 2G``, I`; see '*'). The typical blue coloration in zebrafish was lost completely in FOR (Fig. 2G–G``) and nearly completely in PFA and ZnFA (Fig. 2H–I`). Furthermore, after FOR and ZnFA fixation the trunk segment showed signs of lacerations (Fig. 2G``,H`, see '#').

Influence of different parameters on tissue size

The influence of the fixation chemicals on the trunk segment was analyzed via size measurements before and after fixation. In dogfish, this analysis revealed no size differences in the fixation groups (FOR, PFA and ZnFA) in relation to the age of the larvae (Fig. 3A–C). In the FOR group, the tissue decreased in length around 3–7% and in height 4–14% after fixation (Fig. 3A). In the PFA group, a decrease of 3–13% in length and 10–32% in tissue height could be found, but no significant difference existed between these groups (Fig. 3B). In the ZnFA fixation group the influence on the tissue due to the chemicals followed a similar trend: in all age groups, we measured a decrease of 2–4% in length and 1–7% in height (Fig. 3C).

The examination of the trunk segments in sturgeon revealed no significant difference in the FOR and the ZnFA group in relation to the age of the animals (Fig. 3D,F). In the FOR and ZnFA groups, the highest decrease in tissue length was measured in the 80 DD group: the tissues shrank around 16.1 ± 2.3% (FOR) and 18.5 ± 5.4% (ZnFA). In comparison, concerning tissue height, the 320 DD FOR group showed an average decrease of 11.8 ± 3.1% and in the ZnFA group the 620 DD old larvae had a decrease of their trunk segment of around 15.4 ± 1.8%. The PFA-fixed trunk segments of sturgeon revealed a significant difference ($p < 0.05$) between the tissue shrinkage of the 80 DD (21.5 ± 3.4%) and the 620 DD group (12.1 ± 1.3%) (Fig. 3E). No significant tissue shrinkage was found for height among the 3 developmental stages tested.

The comparison of the fixation methods independent of the age of the animal is shown in Fig. 4. In all 3 species, a significant difference was calculated in the PFA groups. In dogfish, the tissue dimensions were highly significantly different ($p < 0.001$) compared to the ZnFA-fixed tissue (Fig. 4). A decrease of 7.9 ± 1.5% (length) and 14.3 ± 2.3% (height) after PFA fixation and 3.2 ± 0.6% (length) and 5.1 ± 1.2% (height) after ZnFA fixation was measured. In addition, the FOR group had 9.0 ± 1.9% less tissue height shrinkage than the PFA group ($p < 0.05$). Furthermore, there was a significant ($p < 0.001$) difference between the FOR and the PFA group in tissue height for sturgeon (Fig. 4). The tissue after PFA fixation decreased by 18.1 ± 2.9%, and after FOR fixation by 7.6 ± 1.6%. Similar to the dogfish results, in zebrafish there was a significant difference between the ZnFA and the PFA group, well as between the FOR and the PFA group (both $p < 0.05$; Fig. 4). The tissue height decreased by 5.5 ± 0.9% in FOR, 4.8 ± 0.9% in PFA and 3.1 ± 0.4% in ZnFA.

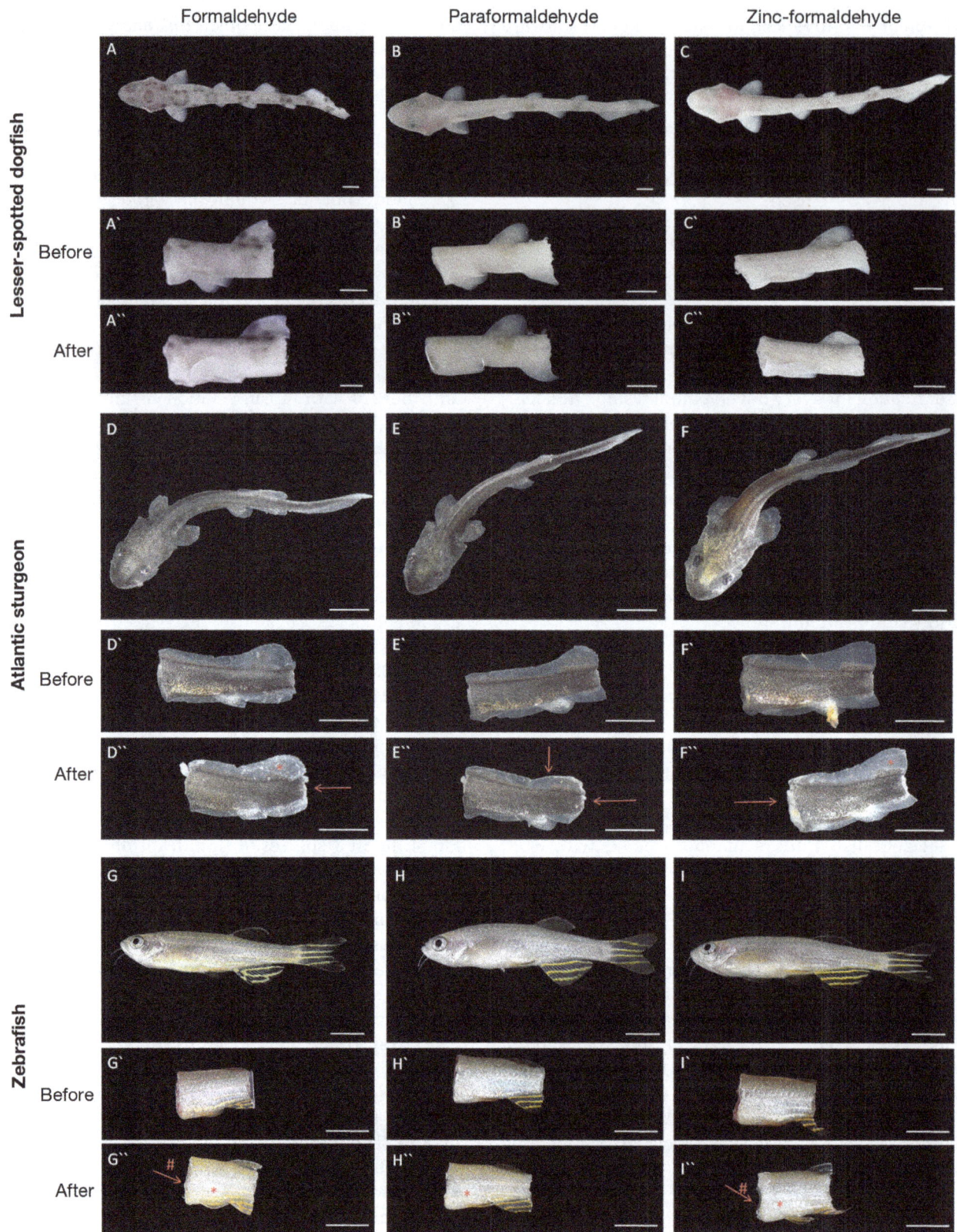

Fig. 2. Effect of different fixation methods on fish tissue. Freshly killed (A–C) lesser-spotted dogfish (2000 degree-day [DD] old; scale 5 mm), (D–F) Atlantic sturgeon (320 DD old; scale 2 mm) and (G–I) adult zebrafish (scale 5 mm). Trunk segments were documented before (`) and after (``) fixation. The pictures show the effect of (A,D,G) formaldehyde, (B,E,H) paraform-aldehyde and (C,F,I) zinc-formaldehyde fixation for 4 h at room temperature (20°C) on the tissues. A loss of pigmentation (*) was present in sturgeon and zebrafish treated with (D``,G``) formaldehyde and (F``,I``) zinc-formaldehyde as well as (H``) para-formaldehyde for zebrafish. (D`–F``) The tissue from sturgeon exhibited a decrease in tissue size length (→/←) as well as height (↓). In zebrafish, after formaldehyde and zinc-formaldehyde treatment, the tissue appeared lacerated (#)

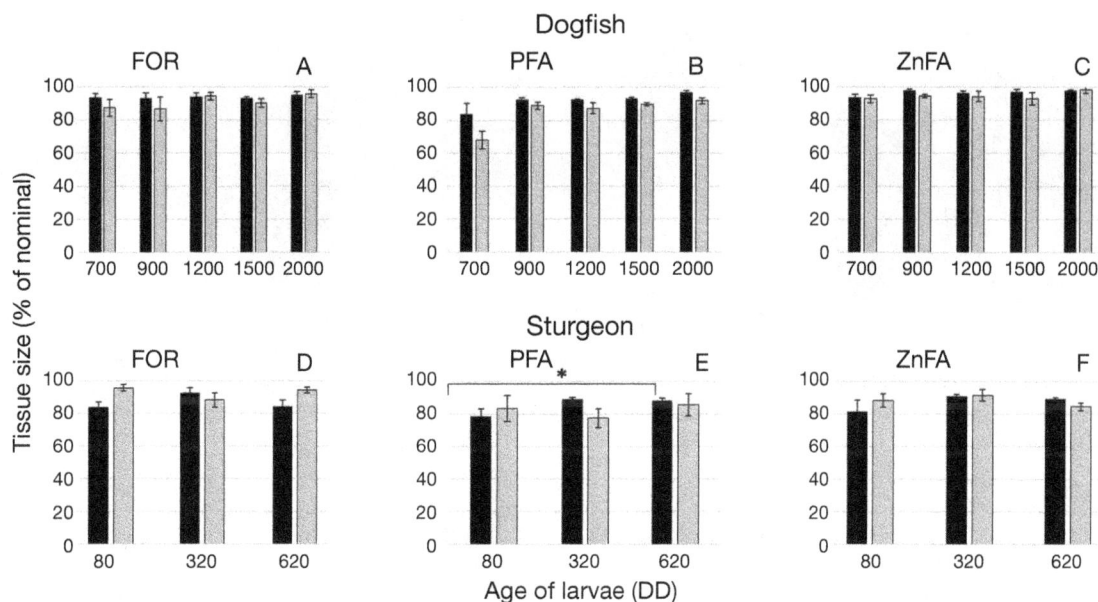

Fig. 3. Changes in tissue size after fixation in relation to age. The influence of a (A,D) 4 % formaldehyde (FOR), (B,E) 4 % paraformaldehyde (PFA), and (C,F) zinc-formaldehyde (ZnFA) fixation was analysed on body tissue length (black columns) and height (grey column) from different developmental stages for (A–C) lesser-spotted dogfish and (D–F) Atlantic sturgeon. In dogfish 700, 900, 1200, 1500 and 2000 degree-days (DD) and in Atlantic sturgeon 80, 320 and 620 DD old larvae were analysed. Between the different fixation regimes, no influence in dependence to the age of larvae in dogfish and sturgeon could be found, except in (E) PFA-fixed sturgeon of 80 and 620 DD, which shows a significant decrease in tissue length in 80 DD compared to 620 DD old larvae ($*p < 0.05$, Wilcoxon rank-sum test)

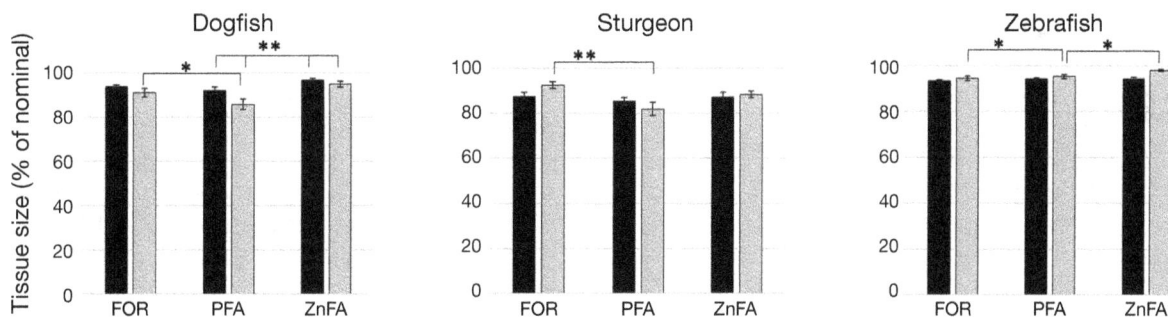

Fig. 4. Change in tissue size in relation to fixation method. The influence of 4 % formaldehyde (FOR), 4 % paraformaldehyde (PFA) and zinc-formaldehyde (ZnFA) fixation on body tissue length (black columns) and height (grey column) was analysed independently of the age of the animals in lesser-spotted dogfish, Atlantic sturgeon and adult zebrafish. In all 3 species a significant difference in tissue shrinkage was detected in the PFA groups in comparison to the FOR groups, and in zebrafish and dogfish samples in the ZnFA groups. ($**p < 0.001$, $*p < 0.05$; Wilcoxon rank-sum test)

Quality of the sections

The optical characteristics of the embedding blocks were diverse. The GEL blocks were completely opaque (Fig. 5A) and therefore a precise orientation of the tissue before sectioning was time-consuming, difficult, or even impossible. Furthermore, the material was hard and bigger trimming steps resulted in breaches of the block. Multiple embedding steps of the isolated fish tissues with the medium were

sometimes necessary. In comparison to GEL, AGA (Fig. 5B) and LMA (Fig. 5C) were softer and the blocks more flexible in their form, which in some cases resulted in problems cutting the embedded tissue due to high forces and pressure on the whole block. Nevertheless, tissue orientation was easier and bigger trimming steps were possible without any damage of tissue and the surrounding embedding material.

The vibratome sections of the animals differed in quality (Table 3). Sections were graded according to

Fig. 5. Optical characteristics of the embedding media: trunk segments of the lesser-spotted dogfish embedded in (A) 4% gelatine/20% albumin (trunk not visible), (B) 4% agarose (trunk visible) and (C) 4% low-temperature melting agarose (trunk highly visible)

Table 3. Quality of vibratome sections (50 µm thick) analysed via microscopic examination. Trunk segments of 2000 degree-day (DD) old lesser-spotted dogfish, 320 DD old Atlantic sturgeon and adult zebrafish were fixed with 4% formaldehyde (FOR), 4% paraformaldehyde (PFA), or 340 mM or 800 mM zinc-formaldehyde (ZnFA), and embedded in 4% gelatine/20% albumin (GEL), 4% agarose (AGA) or 4% low-temperature melting agarose (LMA). Grading: (+) poor, (++) good, (+++) best

| | FOR | | | PFA | | | ZnFA | | |
	GEL	AGA	LMA	GEL	AGA	LMA	GEL	AGA	LMA
Lesser-spotted dogfish	++	++	++	++	+++	++[b]	++[a,b]	++	+++
Atlantic sturgeon	+[a,b]	+++[b]	++	+[a,b]	++[a]	++	+[a,b]	+	+++
Zebrafish	++[a,b]	++	++	+++[b]	+++[b]	++	+	+	++

[a]Deformation and/or extreme shrinkage of the tissue after embedding and sectioning
[b]Sections are still embedded in embedding media

aspects of the general appearance such as form and tissue completeness (no damages or deformation). For each examined fish species, the grading of the sections showed 2 very good results. For dogfish trunk segments, PFA fixation and AGA embedding (Fig. 6A) as well as ZnFA fixation and LMA embedding seem to be preferable methods (Fig. 6B). In sturgeon tissue, best results were obtained by FOR fixation and AGA embedding (Fig. 6C) and the ZnFA fixation and LMA embedding (Fig. 6D). In zebrafish, the highest quality vibratome sections were PFA-fixed tissues embedded in GEL (Fig. 6E) and AGA (Fig. 6F). The quality of all these sections was high, as the morphology of the body cross-section from the trunk segment was well recognizable and no large gap between the muscle fibres or between the muscle and the epidermis was observed. Furthermore, no deformation of the tissue was present. The deformation of the tissues was a particular problem in some GEL-embedded samples. Multiple repeated trials showed the same results. Mainly in sturgeon, all GEL-embedded trunk segments had a high ratio of deformation, restricting identification of anatomical structures (Table 2; see Fig. 6G). In comparison, ZnFA-fixed and GEL-embedded trunk segments from dogfish displayed little shrinkage but showed a curved appearance (Fig. 6H).

Quality of histochemical staining

High-quality sections were used to assess the quality of histochemical labelling with fluorescent probes. The penetration of the labelling reagents was apparently complete in the transverse sections of the trunk segment. Structures labelled with the marker for actin (phalloidin) and labelled nuclei (Hoechst stain) were present throughout the entire vibratome sections. Table 4 summarizes the morphology of the samples after staining as well as the staining intensity with background noise, resolution of details within the sections and exposure time. The staining procedure had no influence on the morphology of the samples in dogfish (Table 4, Fig. 7A–H) and zebrafish (Table 4, Fig. 7M–T). These vibratome sections were of high quality. In the category of the signal-to-noise ratio, the 2 different treatments of dogfish tissues showed no differences. In this category (signal-to-noise ratio), the samples achieved the best grading (Table 4), because the background noise was low and the structures labelled for actin, that is, the entire musculature as well as the nuclei were highly visible at higher magnifications. Furthermore, these samples shared the same short exposure time for very good fluorescence intensities. In zebrafish, the GEL-

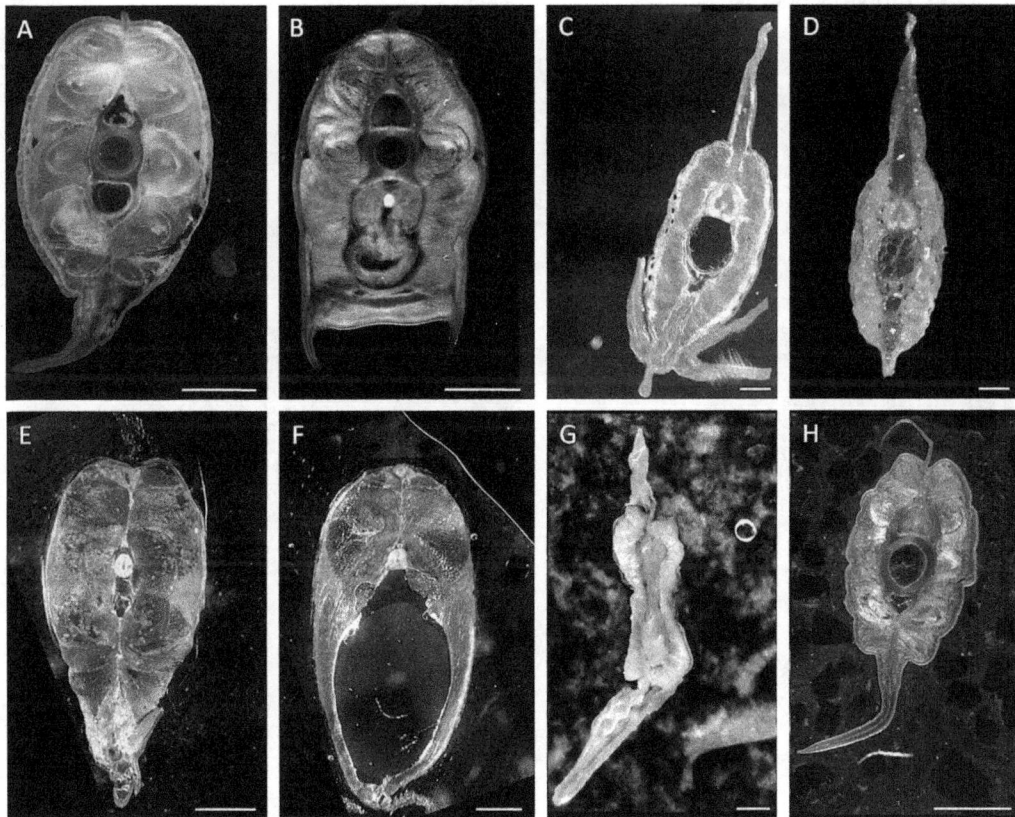

Fig. 6. Vibratome sections (50 μm thick) of trunk sections from 3 fish species with different fixation and embedding techniques which are graded as high quality (+++; A–F) and low quality (+; G, H) (see 'Material and methods: Embedding and sectioning'). Lesser-spotted dogfish: (A) PFA-fixed and embedded in AGA and (B) 800 mM ZnFA-fixed and embedded in LMA. Atlantic sturgeon: (C) FOR-fixed and embedded in AGA and (D) 340 mM ZnFA-fixed and embedded in LMA. Adult zebrafish: PFA-fixed and embedded in (E) GEL and (F) AGA. (G) Trunk segment from sturgeon fixed in FOR and embedded in GEL showed a high level of deformation. (H) Trunk segment from dogfish fixed in ZnFA and embedded in GEL showed a curved appearance due to shrinkage. See Table 3 for abbreviations. Scale bars: 1000 μm for dogfish and zebrafish, 100 μm for sturgeon

Table 4. Sample quality and staining intensity after fluorescence staining for high quality vibratome sections of trunk segments for 3 fish species analysed via microscopic examination. Grading: (–) n/a, (+) poor, (++) good, (+++) best. Fixation: FOR: 4 % formaldehyde, PFA: 4 % paraformaldehyde, ZnFA: 340 mM or 800 mM zinc-formaldehyde. Embedding mediums: GEL: 4 % gelatine/20 % albumin, AGA: 4 % agarose, LMA: 4 % low-temperature melting agarose

| Species | Fixation | Embedding medium | Staining intensity | | Exposure time | Morphology of sections |
| | | | Signal-to-noise ratio | | | |
			Background	Detailed structures		
Lesser-spotted dogfish	PFA	AGA	+++	+++	+++	+++
	ZnFA	LMA	+++	+++	+++	+++
Atlantic sturgeon	FOR	AGA	++	+	++	+
	ZnFA	LMA	–	–	–	–
Zebrafish	PFA	GEL	++	++	+	+++
	PFA	AGA	+++	++	+++	++

embedded samples (Fig. 7M,O,P) needed up to 5 to 6 times longer exposure times to document the fluorescence signal of the phalloidin probe than in the AGA-embedded samples, indicating that the labelling was weaker (Fig. 7Q,S,T, Table 4). The resolution of details in both zebrafish samples achieved only a

'good' grading because, in comparison to the dogfish samples, the structures were not as visible (Table 4). The vibratome sections of the ZnFA-fixed and LMA-embedded trunk segment from sturgeon could not be stained because during the procedure all samples within this approach were completely destroyed (the

Overview	Detail - Hoechst	Detail - phalloidin	Detail - merged

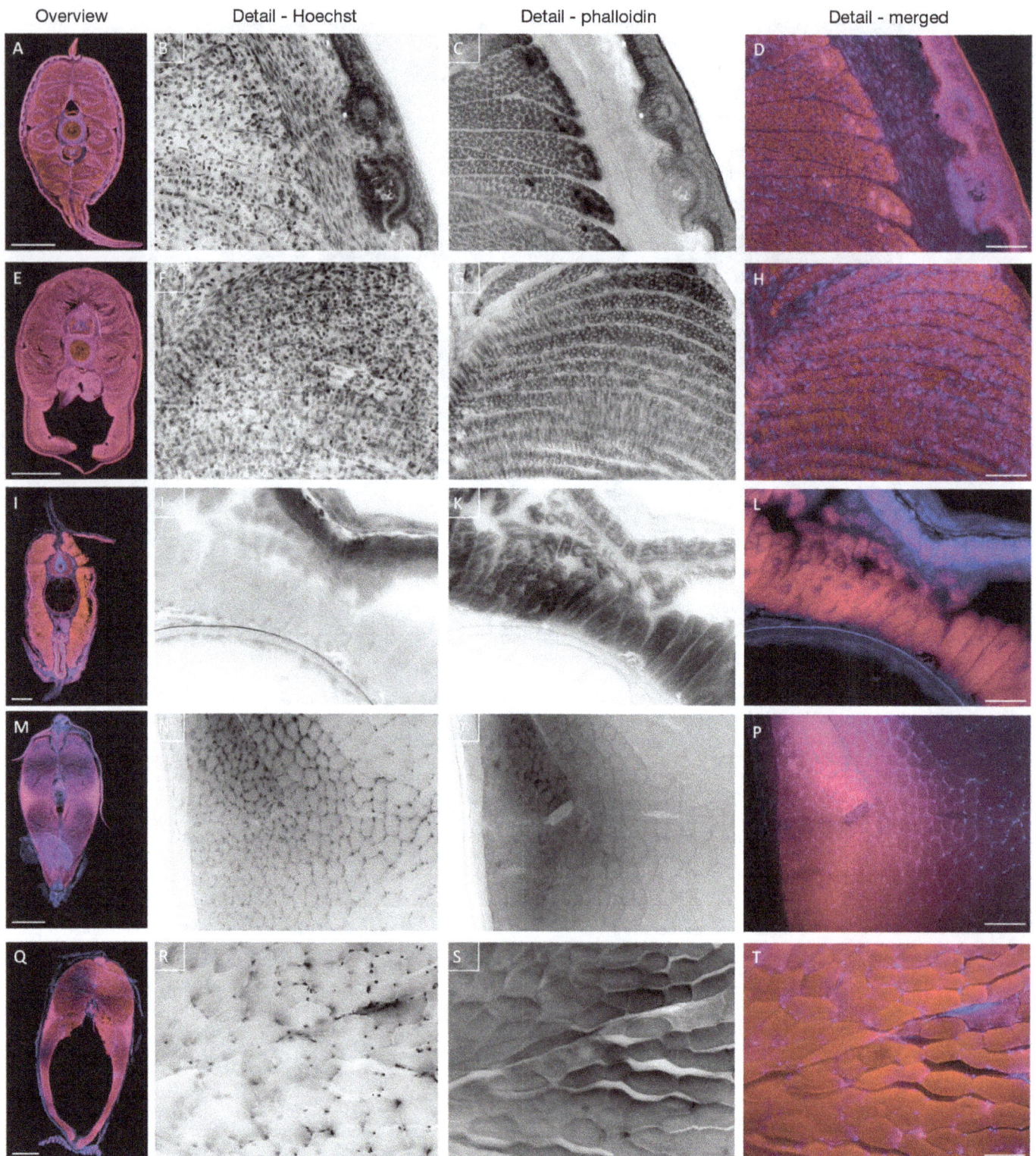

Fig. 7. Double-fluorescence staining for DNA (Hoechst H 33258) and with a probe for actin (phalloidin) in high-quality vibratome sections from 3 fish species. Lesser-spotted dogfish: (A–D) PFA-fixed and embedded in AGA, and (E–H) 800 mM Zn-FA-fixed and embedded in LMA. (I–L) Atlantic sturgeon: FOR-fixed and embedded in AGA. Adult zebrafish: PFA-fixed and embedded in (M–P) GEL and (Q–T) AGA. The first column shows the overview of the whole sections; the other 3 columns show details of the stain with Hoechst, phalloidin, and a merged image of both channels. Scale bar 1000 µm for A, E, M and Q, 100 µm for I, 50 µm for B–D, F–H, N–P and R–T, and 25 µm for J–L. See Table 4 for abbreviations

tissue disintegrated). One reason could be the long washing steps. Also, the FOR-fixed and AGA-embedded sturgeon samples showed a massive degradation in their morphology (Fig. 7I–L). In these sections, structural details were only poorly resolved (Table 4).

DISCUSSION

Pre-analytical variables such as species and age of the animals and therefore structure of the epidermis, skeleton, sample size, type of fixative, fixation time and processing conditions can have a profound influence on the quality of tissue preservation in vertebrate histology (Chung et al. 2008, Verderio 2012). Samples can be under-fixed, so that tissue preservation is poor; samples can be over-fixed so that the sample becomes difficult to infiltrate with other reagents (Wu et al. 2012). Increasing the temperature of fixation will increase the rate of diffusion of the fixative into the tissue and speed up the rate of chemical reaction between the fixative and tissue elements. It can also potentially increase the rate of tissue degeneration in unfixed areas of the specimen. Rolls (2012) noted for light microscopy that the initial fixation is usually carried out at room temperature. Therefore, in this study all 3 chemicals were used for 4 h at 20°C to compare the influence of the fixation process on the trunk segments. Taking into account that PFA is a standard fixation method (e.g. Ward et al. 2008, Kuhlmann 2009), it was surprising that our analysis of the tissue revealed in all 3 examined species that PFA fixation is inferior in comparison to FOR and ZnFA, due to higher shrinkage of the tissue. Similar effects were also described by Wehrl et al. (2014), which analyzed the degree of shrinkage by measuring the volumes of mouse brains. Shrinkage was highest in paraformaldehyde-lysine-periodate and PFA. Furthermore, in Wehrl et al. (2014), a zinc-based fixative caused the smallest degree of brain shrinkage and only small deformations and consequentially was recommended for *in vivo–ex vivo* comparison studies. Especially in morphometric analysis, the shrinkage process has to be considered. In our study, we adapted the FOR and ZnFA fixatives to the sample organisms' osmolarity, and this step was crucial for achieving good tissue preservation. Therefore, in future studies, the use of PFA should be undertaken carefully because proper adaptation to the osmolarity is essential. In conclusion, our findings show that the influence of the fixative must be examined carefully for other tissues in fish, especially when studying marine species. Furthermore, the size of the tissue samples has to be considered because the time of incubation within the fixative has to be prolonged with increasing sample volume, and multiple fixatives should be tested, such as glutaraldehyde, Bouin or Hollande. Benerini Gatta et al. (2012) reported that for different human organs the Hollande fixative (containing cooper actetat, pciric acid, formalin, and acetic acid) proved to be the best for morphology and histochemistry. In comparison, the results obtained with Bouin fixation (containing picric acid, acetic acid, and formalin) was comparable to the one with formalin (Benerini Gatta et al. 2012). Enzymes such as collagenase are frequently used after fixation of whole mount preparations in order to improve the penetration of the labelling reagents. Additionally, Olson (1985) described another technique for microscopic analysis of tissue: the perfusion fixation with specific radionuclides, fluorescent tags, or polymerizable resins can provide a wealth of information on the anatomy, physiology, pharmacology, and biochemistry of fish tissues.

Wu et al. (2012) described the influence of fixatives at different incubation temperatures. Compared to fixations at room temperature (20–25°C); fixation of samples at 4°C tends to reduce extraction of cell contents, to slow down autolytic processes and to reduce tissue shrinkage. However, 4°C will also slow down the process of the fixation itself and therefore a longer incubation time is necessary at lower temperature. This finding should also be analysed in fish trunk segments in future studies.

The sectioning of the fish tissue via a vibratome revealed differences in the quality of the sections. Problems ranging from deformation to breaking of the tissue were encountered in this study. In these cases, the embedding medium was not the perfect choice for this kind of tissue due to the fixation of the tissue structure itself. Our results show that in sturgeon, the GEL embedding method is not optimal because of the deformation or shrinkage of the tissue. The reasons for the extreme shrinkage can be manifold. Kuhlmann (2009) noted that a hypertonic solution will lead to shrinkage of the tissue, caused by an inappropriate pH, buffering capacity or osmolarity. These aspects need to be addressed in future studies. The sectioning of the tissue from dogfish and sturgeon was better in the softer embedding material AGA and LMA because of the compact epidermis. In contrast, in the teleost zebrafish the GEL embedding method was very successful. In this species, the PFA–GEL combination yielded especially high quality vibratome sections. Nevertheless, the PFA fixation caused a significant shrinkage of the tissue. Further-more, most of the trunk segments, mainly in AGA and LMA, broke out of the embedding material after cutting (Table 3). This factor can be difficult to control for smaller samples. In the current study this played a minor role because the tissues were large enough for optical localisation and easy handling. Additional factors affecting tissue preservation need to be examined further, including different con-centrations of the embedding medium or different brands and providers of the embedding medium with different melting temperatures. For example, the quality of low-temperature melting agarose in our study proved to be of prime importance for the per-formance of this medium.

Because of the time-consuming and expensive nature of optimizing protocols for immunostaining in a market full of different antibodies, in the subse-quent next step in this study we only examined the staining properties of the sections using histochemi-cal probes for actin (phalloidin) and DNA (Hoechst stain). The staining procedure yielded good to very good results in dogfish and zebrafish samples, but not in sturgeon. The use of fluorescent probes instead of antibodies might be also the reason for a good staining intensity in FOR-fixed samples be-cause they are much smaller molecules than immuno-globulins. Immunochemical examinations showed poor or no penetration of many antibodies after FOR fixation (Molgaard et al. 2014) due to the extensive cross-linking of proteins and blocking of amino groups (Sternberger 1979). We conclude that for immunochemical examinations of fish tissues with different antibodies, fixation, embedding and section-ing parameters need to be adjusted again according to the specific area of interest.

CONCLUSIONS

The goal of this study was to explore the effects of different fixation and embedding protocols on the quality of vibratome sections and fluorescence histo-chemistry in 3 different fish species. In dogfish, a ZnFA fixation and LMA embedding medium is rec-ommended as there was little tissue shrinkage and very good staining intensities as well as no morpho-logical degradation of the vibratome sections. In stur-geon, we did not obtain good results with any of our methods — other protocols or methods such as cryo-sectioning will have to be tested. For zebrafish, and likely other teleosts, the PFA fixation combined with ARA embedding is recommend for optimal staining

intensity and preservation of the samples, even if this fixation shows a significant decrease in tissue size. Overall, the success of histological examination of vibratome sections from trunk segments will depend on the fish species as well as the development and adjustment of appropriate protocols.

Acknowledgements. The Institute for Fisheries, State Research Center Mecklenburg-Vorpommern for Agricul-ture and Fishery in Born/Darss and the Department of Anatomy and Cell Biology, Universitätsmedizin Greifswald, Greifswald, Germany are gratefully acknowledged for sup-plying the Atlantic sturgeon and the zebrafish. The aquar-ium team of the Ozeaneum in Stralsund helped in supply and keeping of the specimens. We wish to thank all mem-bers of the Department of Cytology and Evolutionary Biol-ogy for their support during this study. This work was sup-ported by the project starter grant of the University of Greifswald (Germany) E404 3100.

LITERATURE CITED

➤ Alvarez-Viejo M, Cernuda-Cernuda R, Alvarez-López C, García-Fernández JM (2004) Identification of extra-retinal photoreceptors in the teleost *Phoxinus phoxinus*. Histol Histopathol 19:487–494

Ballard WW, Mellinger J, Lechenault H (1993) A series of normal stages for development of *Scyliorhinus cani-cula*, the lesser spotted dogfish (Chondrichthyes: Scylio-rhinidae). J Exp Biol 267:318–336

➤ Benerini Gatta L, Cadei M, Balzarini P, Castriciano S and others (2012) Application of alternative fixatives to for-malin in diagnostic pathology. Eur J Histochem 56:e12

➤ Blackiston D, Vandenberg LN, Levin M (2010) High-throughput *Xenopus laevis* immunohistochemistry using agarose sections. Cold Spring Harb Protoc 12:pdb.prot 5532

Bruno DW, Noguera PA, Poppe TT (2013) A colour atlas of salmonid diseases, 2nd edn. Springer, Dordrecht

➤ Bryson JL, Coles MC, Manley NR (2011) A method for labelling vasculature in embryonic mice. J Vis Exp 56: 3267

➤ Cheung ID, Bagnat M, Ma TP, Datta A and others (2012) Regulation of intrahepatic biliary duct morphogenesis by Claudin 15-like b. Dev Biol 361:68–78

➤ Chung JY, Braunschweig T, Williams R, Guerrero N and others (2008) Factors in tissue handling and processing that impact RNA obtained from formalin-fixed, paraffin-embedded tissue. J Histochem Cytochem 56:1033–1042

Dietrich HW III, Westerfield M, Zon LI (2010) The zebrafish: cellular and developmental biology, Part A, 3rd edn. In: Wison L, Tran P (eds) Methods in cell biology, Vol 100. Elsevier, p 103–108

➤ Enoki R, Koizumi A (2013) A method of horizontally sliced preparation of the retina. Methods Mol Biol 935:201–205

➤ Freel CD, Gilliland KO, Mekeel HE, Giblin FJ, Costello MJ (2003) Ultrastructural characterization and Fourier analysis of fiber cell cytoplasm in the hyperbaric oxygen treated guinea pig lens opacification model. Exp Eye Res 76:405–415

➤ Fukuda T, Morooka K, Miyagi Y (2010) A simple but accurate method for histological reconstruction of the large-sized brain tissue of the human that is applicable to construction of digitized brain database. Neurosci Res 67:260–265

▶ Funch PG, Wood MR, Faber DS (1984) Localization of active sites along the myelinated goldfish *Mauthner axon*: morphological and pharmacological evidence for salutatory conduction. J Neurosci 4:2397–2409

▶ Griffioen HA, Van der Beek E, Boer GJ (1992) Gelatine embedding to preserve lesion-damaged hypothalami and intracerebroventricular grafts for vibratome slicing and immunocytochemistry. J Neurosci Methods 43:43–47

▶ Hampton JA, Klaunig JE, Goldblatt PJ (1987) Resident sinusoidal macrophages in the liver of the brown bullhead (*Ictalurus nebulosus*): an ultrastructural, functional and cytochemical study. Anat Rec 219:338–346

▶ Hökfelt T, Ljungdahl A (1972) Modification of the Falck-Hillarp formaldehyde fluorescence method using the vibratome: simple, rapid and sensitive localization of catecholamines in sections of unfixed or formalin fixed brain tissue. Histochemie 29:325–339

▶ Jayachandran P, Hong E, Brewster R (2010) Labelling and imaging cells in the zebrafish hindbrain. J Vis Exp 41: 1976

▶ Kah O, Dubourg P, Onteniente B, Geffard M, Calas A (1986) The dopaminergic innervation of the goldfish pituitary. An immunocytochemical study at the electron-microscope level using antibodies against dopamine. Cell Tissue Res 244:577–582

▶ Kálmán M, Ari C (2002) Distribution of GFAP immunoreactive structures in the rhombencephalon of the sterlet (*Acipenser ruthenus*) and its evolutionary implication. J Exp Zool 293:395–406

Kuhlmann WD (2009) Fixation of biological specimens. www.immunologie-labor.com/cellmarker_files/IET_tissue_02.pdf (accessed 06/10/2014)

▶ Lechtreck KF, Sanderson MJ, Witman GB (2009) High-speed digital imaging of ependymal cilia in the murine brain. Methods Cell Biol 91:255–264

▶ Lieberoth BC, Becker CG, Becker T (2003) Double labelling of neurons by retrograde axonal tracing and non-radioactive in situ hybridization in the CNS of adult zebrafish. Methods Cell Sci 25:65–70

▶ Lindvall O, Björklund A, Hökfelt T, Ljungdahl A (1973) Application of the glyoxylic acid method to vibratome sections for the improved visualization of central catecholamine neurons. Histochemie 35:31–38

▶ Loesel R, Nässel DR, Strausfeld NJ (2002) Common design in a unique midline neuropil in the brains of arthropods. Arthropod Struct Dev 31:77–91

▶ Mahler J, Driever W (2007) Expression of the zebrafish intermediate neurofilament Nestin in the developing nervous system and in neural proliferation zones at postembryonic stages. BMC Dev Biol 7:89

▶ McGrail M, Batz L, Noack K, Pandey S, Huang Y, Gu X, Essner JJ (2010) Expression of the zebrafish *CD133/prominin1* genes in cellular proliferation zones in the embryonic central nervous system and sensory organs. Dev Dyn 239:1849–1857

Molgaard S, Ulrichsen M, Boggild S, Holm ML, Vaegter C, Nyengaard J, Glerup S (2014) Immunofluorescent visualization of mouse interneuron subtypes. Version 2. F1000 Res 3:242

▶ Nyholm MK, Abdellilah-Seyfried S, Grinblat Y (2009) A novel genetic mechanism regulates dorsolateral hinge-point formation during zebrafish cranial neurulation. J Cell Sci 122:2137–2148

▶ Olson KR (1985) Preparation of fish tissues for electron microscopy. J Electron Microsc Tech 2:217–228

▶ Ott SR (2008) Confocal microscopy in large insect brains: zinc–formaldehyde fixation improves synapsin immunostaining and preservation of morphology in wholemounts. J Neurosci Methods 172:220–230

▶ Pilaz LJ, Silver DL (2014) Live imaging of mitosis in the developing mouse embryonic cortex. J Vis Exp 88

▶ Price RL, Haley ST, Bullard T, Davis J, Borg TK, Terracio L (2014) Confocal microscopy of cardiac myocytes. Methods Mol Biol 1075:185–199

▶ Reichmann F, Painsipp E, Holzer P, Kummer D, Bock E, Leitinger G (2015) A novel unbiased counting method for the quantification of synapses in the mouse brain. J Neurosci Methods 240:13–21

Robertson N (2002) Vibratome sectioning of tissues: an alternative to paraffin methods. http://www4.ncsu.edu/~rgfranks/research/protocols/immuno%20histio%20chemistry/robertson%20vbratmimuno.doc (accessed 12/09/2014)

Rolls G (2012) Fixation and fixatives (2) – factors influencing chemical fixation, formaldehyde and glutaraldehyde. www.leicabiosystems.com/pathologyleaders/fixation-and-fixatives-2-factors-influencing-chemical-fixation-formaldehyde-and-glutaraldehyde/ (accessed 24/09/2014)

▶ Satoh K, Takahashi G, Miura T, Hayakari M, Hatayama I (2005) Enzymatic detection of precursor cell populations of preneoplastic foci positive for gamma-glutamyl-transpeptidase in rat liver. Int J Cancer 115:711–716

Sternberger LA (1979) Immunocytochemistry, 2nd edn. John Wiley & Sons, New York, NY

▶ Verderio P (2012) Assessing the clinical relevance of oncogenic pathways in neoadjuvant breast cancer. J Clin Oncol 30:1912–1915

▶ Ward TS, Rosen GD, von Bartheld CS (2008) Optical disector counting in cryosections and vibratome sections underestimates particle numbers: effects of tissue quality. Microsc Res Tech 71:60–68

▶ Wehrl HF, Bezrukov I, Wiehr S, Lehnhoff M and others (2014) Assessment of murine brain tissue shrinkage caused by different histological fixatives using magnetic resonance and computed tomography imaging. Histol Histopathol (in press)

▶ Wu S, Baskin T, Gallagher KL (2012) Mechanical fixation techniques for processing and orienting delicate samples, such as the root of *Arabidopsis thaliana*, for light or electron microscopy. Nat Protoc 7:1113–1124

▶ Yulis CR, Lederis K (1986) The distribution of 'extraurophyseal' urotensin I-immunoreactivity in the central nervous system of *Catostomus commersoni* after urophysectomy. Neurosci Lett 70:75–80

14

Reproduction reduces HSP70 expression capacity in *Argopecten purpuratus* scallops subject to hypoxia and heat stress

Katherina Brokordt*, Hernán Pérez, Catalina Herrera, Alvaro Gallardo

Centro de Estudios Avanzados en Zonas Áridas (CEAZA), Universidad Católica del Norte, Larrondo 1281, Coquimbo, Chile

ABSTRACT: In scallops, gonad production is highly demanding energetically, and reproduction usually occurs during spring–summer, a period of strong environmental changes. The synthesis of heat-shock proteins (HSPs) is a major mechanism of stress tolerance in animals, including scallops, and HSP expression contributes considerably to cellular energy demand. Therefore, reproductive investment may limit the availability of energy (in terms of ATP) for the expression of HSP in organisms exposed to environmental stress. We evaluated the stress response capacity of adult *Argopecten purpuratus* scallops to high temperature and hypoxia. Stress response capacity was assessed through gene expression (for temperature stress) and protein induction of 70 kD HSP at 3 reproductive stages: immature, mature and spawned. We also evaluated the effect of reproductive status on the cellular ATP provisioning capacity through citrate synthase activity. Immature scallops exposed to thermal stress showed 1.3- and 1.5-fold increases in *hsp70* mRNA and HSP70 protein levels, respectively, and those exposed to hypoxia doubled their level of HSP70 compared to non-stressed immature scallops. However, following gonad maturation and spawning, *hsp70* mRNA increased by only 0.49- and 0.65-fold, respectively, after thermal stress and HSP70 protein levels of scallops exposed to thermal and hypoxia stressors did not differ from those of non-stressed animals. In parallel, citrate synthase showed its highest level in immature scallops, declined with gonad maturation, and was lowest in spawned scallops. These results suggest that reproductive investment reduces the stress response capacity of *A. purpuratus* and that mature and spawned scallops could be more vulnerable to environmental stressors than immature individuals.

KEY WORDS: Reproductive cost · HSP70 · *hsp70* mRNA · Stress response · Thermal stress · Hypoxia stress · Scallops · *Argopecten purpuratus*

INTRODUCTION

In most organisms, reproduction requires high levels of energy investment. The amount of energy invested and the stage during which this energy is needed depend, in part, on the organism's reproductive strategy. In broadcast-spawning mollusks such as scallops, the majority of energy investment into gamete production occurs during gonad maturation (Sastry 1968, Barber & Blake 1983, Martínez 1991). In scallops, this process occurs even when food is limited, and as much as 50% of the energy reserves stored in somatic tissues can be mobilized to support reproductive maturation (Martínez & Mettifogo 1998, Lodeiros et al. 2001, Brokordt & Guderley 2004). The mobilization of energy reserves and re-channeling of consumed energy to support gonad maturation limit the amount of energy available or decrease the metabolic capacity to support other vital processes. For example, in mollusks, reproductive investment has

*Corresponding author: kbrokord@ucn.cl

been shown to decrease the amount of energy available for growth (Iglesias & Navarro 1991), increase maintenance metabolic demands (Kraffe et al. 2008), decrease aerobic metabolic capacity (Brokordt et al. 2000a,b) and decrease swimming ability and escape response capacity (Brokordt et al. 2000a,b, 2003, 2006, Kraffe et al. 2008). Additionally, in several bivalve species (e.g. mussels, scallops and oysters), periods of reproductive activity and mass mortality have been observed to coincide (Tremblay et al. 1998, Xiao et al. 2005, Samain et al. 2007), possibly due to the presence of several stressful environmental factors. In several bivalve species, reproductive maturation occurs during the spring–summer season, characterized by strong environmental fluctuations in conditions such as temperature and oxygen level (Cheney et al. 2000, Tomaru et al. 2001, Cabello et al. 2002, Xiao et al. 2005, Li et al. 2007, Zhang et al. 2010).

When environmental conditions exceed an organism's ability to adapt physiologically, this produces physiological stress. Organisms have evolved several physiological responses to tolerate environmental stress. Among cellular stress responses, one of the most important is the production of stress proteins, commonly known as heat-shock proteins (HSPs) (Morris et al. 2013). HSPs are present in all organisms; they are among the most abundant soluble proteins in the body and are induced by most stressors (Calderwood 2007). HSPs are molecular chaperones that decrease the aggregation of unfolded proteins, assist in protein refolding, and facilitate the channeling of irreversibly denatured proteins towards proteolytic degradation (Parsell & Lindquist 1993). Among the HSPs, the HSP70 family (so denoted due to its 70 kD mass) is the most abundant, and, in many organisms, is considered the most important (Feder & Hofmann 1999, Sørensen 2010). HSP70 activity augments tolerance to several stressful conditions, including extreme temperature, hypoxia, UV radiation, and the presence of toxins; it also participates in the immune response (reviewed by Feder & Hofmann 1999, Calderwood 2007). Under 'non-stress' conditions, HSPs also play an important role in protein biogenesis by preventing the premature folding and aggregation of emerging polypeptides (Frydman et al. 1994, Hartl & Hayer-Hartl 2002). Therefore, HSPs not only increase the organism's capacity to tolerate stress conditions but also enhance the efficiency of protein synthesis. However, gene expression and protein synthesis processes, as well as the chaperoning activity of these proteins, generate high cellular demands for energy in terms of ATP (Hofmann & Somero 1995, Somero 2002, Sharma et al. 2010). Therefore, we propose that upon exposure to a stress factor, the observed level of induced HSP70 may reflect not only the level of molecular damage, but also the capacity to express this protein, which could be energetically limited or affected by the physiological status of the animals. There is some evidence that supports this idea; for example, a study made in blood cells from rainbow trout showed that the *in vitro* inhibition of energetic metabolic pathways decreased ATP by 79%, and this reduced *hsp70* mRNA levels after heat shock (Currie et al. 1999). Also, Mizrahi et al. (2011) observed a negative correlation between endogenous levels of the HSP70 isoform of the foot tissue and albumen gland mass (larger albumen glands indicative of more mature animals) of land snails (*Shincterochila cariosa* and *S. zonata*). Finally, we have recently shown that the marine snail *Concholepas concholepas* under reduced nutritional status showed lower levels of HSP70 induction upon exposure to stress factors during low tide compared with snails in good nutritional status (Jeno & Brokordt 2014).

As both reproduction and the synthesis of stress proteins are energetically expensive, there could be an energetic compromise between these 2 processes. We hypothesized that reproductive investment may limit the availability of energy (in terms of ATP) for HSP70 expression in organisms exposed to environmental stress. To test this hypothesis, we compared the *hsp70* mRNA levels and HSP70 abundance among *Argopecten purpuratus* scallops at different reproductive stages (immature, mature and spawned) exposed to 2 different stress factors (hypoxia and high temperature). *Hsp70* mRNA levels were measured only upon exposure to thermal stress. In parallel, we evaluated the effect of reproductive status on the aerobic metabolic capacity of these scallops, measured through citrate synthase (CS) enzymatic activity. Through the activity of this key enzyme we aimed to assess the effect of reproduction upon one of the main ATP provisioning metabolic pathways. We used *A. purpuratus* as a model because as a broadcast spawner this species has previously been shown to have high reproductive investment (Martínez 1991, Martínez & Mettifogo 1998, Martínez et al. 2000). Also, most natural beds and cultures of *A. purpuratus* are located in bays near upwelling zones and are thus exposed to frequent environmental changes, especially during the reproductive period (Zhang et al. 2010, CEAZA Oceanographic Monitoring System unpubl. data).

MATERIALS AND METHODS

Animal procurement and holding conditions

Adult *Argopecten purpuratus* (70–80 mm shell height) with immature (n = 60) and mature (n = 120) gonads were obtained from aquaculture centers located in Coquimbo, northern Chile (30° 16′ S, 71° 35′ W), during summer 2011. Summer is the most active reproductive season for *A. purpuratus* at Coquimbo, and because reproduction in this species is somewhat asynchronous (Cantillanez et al. 2005), it was possible to find animals at different stages of maturation. Reproductive stage was initially determined using a visual scale following Disalvo et al. (1984) and Martínez & Pérez (2003), where immature individuals show flaccid and pale gonads, and mature individuals show intensely coloured, turgid gonads. In the case of spawned scallops, individuals were chosen from animals that were induced to spawn and subsequently showed empty gonads. Visual observations were thereafter verified via gonad mass measurements, as described below. The scallops were transported to the Universidad Católica del Norte's laboratory in Coquimbo. In order to reduce stress arising from the transport process, individuals were acclimated to laboratory conditions for 4 d in 1000 l tanks supplied with filtered, aerated, running seawater (~18°C) and fed a diet of 50 % *Isochrysis galbana* and 50 % *Chaetoceros calcitrans*. Following acclimation, 60 mature scallops were stimulated to spawn by adding excess microalgae.

Experimental design

We first performed the thermal stress trial. Fifteen scallops from each reproductive stage (immature, mature and spawned) were subjected to a rapid temperature increase from 18 to 24°C and then maintained at 24°C for 6 h. This increase of temperature was previously tested in a preliminary experiment to ensure a significant stress response in the scallops. An additional 15 scallops from each reproductive stage were maintained at baseline temperature (18°C) over the same 6 h and served as controls (i.e. unstressed scallops).

For the hypoxia stress trial, 15 scallops from each reproductive stage were subjected to a rapid decrease in seawater oxygen content from saturation levels (~8.0 mg O_2 l^{-1}) to hypoxic conditions (2.0–1.5 mg O_2 l^{-1}) by adding gaseous nitrogen. Scallops were maintained under this hypoxic condition for 6 h.

As controls (i.e. unstressed scallops), an additional 15 scallops from each reproductive stage were maintained under baseline oxygen conditions over the same 6 h. During this trial, scallops were maintained at 18°C.

Hypoxia treatments were based upon the oxygen fluctuations measured in Tongoy Bay during the spring–summer season by the CEAZA Oceanographic Monitoring System. Therefore, both the hypoxia level (2.0–1.5 mg O_2 l^{-1}) and the duration of the hypoxia cycles followed real environmental conditions to which the scallops are exposed.

Following each experiment, individuals were measured and their gonads were dissected and weighed. Each individual's gills were then extracted, deep frozen in liquid nitrogen and stored at −80°C for later HSP70 quantification and CS enzymatic activity determination. Gill tissues from each scallop exposed to thermal stress trials were stored in RNAlater (Ambion) at −20°C until processing for *hsp70* gene transcription determination.

In preliminary assays, we evaluated HSP70 levels after stress exposure in different tissues: muscle, mantle and gills. We did not measure HSP70 in gonads or digestive glands because these are very unstable tissues, i.e. they experience large changes in short time periods associated with reproductive status or food availability, respectively. The preliminary assays showed that after stress, HSP70 levels increased 1.5 times in muscle tissue, 2.0 times in mantle tissue and 3 times in gill tissue, compared with their respective control tissue from unstressed scallops. Based on these results, we performed the complete study in the gill tissue, which represents a large proportion of the scallops' soft tissues. Because we were looking for the association between HSP70 induction capacity and energetic metabolism capacity, CS was also measured in gill tissue.

Total RNA extraction, cDNA synthesis and mRNA transcription analysis with quantitative real-time PCR

Total RNA was isolated from the gill tissues using Trizol reagent (Invitrogen) according to the manufacturer's instructions. RQ1 RNase-Free DNase (Promega) was used to eliminate DNA contamination. Equal amounts of RNA from 3 individuals per treatment were pooled (thus the 15 individuals per treatment became n = 5 per treatment). Each RNA pool was reverse transcribed using the Revertid H Minus First Strand cDNA Synthesis Kit (Fermentas) according to the manufacturer's instructions. We used 200 ng of total RNA for the real-time PCR analysis. *Hsp70* gene transcription was performed using specific *A. purpuratus* primers (*hsp70*F 5'GAG GCC GTC GCC TAT GGT GC3'; *hsp70*R 5'GCG GTC TCG ATA-CCC AGG GAC A3') (GenBank accession number FJ839890), for which PCR efficiency was previously verified through the standard curve method. The designed primers were selected from a conserved region that does not discriminate between genes encoding for different *hsp70* isoforms and thus between constitutive and inducible ones. However, in preliminary studies using these primers, *hsp70* mRNA showed a strong increase after stress, which indicates that we are measuring the *hsp70* inducible isoforms.

All real-time PCR reactions were performed in triplicate in a 20 µl reaction mixture containing 5 µl cDNA, 0.2 mM of each primer, SYBR Green qPCR master mix 2X (Fermentas) and 50 nM RoX solution. Real-time PCR reactions were run in a StepOne Plus Real-Time PCR System (Applied Biosystems) using the comparative $\Delta\Delta C_T$ method (Livak & Schmittgen 2001) and EF1α (GenBank accession number ES469321.1) as an endogen control. In preliminary studies, the stability of this endogenous gene was tested for our species and tissues. Initial denaturing time was 10 min at 95°C, followed by 40 cycles of denaturing at 95°C for 30 s, annealing at 60°C for 30 s, and extension at 72°C for 30 s, with a ramp rate of the melt curve of 95°C (15 s), 55°C (15 s) and 95°C (15 s).

Extraction and quantification of total protein for HSP70 determination

Total protein was quantified for 0.03 g of gill from each individual. Gill tissue was homogenized in 150 µl of homogenization buffer (32 mM Tris-HCl at pH 7.5, 2% SDS, 1 mM EDTA, 1 mM Pefabloc and 1 mM protease inhibitor cocktail). The homogenate was incubated for 5 min at 100°C, then resuspended in 100 µl of homogenization buffer and re-incubated at 100°C for 5 min. The homogenate was centrifuged at 10 600 × *g* for 20 min. Total protein was quantified in an aliquot of the supernatant with a Micro-BCA kit using a microplate spectrophotometer EPOCH (BioTek).

Quantification of HSP70 protein levels

HSP70 was measured in the gill tissue of each individual by enzyme-linked immunosorbent assay (ELISA), which was validated in previous assays by comparing ELISA results with immunoprobing of western blots from western blot analyses (using the antibodies described below), which showed only one band at the level of 70 kD HSP. Total protein (30 µg ml^{-1}) was diluted in 0.05 M carbonate-bicarbonate buffer at pH 9.6, and 50 µl of sample per well was incubated in an ELISA plate overnight at 4°C with 3 blanks (containing buffer only). The plate was then washed twice with phosphate-buffered saline (PBS) (200 µl per well). Next, 200 µl of blocking buffer (PBS + 5% skim milk) was added to each well and incubated for 2 h. The wells were washed again with PBS. Subsequently, 100 µl of the primary antibody — polyclonal mono-specific anti-epitope that recognize the inducible and constitutive forms of HSP70 specific for *Argopecten purpuratus*, developed in immunized mice with a synthetic peptide epitope (Group of Immunological Markers on Aquatic Organisms, Catholic University of Valparaiso) — diluted 1:400 in blocking buffer + 0.05% Tween-20 was added to each well, and the plate was incubated overnight at 4°C. The plate was then washed 4 times with PBS, incubated with goat anti-mouse IgG (Thermo Scientific) secondary antibody, diluted in blocking buffer + 0.05% Tween-20 for 2 h at 25°C, and washed again 4 times with PBS. Next, 100 µl of substrate solution (10 mg *o*-phenylenediamine dihydrochloride in 25 ml of 0.05 M citrate phosphate buffer) was added, followed by incubation of the plate for 30 min at 25°C. Finally, the plate was read at 450 nm in a microplate spectrophotometer. The absorbance of the sample was corrected by the mean absorbance of the blanks. HSP70 levels were expressed relative to the respective protein level of the unstressed immature group.

CS activity

We homogenized the samples of gill tissue on ice (n = 7 per reproductive status), in 10 volumes of 50 mM imidazole-HCl, 2 mM EDTA-Na$_2$, 5 mM EGTA, 1 mM dithiothreitol and 0.1% Triton X-100, pH 6.6. The homogenates were centrifuged at 4°C for 15 min at 600 ×*g*. We measured enzyme activity at controlled room temperature (20°C) using a UV/Vis spectrophotometer (Cary 50 Bio, Varian). Conditions for enzyme assays were adapted from conditions used by Brokordt et al. (2000a) for *Chlamys islandica*, as follows (all concentrations in mmol l^{-1}): TRIS-HCl 75, oxaloacetate 0.3 (omitted for the control), DTNB 0.1 and acetyl CoA 0.2, pH 8.0. Enzyme activity was examined by following the absorbance changes at

412 nm to detect the transfer of sulphydryl groups from CoASH to DTNB. The extinction coefficient used for DTNB was 13.6 cm^{-1} μmol^{-1}. All assays were run in duplicate and the specific activities are expressed in international units (μmol of substrate converted to product per minute) per gram of gill mass.

Statistical analyses

To evaluate the effect of reproductive stage on *hsp70* mRNA levels and HSP70 abundance in *A. purpuratus* exposed to thermal and hypoxia stress factors, we performed 2-way ANOVAs for each stress factor. Model predictors were reproductive stage (with 3 levels: immature, mature and spawned) and presence/absence of the stressor (high temperature or hypoxia). To evaluate the effect of reproductive stage on CS enzyme activity, we performed a 1-way ANOVA. For both ANOVAs, normality of the dependent variable was tested using the Shapiro-Wilks test (SAS Institute 1999) and homogeneity of variances using Levene's test (Snedecor & Cochran 1989) to verify that the data met model assumptions. A posteriori tests for specific differences were conducted via the multiple pairwise comparisons least-square means (Lenth & Hervé 2015), with significance evaluated at $p \leq 0.05$.

RESULTS

Changes in gonad mass and metabolic capacity with reproductive status

The mean gonad mass was approximately 170% greater in mature *Argopecten purpuratus* scallops than in immature scallops, indicating substantial reproductive investment (Fig. 1). After spawning, the mean gonad mass was similar to that of immature animals.

CS activity varied with *A. purpuratus* reproductive status (Fig. 2). This enzyme showed its highest levels in immature scallops, tended to decline with gonadal maturation, and attained its lowest level in spawned scallops (Fig. 2). CS activity was significantly different between immature and spawned scallops ($p = 0.044$).

Hsp70 mRNA levels and HSP70 abundance in scallops of different reproductive status following different stressors

Hsp70 mRNA relative levels increased significantly after exposure to thermal stress in scallops with each of the different reproductive statuses (Table 1, Fig. 3). However, in immature scallops, this increase in *hsp70* mRNA levels was more than 2-fold higher than in mature and spawned scallops. Immature stressed

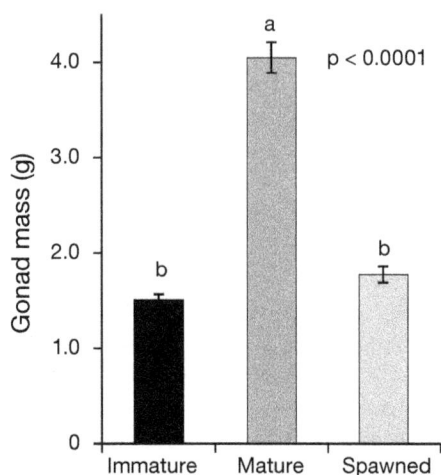

Fig. 1. Changes in gonad mass in *Argopecten purpuratus* scallops in 3 different reproductive stages (immature, mature and spawned). Immature and mature scallops were obtained from a cultured population. Spawning was induced in the laboratory. Values represent means ± SE (*n* = 60 per reproductive stage). Means sharing the same letter are not significantly different (p ≥ 0.05) from one another as indicated by *a posteriori* multiple comparisons (least-square means)

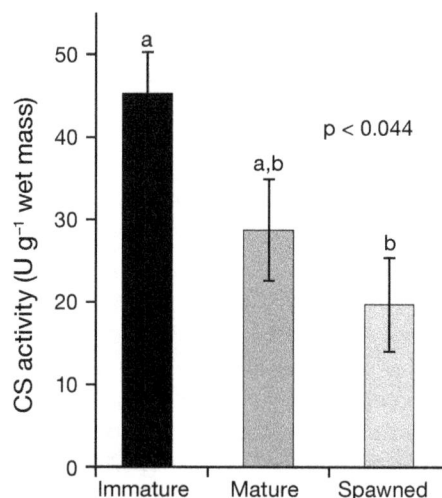

Fig. 2. Citrate synthase (CS) activities in gill tissue of *Argopecten purpuratus* scallops in 3 different reproductive stages (immature, mature and spawned) as an indicator of the tissue capacity for ATP generation via mitochondrial oxygen-dependent metabolism. Values represent means ± SE (*n* = 7 per reproductive stage). Means sharing the same letter are not significantly different (p ≥ 0.05) from one another as indicated by *a posteriori* multiple comparisons (least-square means)

Table 1. Results of 2-way ANOVAs comparing *hsp70* gene transcription and protein induction levels between *Argopecten purpuratus* scallops with different reproductive status, exposed to stress by temperature or hypoxia. Reproductive status: immature, mature and spawned; temperature: 24°C (stress temperature) or 18°C (control temperature); oxygen: hypoxia (~2.0–1.5 mg O_2 l^{-1}, stress) or normoxia (~8.0 mg O_2 l^{-1}, control). For gene expression, n = 5 replicates per condition (each replicate includes 3 individuals' total RNA). For protein expression, n = 12–14 individuals per condition

Source	df	F	p
Hsp70 gene transcription after thermal stress			
Model	1	837	0.000000
Reproductive status (RS)	2	31.8	0.000000
Stress level (SL)	1	82.7	0.000000
RS × SL	2	14.1	0.000057
Error	b		
HSP70 protein levels after thermal stress			
Model	1	466	0.000000
RS	2	8.31	0.000611
SL	1	2.09	0.152843
RS × SL	2	18.9	0.000000
Error	71		
HSP70 protein levels after hypoxia stress			
Model	1	128	0.000000
RS	2	3.16	0.047551
SL	1	13.5	0.000433
RS × SL	2	6.55	0.002321
Error	83		

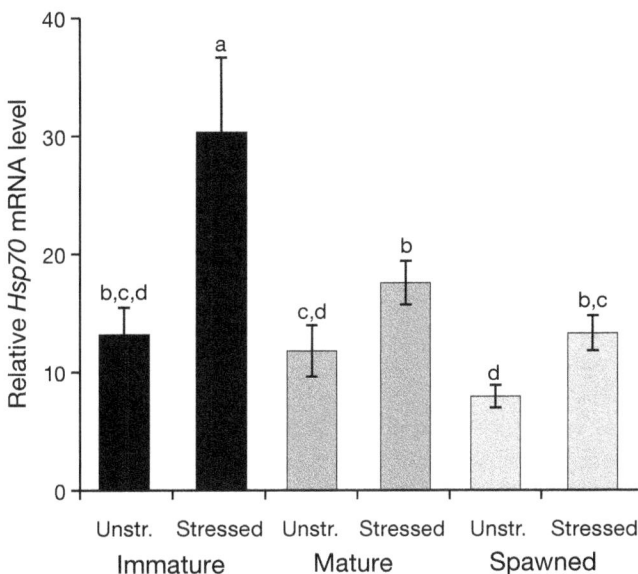

Fig. 3. Relative *Hsp70* gene transcription (mRNA) levels measured in gill tissue of *Argopecten purpuratus* scallops exposed (stressed) and not exposed (unstressed, controls) to thermal stress in 3 different reproductive stages (immature, mature and spawned). Values represent means ± SE (n = 5 replicates per condition; each replicate includes 3 individuals' total RNA). Means sharing the same letter are not significantly different (p ≥ 0.05) from one another as indicated by *a posteriori* multiple comparisons (least-square means)

scallops showed 130% more *hsp70* mRNA than unstressed scallops, whereas stressed mature individuals showed 49% and spawned scallops 65% more *hsp70* mRNA than unstressed scallops.

Following exposure to thermal or hypoxia stress, immature *A. purpuratus* had significantly higher HSP70 protein levels (1.5- and 2-fold higher, respectively) than non-stressed immature scallops (Table 1, Fig. 4). However, following gonad maturation and spawning, there were not statistically significant differences (p > 0.05) in the HSP70 levels between scallops exposed and not exposed to thermal or hypoxia stress. Spawned scallops exposed to thermal stress and hypoxia had slightly higher levels of HSP70 relative to control animals, but this difference was not significant (Table 1, Fig. 4).

DISCUSSION

Our results show that the capacity of mature and spawned *Argopecten purpuratus* to increase *hsp70* mRNA and HSP70 levels following exposure to thermal and hypoxia stressors was markedly reduced relative to that of immature individuals. Interestingly, the reduction in HSP70 induction capacity after gonad maturation was similar for both stressors. In parallel to these reductions, we observed a decrease in the activity of the mitochondrial enzyme CS. CS is a key enzyme regulating the tricarboxylic acid cycle and thus the capacity for ATP generation via mitochondrial oxygen-dependent metabolism of cells and/or tissue (Storey 2004). Therefore, reduced *hsp70* mRNA and HSP70 levels following reproductive processes could reflect an energetic compromise between the production of these proteins and gametogenic activity.

In several scallop species, available energy becomes limited during reproduction, as revealed by the sharp decrease in energy reserves following gonadal maturation and spawning (Martínez 1991, Martínez et al. 2000, Brokordt & Guderley 2004). In *A. purpuratus* from the same studied population, an approximately 50% decrease in adductor muscle carbohydrate stores has been observed following gonadal maturation, persisting until after spawning (K. Brokordt unpubl. data). We have also observed reductions in the quality and oxidative capacity of muscle mitochondria following gonadal maturation and spawning in *Chlamys islandica* scallops (Brokordt et al. 2000a). Decreased metabolic capacity of the anaerobic and aerobic metabolic pathways (through reduction of glycogen phosphorylase, pyru-

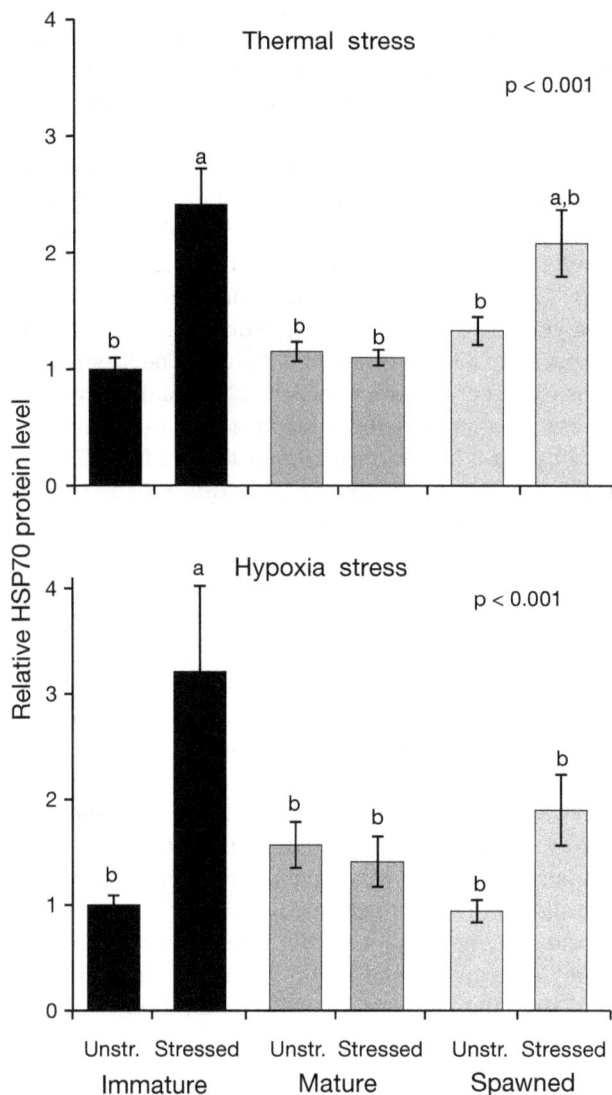

Fig. 4. Relative HSP70 protein levels measured in gill tissue of *Argopecten purpuratus* scallops exposed (stressed) and not exposed (unstressed, controls) to thermal and hypoxia stress in 3 different reproductive stages (immature, mature and spawned). Values represent means ± SE (n = 15 per condition). Means sharing the same letter are not significantly different (p ≥ 0.05) from one another as indicated by *a posteriori* multiple comparisons (least-square means)

vate kinase, phosphofructokinase, octopine dehydrogenase and CS) following gonadal maturation and spawning has also been observed in *C. islandica* (Brokordt et al. 2000a) and *Euvola ziczac* (Brokordt et al. 2000b). Moreover, maintenance metabolic requirements were higher after gonad maturation in *Placopecten magellanicus* (Kraffe et al. 2008). Therefore, the declines in energy availability and metabolic capacity along with the increased metabolic demands of gonadal maturation, which last until spawning, could limit the amount of energy available

for the gene transcription and protein synthesis of HSP70, explaining our results.

The actual energetic cost of mounting a cellular stress response through HSPs is poorly understood. Stress proteins play an important role in energy-dependent processes, such as protein translocation into cellular organelles (Beckmann et al. 1990) and the binding and release of denatured proteins and short peptides to HSP70 (McKay et al. 1994, Mayer & Bukau 2005). Recently, it was observed that under *in vitro* conditions, one Hsp70 molecule consumed 5 ATPs to unfold a single misfolded protein into an intermediate that, upon chaperone dissociation, spontaneously refolded to its native state (Sharma et al. 2010). Thus, ATP is considered indispensable for correct HSP70 functioning (Morimoto 1993, Currie et al. 1999). Gene transcription and translation are also energetically expensive. *Hsp70* mRNA transcription appears to be energetically limited in the red blood cells of rainbow trout subjected to extreme energetic stress, such as that occurring during combined heat shock and metabolic inhibition (i.e. decreased ATP by 79%; Currie et al. 1999). Although the exact amount of ATP necessary for HSP protein synthesis is unknown, the entirety of protein synthesis is estimated to cost between 25 and 40% (in mussels and fish, respectively) of an organism's total oxidative metabolism (Hawkins 1985, Lyndon et al. 1992). A study evaluating the cost of maintaining high HSP levels upon exposure to long-term stress in the trout *Oncorhynchus mykiss* found that chronic stress (i.e. elevated HSP72 and HSP89) decreased metabolic condition (i.e. reduced muscle phosphocreatine, ATP and glycogen) (Viant et al. 2003). Similarly, individual *Drosophila melanogaster* that express additional copies of the *hsp70* gene increase their metabolic rate by as much as 35% following exposure to heat stress (Hoekstra & Montooth 2013). We observed different effects of the reproduction on the *hsp70* mRNA transcription and HSP70 protein levels in scallops exposed to thermal stress. After gonad maturation and spawning, *hsp70* mRNA significantly increased (but less than before gonad maturation) in stressed scallops; however, there were no such increases in protein level in the same individuals. Although post-transcriptional regulation cannot be dismissed, these differences may reflect how the energetic restriction affected each of these processes. Given that we measured the expression of only one form of HSP70, and we cannot discard the potential existence of other forms of HSP70 in *A. purpuratus*, as have been reported for other bivalve species (Zhang et al. 2012), we may have evaluated only part of the response due to HSP70. Therefore, the effect of the energetic restriction may be even higher than observed in this study. Our results and the previous examples show that HSP synthesis and activity are energetically costly; however, the processes of degrading a misfolded protein and re-synthesizing another cost one thousand times more (Sharma et al. 2010). Therefore, this cellular stress response is considered a key physiological adaptation (Sharma et al. 2010, Morris et al. 2013).

HSP70 levels have traditionally been used as biomarkers of stress; i.e. the result of individuals attempting to cope with the damage caused by environmental assaults (Sørensen 2010, Morris et al. 2013). However, considering the energetic cost of this cellular response, variation in HSP70 levels among individuals exposed to the same stressor could indicate variation in the capacity to mount this cellular stress response (Sørensen 2010). This latter approach, i.e. HSP70-induced levels as an indicator of stress response capacity, although suggested in the literature (Sørensen 2010), is supported by almost no empirical evidence. However, a study in 2 land snails has shown a negative correlation between the endogenous HSP70 isoform levels in foot tissue and albumen gland mass (indicative of reproductive maturation) (Mizrahi et al. 2011). These authors indicate that reduced levels of HSP70 isoforms, preceding egg laying, may reduce the snail's ability to cope with external stress (Mizrahi et al. 2011). Nevertheless, this study did not evaluate the changes in HSP induction in these snails upon exposure to stress. Additionally, in a recent study in the intertidal marine snail Concholepas concholepas, we found that individuals in poor energetic condition showed reduced HSP70 induction in response to hypoxia and/ or heat and cold thermal stress relative to individuals in good energetic condition (Jeno & Brokordt 2014). Finally, in Crassostrea gigas oysters, levels of the 72 and 69 kDa HSPs were lower in post-spawning than in pre-spawning individuals after being stimulated by heat shock, indicating that spawning reduced HSP synthesis (Li et al. 2007). Also, the post-spawning oysters had depleted glycogen stores and reduced adenylate energy charge compared with pre-spawning individuals, indicative of lower energy availability for metabolic activity (Li et al. 2007). These studies and the present study support the 'capacity approach' for observed HSP levels.

In life-history evolution, trade-offs represent the cost paid in the currency of fitness (i.e. the ability to produce fertile progeny), because an improvement in a fitness-related trait is associated with a detrimental change in another fitness-associated trait (Reznick 1985, Stearns 1989). One prominent trade-off is the cost of reproduction, in which current reproductive parental investment reduces parental longevity (e.g. reducing survival probability) or future reproductive capacity (Reznick 1985). Several energetic trade-offs between fitness-associated traits have been observed in mollusks. For example, in the cockle Cerastoderma edule, gonadal maturation has been associated with a decrease in the amount of energy available for growth (Iglesias & Navarro 1991), and for the scallops C. islandica (Brokordt et al. 2000a,b), A. purpuratus (Brokordt et al. 2006, Pérez et al. 2009) and Placopecten magellanicus (Kraffe et al. 2008), it is associated with a decrease in metabolic support for the escape response. In D. melanogaster and the abalone Haliotis discus hannai, induction of HSP70 has been associated with fitness (Krebs & Feder 1997, Sørensen et al. 2003, Cheng et al. 2006), because it increases tolerance to physiological stressors and may be critical for survival in some circumstances. Our results suggest that the observed reduction in HSP70 synthesis capacity following reproductive investment is a reproductive cost, as the decreased stress response capacity may limit future survival probability, and may reflect the existence of an evolutionary trade-off between these traits.

Energetic trade-offs between reproduction and other physiological functions can be sufficiently strong that mortality rates increase during post-maturation and spawning processes (Perdue et al. 1981, Barber & Blake 1983, Rocha et al. 2001). Mass mortality after maturation and spawning has been reported in several species of bivalves (Perdue et al. 1981, Barber & Blake 1983) and often coincides with strong environmental change (Cheney et al. 2000, Tomaru et al. 2001, Xiao et al. 2005). Argopecten purpuratus beds and artificial culture systems typically occur in bays near upwelling zones and are therefore exposed to fluctuations in dissolved oxygen levels. Periods of decreased oxygen levels can persist for several days during the spring–summer season (Zhang et al. 2010, CEAZA Oceanographic Monitoring System unpubl. data). Similarly, increased mortality rates of mature A. purpuratus have been observed in cultures in northern Chile (Bahía Inglesa and Tongoy Bay; Camanchaca Ostimar and INVERTEC growing companies, unpubl. data) and in natural beds in Peru (Cabello et al. 2002), most likely in association with periods of hypoxia. During the spring–summer season, scallops are not only more subject to hypoxia (<2.0 mg O_2 l^{-1}), but have also reached gonadal maturation and are initiating spawning. In culture, these processes are compounded with increased manipulation (e.g. sorting and cleaning of culture systems). Increased energy demands resulting from reproduction and environmental stressors, in combination with reduced HSP70 synthesis capacity, could partially explain the increase in mortality rates observed in natural and cultured *A. purpuratus* populations.

In conclusion, our results suggest that reproductive investment may reduce the stress response capacity of *A. purpuratus*, such that mature and spawned scallops should be more vulnerable to environmental stress effects than immature individuals. The similar observed effects of the 2 different types of stressor (hypoxia and heat stress, both of which occur frequently in natural and cultured scallops populations) suggest a consistent pattern of response in this species.

Acknowledgements. We are grateful to Felipe Campos and Igor Vidal for their help with animal manipulation during the experiments. This study was supported by FONDECYT 3110101 funding to H.P. and K.B.

LITERATURE CITED

▶ Barber BJ, Blake NJ (1983) Growth and reproduction of the bay scallop, *Argopecten irradians* (Lamarck) at its southern distributional limit. J Exp Mar Biol Ecol 66:247–256

▶ Beckmann RP, Welch WJ, Mizzen LA (1990) Interaction of HSP70 with newly synthesized proteins: implication for protein folding and assembly. Science 248:850–854

Brokordt K, Guderley H (2004) Energetic requirements during gonad maturation and spawning in scallops: sex differences in *Chlamys islandica* (Muller 1776). J Shellfish Res 23:25–32

▶ Brokordt KB, Himmelman JH, Guderley H (2000a) Effect of reproduction on escape responses and muscle metabolic in the scallop *Chlamys islandica* Muller 1776. J Exp Mar Biol Ecol 251:205–225

▶ Brokordt KB, Himmelman JH, Nusetti OA, Guderley H (2000b) Reproductive investment reduces recuperation from exhaustive escape activity in the tropical scallop *Euvola ziczac*. Mar Biol 137:857–865

▶ Brokordt K, Guderley H, Guay M, Gaymer C, Himmelman J (2003) Sex differences investment: maternal care reduces escape responses capacity in the whelk *Buccinum undatum*. J Exp Mar Biol Ecol 291:161–180

▶ Brokordt K, Fernandez M, Gaymer C (2006) Domestication reduces the capacity to escape from predators. J Exp Mar Biol Ecol 329:11–19

Cabello R, Tam J, Jacinto ME (2002) Procesos naturales y antropogénicos asociados al evento de mortalidad de conchas de abanico ocurrido en la bahía de Paracas (Pisco, Perú) en junio del 2000. Rev Peruana Biol 9:49–65

Calderwood SK (2007) Cell stress proteins. In: Zouhair Atassi M (ed) Protein reviews, Vol 7. Springer, New York, NY

▶ Cantillanez M, Avendaño M, Thouzeau G, Le Pennec M (2005) Reproductive cycle of *Argopecten purpuratus* (Bivalvia: Pectinidae) in La Rinconada marine reserve

(Antofagasta, Chile): response to environmental effects of El Niño and La Niña. Aquaculture 246:181–195

Cheney DP, MacDonald BF, Elston RA (2000) Summer mortality of Pacific oysters, *Crassostrea gigas* (Thunberg): initial findings on multiple environmental stressors in Puget Sound, Washington, 1998. J Shellfish Res 19: 353–359

▶ Cheng P, Liu X, Zhang G, Deng Y (2006) Heat-shock protein 70 gene expression in four hatchery pacific abalone *Haliotis discus hannai* Ino populations using for marker-assisted selection. Aquac Res 37:1290–1296

▶ Currie S, Tufts BL, Moyes CD (1999) Influence of bioenergetic stress on heat shock protein gene expression in nucleated red blood cells of fish. Am J Physiol 276: R990–R996

Disalvo L, Alarcon E, Martínez E, Uribe E (1984) Progress in mass culture of *Chlamys* (*Argopecten*) *purpurata* Lamarck (1819) with notes of its natural history. Rev Chil Hist Nat 57:35–45

▶ Feder ME, Hofmann GE (1999) Heat-shock proteins, molecular chaperones, and the stress response. Annu Rev Physiol 61:243–282

▶ Frydman J, Nimmesgern E, Ohtsuka K, Hartl FU (1994) Folding of nascent polypeptide chains in a high molecular mass assembly with molecular chaperones. Nature 370:111–117

▶ Hartl FU, Hayer-Hartl M (2002) Molecular chaperones in the cytosol: from nascent chain to folded protein. Science 295:1852–1858

▶ Hawkins AJS (1985) Relationships between the synthesis and breakdown of protein, dietary absorption and turnovers of nitrogen and carbon in the blue mussel, *Mytilus edulis*. Oecologia 66:42–49

▶ Hoekstra LA, Montooth KL (2013) Inducing extra copies of the Hsp70 gene in *Drosophila melanogaster* increases energetic demand. BMC Evol Biol 13:68

▶ Hofmann G, Somero G (1995) Evidence for protein damage at environmental temperatures: seasonal changes in levels of ubiquitin conjugates and hsp70 in the intertidal mussel *Mytilus trossulus*. J Exp Biol 198:1509–1518

▶ Iglesias JIP, Navarro E (1991) Energetics of growth and reproduction in cockles (*Cerastoderma edule*): seasonal and age-dependent variations. Mar Biol 111:359–368

▶ Jeno K, Brokordt K (2014) Nutritional status affects the capacity of the snail *Concholepas concholepas* to synthesize Hsp70 when exposed to stressors associated with tidal regimes in the intertidal zone. Mar Biol 161: 1039–1049

▶ Kraffe E, Tremblay R, Belvin S, LeCoz J, Marty Y, Guderley H (2008) Effect of reproduction on escape responses, metabolic rates and muscle mitochondrial properties in the scallop *Placopecten magellanicus*. Mar Biol 156:25–38

▶ Krebs RA, Feder ME (1997) Deleterious consequences of Hsp70 overexpression in *Drosophila melanogaster* larvae. Cell Stress Chaperones 2:60–71

Lenth RV, Hervé M (2015) Least-squares means: R package version 2.14. http://cran.r-project.org

▶ Li Y, Qin JG, Abbott C, Li X, Bekendorff K (2007) Synergistic impacts of heat shock and spawning on the physiology and immune health of *Crassostrea gigas*: an explanation for summer mortality in Pacific oysters. Am J Physiol 293: R2353–R2362

▶ Livak KJ, Schmittgen TD (2001) Analysis of relative gene expression data using real-time quantitative PCR and the $2^{-\Delta\Delta C_T}$ method. Methods 25:402–408

Lodeiros C, Maeda-Martinez AN, Freites L, Uribe E, Lluch-Cota DB, Sicard MT (2001) Ecophysiology of scallops from Iberoamerica. In: Maeda-Martínez AN (ed) Los moluscos pectinidos de Iberoamerica: ciencia y acuicultura. Editorial Limusa, Mexico City, p 77–88

Lyndon AR, Houlihan DF, Hall SJ (1992) The effect of short-term fasting and a single meal on protein synthesis and oxygen consumption in cod, Gadus morhua. J Comp Physiol B 162:209–215

Martínez G (1991) Seasonal variation in biochemical composition of three size classes of the Chilean scallop Argopecten purpuratus Lamarck, 1819. Veliger 34:335–343

Martínez G, Mettifogo L (1998) Mobilization of energy from adductor muscle for gametogenesis of the escallops, Argopecten purpuratus Lamarck. J Shellfish Res 17:113–116

Martínez G, Pérez H (2003) Effect of different temperature regimes on reproductive conditioning in the scallop Argopecten purpuratus. Aquaculture 228:153–167

Martínez G, Brokordt K, Aguilera C, Soto V, Guderley H (2000) Effect of diet and temperature upon muscle metabolic capacities and biochemical composition of gonad and muscle in Argopecten purpuratus Lamarck 1819. J Exp Mar Biol Ecol 247:29–49

Mayer MP, Bukau B (2005) Hsp70 chaperones: Cellular functions and molecular mechanisms. Cell Mol Life Sci 62:670–684

McKay DB, Wilbanks SM, Flaherty KM, Ha JH, O'Brien MC, Shirvanee LL (1994) Stress-70 proteins and their interaction with nucleotides. In: Morimoto RI, Tissieres A, Georgopoulos C (eds) The biology of heat shock proteins and molecular chaperones. Cold Spring Harbor Laboratory Press, Cold Spring Harbor, New York, NY, p 153–177

Mizrahi T, Heller J, Goldberg S, Arad Z (2011) Heat shock protein expression in relation to reproductive cycle in land snails: implication for survival. Comp Biochem Physiol A 160:149–155

Morimoto RI (1993) Cells in stress: transcriptional activation of heat shock genes. Science 259:1409–1410

Morris JP, Thatje S, Hauton C (2013) The use of stress-70 proteins in physiology: a re-appraisal. Mol Ecol 22:1494–1502

Parsell DA, Lindquist S (1993) The function of heat-shock proteins in stress tolerance: degradation and reactivation of damaged proteins. Annu Rev Genet 27:437–496

Perdue A, Beattie JH, Chew KK (1981) Some relationships between gametogenic cycle and summer mortality phenomenon in the Pacific oyster (Crassostrea gigas) in Washington state. J Shellfish Res 1:9–16

Pérez HM, Brokordt K, Martínez G, Guderley H (2009) Effect of spawning on force production during escape responses of the scallop Argopecten purpuratus. Mar Biol 156:1585–1593

Reznick D (1985) Costs of reproduction: an evaluation of the empirical evidence. Oikos 44:257–267

Rocha F, Guerra A, Gonzalez AF (2001) A review of reproductive strategies in cephalopods. Biol Rev Camb Philos Soc 76:291–304

Samain JF, Dégremont L, Solechnik P, Haure J and others (2007) Genetically based resistance to summer mortality in the Pacific oyster (Crassostrea gigas) and its relationship with physiological, immunological characteristics and infection processes. Aquaculture 268:227–243

SAS Institute (1999) The SAS system for Windows, version 8.0. SAS Institute, Cary, NC

Sastry AN (1968) The relationships among food, temperature and gonad development of the bay scallop Aequipecten irradians Lamarck. Physiol Zool 41:44–53

Sharma S, Singh R, Kaur M, Kaur G (2010) Late-onset dietary restriction compensates for age-related increase in oxidative stress and alterations of HSP 70 and synapsin1 protein levels in male Wistar rats. Biogerontology 11:197–209

Snedecor GW, Cochran WG (1989) Statistical methods, 8th edn. Iowa State University Press, Ames, IA

Somero GN (2002) Thermal physiology and vertical zonation of intertidal animals: optima, limits, and costs of living. Integr Comp Biol 42:780–789

Sørensen JG (2010) Application of heat shock protein expression for detecting natural adaptation and exposure to stress in natural populations. Curr Zool 56:703–713

Sørensen JG, Kristensen TN, Loeschcke V (2003) The evolutionary and ecological role of heat shock proteins. Ecol Lett 6:1025–1037

Stearns SC (1989) Trade-offs in life-history evolution. Funct Ecol 3:259–268

Storey KB (2004) Functional metabolism: regulation and adaptation. Wiley-Liss, Hoboken, NJ

Tomaru Y, Kawabata Z, Nakano S (2001) Mass mortality of the Japanese pearl oyster Pinctada fucata martensii in relation to water temperature, chlorophyll a and phytoplankton composition. Dis Aquat Org 44:61–68

Tremblay R, Myrand B, Sevigny JM, Blier P, Guderley H (1998) Bioenergetic and genetic parameters in relation to susceptibility of blue mussels, Mytilus edulis (L.) to summer mortality. J Exp Mar Biol Ecol 221:27–58

Viant MR, Werner I, Rosenblum ES, Gantner AS, Tjeerdema RS, Johnson ML (2003) Correlation between heat-shock protein induction and reduced metabolic condition in juvenile steelhead trout (Oncorhynchus mykiss) chronically exposed to elevated temperature. Fish Physiol Biochem 29:159–171

Xiao J, Ford SE, Yang H, Zhang G, Zhang F, Guo X (2005) Studies on mass summer mortality of cultured zhikong scallops (Chlamys farreri Jones et Preston) in China. Aquaculture 250:602–615

Zhang J, Gilbert D, Gooday AJ, Levin L and others (2010) Natural and human-induced hypoxia and consequences for coastal areas: synthesis and future development. Biogeosciences 7:1443–1467

Zhang G, Fang X, Guo X, Li L and others (2012) The oyster genome reveals stress adaptation and complexity of shell formation. Nature 490:49–54

15

Lipid extraction in stable isotope analyses of juvenile sea turtle skin and muscle

Thaisa F. Bergamo[1,2], Silvina Botta[2,3,*], Margareth Copertino[1,2]

[1]Laboratório de Ecologia Vegetal Costeira, Instituto de Oceanografia, Universidade Federal do Rio Grande - FURG, CP 474, Rio Grande, RS 96203-900, Brazil

[2]Programa de Pós-Graduação em Oceanografia Biológica, Instituto de Oceanografia, Universidade Federal do Rio Grande - FURG, CP 474, Rio Grande, RS 96203-900, Brazil

[3]Laboratório de Ecologia e Conservação da Megafauna Marinha, Universidade Federal do Rio Grande - FURG, CP 474, Rio Grande, RS 96203-900, Brazil

ABSTRACT: Studies involving various aspects of the biology and ecology of sea turtles have successfully applied stable isotope analysis. In many of these studies, the chemical extraction of ^{13}C-depleted lipids of sea turtle tissues has been used as a standard protocol, often without testing whether the time-consuming lipid removal is required. Furthermore, this chemical procedure may unpredictably modify δ^{15}N values, probably due to the loss of proteins associated with lipid structures, thus reinforcing the need for testing the isotopic consequence of the chemical removal of lipids. This study aimed to evaluate the effects of lipid extraction on skin and muscle C and N isotopic values of juvenile green turtles *Chelonia mydas*. We analyzed paired δ^{13}C and δ^{15}N values from lipid-extracted and non-lipid-extracted samples of skin and muscle of 15 juvenile green turtles. Lipid extraction was performed using a mixture of chloroform and methanol. A significant increase was found in δ^{13}C values after lipid extraction for muscle (~0.5‰), but not for skin. C:N ratios were not correlated with the change in δ^{13}C values in either tissue. δ^{15}N values were not affected by lipid extraction in either tissue. The difference found in δ^{13}C values between control and lipid-extracted muscle samples may be biologically significant. On the other hand, the lipid extraction of skin samples does not appear to be necessary in the case of juvenile green turtles. This procedure needs to be tested in green turtles in other life stages.

KEY WORDS: Green turtle · *Chelonia mydas* · Stable isotope analysis · Carbon · Nitrogen · δ^{13}C · δ^{15}N · Lipid extraction

INTRODUCTION

Stable isotope analysis (SIA) of animal tissue carbon (δ^{13}C values) and nitrogen (δ^{15}N values) has been largely used to study the trophic ecology and habitat use of several taxa, including marine megafauna (e.g. seabirds, Cherel et al. 2005; sea turtles, Arthur et al. 2008; sharks, Carlisle et al. 2015; marine mammals, Drago et al. 2015). The method is based on the principle that the isotopic composition in the tissues of consumers reflects the mixture of the isotopic composition present in the different dietary items consumed (DeNiro & Epstein 1978, 1981). The δ^{15}N values predictably increase with increasing trophic level (DeNiro & Epstein 1981, Minagawa & Wada 1984) and consequently, they are used as indicators of a consumer's trophic position (Post 2002, McCutchan et al. 2003, Vanderklift

*Corresponding author: silbotta@gmail.com

& Ponsard 2003). On the other hand, $\delta^{13}C$ values vary little along the food chain and are mainly used to infer primary sources in a food web (McCutchan et al. 2003).

In recent years, several studies using SIA have addressed the issue of the effects of lipids on the isotopic composition of tissue carbon (e.g. Lesage et al. 2010, Ruiz-Cooley et al. 2011, Ryan et al. 2012). Due to differences in the biochemical pathways during synthesis, lipids may be depleted in ^{13}C relative to other major constituents, such as proteins (DeNiro & Epstein 1977). Such differences among protein and lipid metabolism (Martínez del Rio et al. 2009) can bias the interpretation of stable isotope values and result in misleading diet reconstructions, especially when based on tissues with a significant amount of fat (Ricca et al. 2007). Therefore, lipids are usually removed from tissues by either chemical lipid extraction or mathematical correction using species- and tissue-specific equations (e.g. Logan et al. 2008, Elliott et al. 2014). Chemical extraction of lipids is generally effective, yielding isotopic values of carbon that can be considered free of the influence of these biomolecules (Post et al. 2007, Logan et al. 2008, Medeiros et al. 2015), resulting in increased values of $\delta^{13}C$ compared to samples without extraction. However, several studies have reported unexpected changes in $\delta^{15}N$ values after lipid extraction (Sotiropoulos et al. 2004, Ricca et al. 2007, Logan et al. 2008), which is believed to be a consequence of the loss of proteins associated with lipid structures, such as lipid-membrane proteins (Sweeting et al. 2006). Therefore, testing the effects of lipid removal on both $\delta^{13}C$ and $\delta^{15}N$ values is highly relevant before applying SIA to address questions about diet and trophic ecology.

Due to the high correlation between the values of the carbon to nitrogen (C:N) ratio and the percentage of lipids, the C:N ratio itself has been used as an indicator of tissue lipid content for numerous tissues and organisms (e.g. Post et al. 2007). Following this, samples with C:N > 3.5 are considered to have a high lipid content (Post et al. 2007), making chemical extraction, or the use of normalization equations, necessary (Sweeting et al. 2006, Post et al. 2007). However, some studies found no support for the relationship between the bulk C:N ratio and the lipid percentage in organisms such as fish (Fagan et al. 2011), marine mammals (Wilson et al. 2014, Yurkowski et al. 2015), sea turtles (Medeiros et al. 2015), and invertebrates (Kiljunen et al. 2006). Therefore, the need for lipid extraction should be tested empirically before con-

sidering the tissue as free of lipid interference, even for tissues with C:N < 3.5 (Fagan et al. 2011, Yurkowski et al. 2015).

Studies involving various aspects of the biology and ecology of sea turtles have successfully applied SIA (Arthur et al. 2008, Lemons et al. 2011, Thomson et al. 2012, González Carman et al. 2014). In many of these studies, the lipid extraction of sea turtle tissues has been used as a standard protocol, often without testing the extraction efficiency (i.e. lipid removal) and interference on $\delta^{15}N$ values (Revelles et al. 2007, Cardona et al. 2009). On the other hand, other authors decided not to extract lipids mainly based on C:N < 3.5 values (Tomaszewicz et al. 2015, Prior et al. 2016).

Considering that the methodology used in preparing the samples can influence the results and, consequently, the interpretation of isotopic data, we analyzed the effect of lipid extraction on the isotope values of C and N in skin and muscle of juvenile green turtles *Chelonia mydas*. We determined (1) whether the chemical removal of lipids from these tissues increases their $\delta^{13}C$ values, thus evidencing the presence of lipids in the sample, and (2) whether this procedure significantly affects the $\delta^{15}N$ values, either by increasing or decreasing them. The goal of the study was to assess the need for extracting lipids from skin and muscle samples of juvenile sea turtles.

MATERIALS AND METHODS

Collection and sample preparation

The samples of green turtles (n = 15) used in this study were from dead, stranded animals found along the southern coast of Rio Grande do Sul, Brazil (31° 20′ S to 33° 45′ S) between May 2013 and June 2014. The decomposition stage of the specimens was assessed and classified according to Duarte et al. (2011). All animals used were classified into Stages 1 or 2 (fresh carcass or initial decomposition, respectively). Stable isotope composition was considered not to be affected by these decomposition stages as evidenced by previous work (Payo-Payo et al. 2013). For each turtle, curved carapace length (CCL) was measured from the nuchal notch to the posteriormost tip of the carapace with a measuring tape (Bolten 1999). Individuals with CCL < 115 cm were considered to be juveniles according the mean size of nesting females from Brazil (Almeida et al. 2011). Samples of skin and muscle tissue of the right fore flipper (Revelles et al. 2007) were collected with forceps and stainless steel razor blades, placed in labeled plastic bags, and then stored frozen at −20°C.

Paired skin and muscle samples were thawed, rinsed with distilled water, and dried in an oven at 60°C for 24 h. Each tissue sample (15 skin and 15 muscle) was divided into 2 parts and subjected to 2 treatments: with and without lipid extraction. Samples of the first group were extracted by a Soxhlet reflux using chloroform:methanol (2:1) for two 8 h cycles (Folch et al. 1957, Revelles et al. 2007). After the extraction, samples were washed with distilled water and dried in an oven at 60°C for 48 h. All samples (with and without extraction) were then ground to a fine powder with a mortar and pestle, weighed (between 0.5 and 0.9 mg) on a precision scale (AUY 220, Shimadzu), and stored in 5 × 9 mm tin capsules.

For the determination of the $\delta^{13}C$ and $\delta^{15}N$ ratios, all samples were sent to the Stable Isotope Core Laboratory at Washington State University (USA), and analyzed by a continuous flow isotope ratio mass spectrometer (Delta PlusXP, Thermo Finnigan). A conventional delta notation (δ) in parts per thousand (‰) was used to express the stable isotope ratios of the samples in relation to that of standards:

$$\delta X = [(R_{sample}/R_{standard}) - 1] \qquad (1)$$

where X is ^{13}C or ^{15}N, and R denotes the heavier: lighter isotope ratio (^{13}C:^{12}C or ^{15}N:^{14}N) in the sample (R_{sample}) and standard ($R_{standard}$). The standard was Vienna Pee Dee Belemnite for $\delta^{13}C$ values and atmospheric N_2 for $\delta^{15}N$ values. No replicates of the samples were analyzed; however, the analytical precision based on the standard deviations around the means for internal laboratory standards run at set intervals was high (≤ 0.1‰, both for $\delta^{13}C$ and $\delta^{15}N$), thus results are still expected to be significant. The C:N ratio was calculated by dividing the %C by %N.

Statistical analysis

The assumptions of data normality and homoscedasticity were verified using the Shapiro-Wilk and Bartlett tests, respectively. To test whether there was a significant difference in the values of $\delta^{13}C$, $\delta^{15}N$, and the C:N ratio between samples of skin and muscle, and within lipid-extracted and control (i.e. without lipid extraction) samples within the same tissue, data were analyzed using paired t-tests, with a significance level of $\alpha = 0.05$. Simple linear regression models were used to test the relationship between $\Delta\delta^{13}C$ (difference between $\delta^{13}C$ values of con-

trol and lipid-extracted samples) and the C:N ratio of control samples for muscle and skin in order to test whether the C:N ratio of untreated samples could be an indicator of the need for lipid removal. All analyses were performed in the statistical environment R v. 3.1.3 (R Development Core Team 2013).

RESULTS

The CCL of the green turtles analyzed varied between 35 and 47 cm (mean ± SD: 39.8 ± 3.6 cm), thus all were considered juveniles. The $\delta^{13}C$ values were significantly lower in muscle than in skin samples (paired t-tests, $t = 0.14$, p < 0.001 for lipid-extracted and control samples). Nitrogen isotope values, on the other side, did not differ between tissues ($t = 0.14$, p = 0.13 and $t = 0.14$, p = 0.14, for lipid-extracted and control samples, respectively). C:N ratios were significantly higher in muscle samples ($t = 0.14$, p < 0.001, for lipid-extracted and control samples). The values of $\delta^{13}C$ in muscle were significantly lower in control samples than in lipid-extracted samples ($t = 6.2$, p < 0.0001). Mean $\delta^{13}C$ differences between control and lipid-extracted samples ($\Delta\delta^{13}C$) was 0.51‰. However, this difference was mainly driven by 3 samples that showed an increase of >0.5‰ after lipid extraction, with all other samples being less modified by the treatment (i.e. $\Delta\delta^{13}C$ < 0.5‰; Fig. 1). Skin samples did not show differences between treatments ($t = 0.88$, p = 0.39), with a mean $\Delta\delta^{13}C$ of 0.09‰ (Fig. 1, Table 1). Neither the regression between the C:N ratio of control samples and $\Delta\delta^{13}C$ for muscle or for skin were significant ($r^2 = 0.06$, p = 0.38 and $r^2 = 0.05$, p = 0.40, for muscle and skin, respectively).

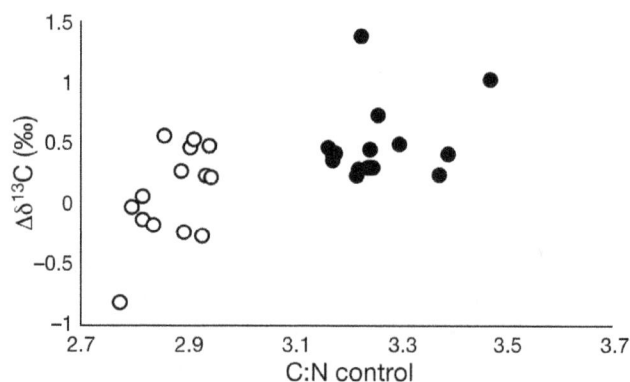

Fig. 1. Relationship between $\Delta\delta^{13}C$ (the difference between carbon isotope values) of control and lipid-extracted samples of skin (open circles) and muscle (black circles) of juvenile green turtles *Chelonia mydas* and the C:N ratio of control samples

Table 1. Carbon (δ^{13}C) and nitrogen (δ^{15}N) isotope values and C:N ratios of muscle and skin of juvenile green turtles *Chelonia mydas*. Mean ± SD, minimum, and maximum values for the control (without lipid extraction) and lipid-extracted samples are reported

Treatment	δ^{13}C (‰)			δ^{15}N (‰)			C:N		
	Mean ± SD	Min	Max	Mean ± SD	Min	Max	Mean ± SD	Min	Max
With extraction									
Muscle	−17.2 ± 0.9	−18.4	−15.5	11.7 ± 1.2	9.2	14.3	3.1 ± 0.1	2.7	3.2
Skin	−15.2 ± 0.6	−16.2	−14.4	12.0 ± 0.9	10.3	13.6	2.8 ± 0.1	2.6	3.0
Control									
Muscle	−17.7 ± 0.8	−18.7	−16.2	11.3 ± 1.4	8.4	13.5	3.2 ± 0.1	3.2	3.5
Skin	−15.3 ± 0.6	−16.1	−14.2	11.9 ± 1.1	9.4	13.2	2.9 ± 0.1	2.8	3.2

Lipid extraction had no significant effect on the average values of δ^{15}N found for both tissues ($t = 1.7$, $p = 0.11$ and $t = 1.05$, $p = 0.31$ for muscle and skin, respectively; Table 1). The mean difference between δ^{15}N values of control and lipid-extracted samples ($\Delta\delta^{15}$N) was 0.38 and 0.16‰ for muscle and skin, respectively. The extraction of lipids significantly modified the values of the C:N ratios of muscle samples ($t = 3.46$, $p < 0.05$) but not C:N ratios of skin ($t = 2.09$, $p = 0.05$; Table 1).

DISCUSSION

Our results show that the effect of lipid extraction (chloroform:methanol solution) on the δ^{13}C values was significant when analyzing the muscle (increase in ~0.5‰) of juvenile *Chelonia mydas*, but did not affect the δ^{13}C values of the skin. Further, significantly lower δ^{13}C and higher C:N ratios were found in muscle samples in relation to those found in skin samples. This difference could be interpreted as evidence of a higher lipid content of muscle samples (Post et al. 2007). However, this difference was maintained when δ^{13}C values in lipid-extracted samples were compared. Similar differences between these 2 tissues were also reported for loggerhead turtles *Caretta caretta* (Revelles et al. 2007) and for pinnipeds (Todd et al. 2010). Therefore, other factors such as the carbon isotope composition of individual amino acids within tissues (Martínez del Rio et al. 2009, Fagan et al. 2011) may play a role in the differences found between δ^{13}C values of both tissues.

We found a mean difference of ~0.5‰ in δ^{13}C values of muscle, where some of the samples (see Fig. 1) showed increases in carbon isotope values higher than 0.5‰ after lipid extraction; thus, caution is required when deciding on the need for lipid extraction, especially when dealing with low sample sizes

that may include such outliers. Moreover, Lesage et al. (2010) calculated the predicted error in prey contribution using Bayesian stable isotope mixing models and found that an error of 0.5‰ due to sample treatment of consumer tissues would introduce a considerable bias depending on the degree of distinctiveness of the end members of the mixing model. Thus, although small $\Delta\delta^{13}$C values were found for juvenile sea turtle muscle samples, we recommended testing for the effect of lipid extraction when using δ^{13}C, before analyzing dietary ratios by Bayesian mixing models.

In the present study, although the mean C:N values of skin and muscle were lower than 3.5, lipid extraction significantly increased the δ^{13}C values and lowered the C:N ratios of muscle. Similar results were found for bone tissue of loggerhead turtles, where although samples showed mean C:N ratios of 3.1, changes in δ^{13}C values of ~1.5‰ were found after lipid extraction (Medeiros et al. 2015), thus concluding that the C:N ratio may not accurately predict the lipid content of samples. Likewise, the reason for the use of C:N as an indicator of the amount of lipids has been discussed in several other studies where tissues with different lipid proportions showed similar C:N ratios (Kiljunen et al. 2006, Fagan et al. 2011, Ruiz-Cooley et al. 2011, Ryan et al. 2012). This finding is also important, as most normalization models of δ^{13}C values (e.g. McConnaughey & McRoy 1979, Post et al. 2007) rely on the C:N ratio as a proxy for the lipid content of the sample. Therefore, caution is required when applying these mathematical methods that use the C:N ratio as a model parameter for calculating lipid-free δ^{13}C values.

δ^{15}N values showed no significant differences after lipid extraction. This result was similar to that found for loggerhead sea turtle bone (Medeiros et al. 2015) and a variety of other aquatic consumers (Ingram et al. 2007, Ricca et al. 2007). However, in other studies, a change in δ^{15}N values was observed (Bodin et al. 2007, Logan et al. 2008, Lesage et al. 2010, Ruiz-Cooley et al. 2011, Wilson et al. 2014) and was associated with the extraction of lipoprotein components that have low nitrogen and the binding of proteins with structural polar lipid components (Bodin et al. 2007, Ruiz-Cooley et al. 2011). Due to the inconsistency of lipid extraction effects on δ^{15}N values, separate runs for carbon and nitrogen SIA are recommended (e.g. Post et al. 2007, Kojadinovic et al.

2008), unfortunately doubling processing time and laboratory costs. However, our findings showed no undesired effects of lipid extraction on $\delta^{15}N$ values when analyzing skin or muscle of young sea turtles using chloroform:methanol as the solvent.

Both the low C:N ratios as well as the lack of lipid extraction effects on carbon isotope values of skin reported in this study showed that there is no need for lipid extraction of this tissue in young green turtles. However, this recommendation may not be valid for tissues from turtles that are in other life cycle phases. During the reproductive phase, the lipid content in adult females increases in the subcutaneous layers and organs to maintain metabolism during vitellogenesis (Hamann et al. 2002). Those authors observed that the concentration of triglycerides in the plasma is lower in non-breeding females and increases at the beginning of vitellogenesis and courting. Thus, the results presented here may not be applicable for SIA of the skin of sea turtles in other development phases, and testing the need for lipid extraction in tissues in these groups is recommended.

CONCLUSION

Given the observed changes in the $\delta^{13}C$ values between control and lipid-extracted muscle samples that may be biologically significant, it is recommended that the lipid content of juvenile green turtle muscle should be accounted for through chemical lipid removal. On the other hand, the lipid extraction of skin samples seems not to be a necessary step in the case of juvenile green turtles, but this procedure needs to be tested in other life stages of this species. Finally, we advise caution when using the C:N ratio of muscle of juvenile green turtles in lipid correction models, as it may not accurately predict lipid-free $\delta^{13}C$ values, at least for samples in the range of C:N ratios observed here (3.2 to 3.5). Future studies should include green turtles from other age classes/ developmental stages in order to encompass a broad range of lipid contents in these tissues.

Acknowledgements. We are indebted to the nongovernmental organization Núcleo de Educação e Monitoramento Ambiental (NEMA) for providing the samples used in this study. Analysis of the data was supported financially by the Coordenação de Aperfeiçoamento Pessoal de Nível Superior (CAPES/AuxPe). CAPES also provided a scholarship to T.F.B. S.B. is currently a postdoctoral fellow (CAPES-PNPD Institucional 2931/2011). This article is part of T.F.B.'s MSc thesis in Biological Oceanography (Graduation Course in Biological Oceanography - IO - FURG, RS, Brazil) under the supervision of M.C. and S.B. and is a contribution of the research groups 'Ecologia e Conservação da Megafauna Marinha-EcoMega/CNPq' and 'Grupo de Análises de Isótopos Estáveis em Ambientes Aquáticos (GAIA-FURG)'.

LITERATURE CITED

▶ Almeida AP, Moreira LMP, Bruno SC, Thomé JCA, Martins AS, Bolten AB, Bjorndal KA (2011) Green turtle nesting on Trindade Island, Brazil: abundance, trends, and biometrics. Endang Species Res 14:193–201

▶ Arthur KE, Boyle MC, Limpus CJ (2008) Ontogenetic changes in diet and habitat use in green sea turtle (*Chelonia mydas*) life history. Mar Ecol Prog Ser 362:303–311

▶ Bodin F, Loc'h LE, Hily C (2007) Effect of lipid removal on carbon and nitrogen stable isotope ratios in crustacean tissues. J Exp Mar Biol Ecol 341:168–175

Bolten AB (1999) Techniques for measuring sea turtles. In: Eckert KL, Bjorndal KA, Abreu-Grobois FA, Donnelly M (eds) Research and management techniques for the conservation of sea turtles. Publication No. 4. IUCN/SSC Marine Turtle Specialist Group, Washington, DC, p 1–5

▶ Cardona L, Aguilar A, Pazos L (2009) Delayed ontogenic dietary shift and high levels of omnivory in green turtles (*Chelonia mydas*) from the NW coast of Africa. Mar Biol 156:1487–1495

▶ Carlisle AB, Goldman KJ, Litvin SY, Madigan DJ and others (2015) Stable isotope analysis of vertebrae reveals ontogenetic changes in habitat in an endothermic pelagic shark. Proc R Soc B 282:20141446

▶ Cherel Y, Hobson KA, Weimerskirch H (2005) Using stable isotopes to study resource acquisition and allocation in procellariiform seabirds. Oecologia 145:533–540

▶ DeNiro MJ, Epstein S (1977) Mechanism of carbon isotope fractionation associated with lipid synthesis. Science 197: 261–263

▶ DeNiro MJ, Epstein S (1978) Influence of diet on the distribution of carbon isotopes in animals. Geochim Cosmochim Acta 42:495–506

▶ DeNiro MJ, Epstein S (1981) Influence of diet on the distribution of nitrogen isotopes in animals. Geochim Cosmochim Acta 45:341–351

▶ Drago M, Franco-Trecu V, Zenteno L, Szteren D and others (2015) Sexual foraging segregation in South American sea lions increases during the pre-breeding period in the Río de la Plata plume. Mar Ecol Prog Ser 525:261–272

Duarte DLV, Monteiro DS, Jardim RD, Soares JCM, Varela AS Jr (2011) Sex determination and gonadal maturation of females of green turtle (*Chelonia mydas*) and loggerhead sea turtle (*Caretta caretta*) in extreme southern of Brazil. Acta Biol Parana 40:87–103

▶ Elliott KH, Davis M, Elliott JE (2014) Equations for lipid normalization of carbon stable isotope ratios in aquatic bird eggs. PLoS One 9:e83597

Fagan KA, Koops MA, Arts MT, Power M (2011) Assessing the utility of C:N ratios for predicting lipid content in fishes. Can J Fish Aquat Sci 68:374–385

▶ Folch J, Lees M, Stanley GHS (1957) A simple method for the isolation and purification of total lipids from animal tissues. J Biol Chem 226:497–509

González Carman V, Botto F, Gaitán E, Albareda D, Campagna C, Mianzan H (2014) A jellyfish diet for the herbivorous green turtle *Chelonia mydas* in the temperate SW Atlantic. Mar Biol 161:339–349

Hamann M, Limpus CJ, Whittier JM (2002) Patterns of lipid storage and mobilisation in the female green sea turtle (*Chelonia mydas*). J Comp Physiol B 172:485–493

Ingram T, Matthews B, Harrod C, Stephens T, Grey J, Markel R, Mazumder A (2007) Lipid extraction has little effect on the $\delta^{15}N$ of aquatic consumers. Limnol Oceanogr Methods 5:338–342

Kiljunen M, Grey J, Sinisalo T, Harrod C, Immonen H, Jones RI (2006) A revised model for lipid-normalizing $\delta^{13}C$ values from aquatic organisms, with implications for isotope mixing models. J Appl Ecol 43:1213–1222

Kojadinovic J, Ricjard P, Le Corre M, Bustamante P (2008) Effects of lipid extraction on $\delta^{13}C$ and $\delta^{15}N$ values in seabird muscle, liver and feathers. Waterbirds 31:169–178

Lemons G, Lewison R, Komoroske L, Gaos A and others (2011) Trophic ecology of green sea turtles in a highly urbanized bay: insights from stable isotopes and mixing models. J Exp Mar Biol Ecol 405:25–32

Lesage V, Morin Y, Rioux È, Pomerleau C, Ferguson SH, Pelletier É (2010) Stable isotopes and trace elements as indicators of diet and habitat use in cetaceans: predicting errors related to preservation, lipid extraction, and lipid normalization. Mar Ecol Prog Ser 419:249–265

Logan JM, Jardine TD, Miller TJ, Bunn SE, Cunjak RA, Lutcavage ME (2008) Lipid corrections in carbon and nitrogen stable isotope analyses: comparison of chemical extraction and modelling methods. J Anim Ecol 77:838–846

Martínez del Rio C, Wolf N, Carleton SA, Gannes LZ (2009) Isotopic ecology ten years after a call for more laboratory experiments. Biol Rev (Camb) 84:91–111

McConnaughey T, McRoy CP (1979) Food-web structure and the fractionation of carbon isotopes in the Bering Sea. Mar Biol 53:257–262

McCutchan JH, Lewis WM, Kendall C, McGrath CC (2003) Variation in trophic shift for stable isotope ratios of carbon, nitrogen, and sulfur. Oikos 102:378–390

Medeiros L, da Silveira Monteiro D, Petitet R, Bugoni L (2015) Effects of lipid extraction on the isotopic values of sea turtle bone collagen. Aquat Biol 23:191–199

Minagawa M, Wada E (1984) Stepwise enrichment of ^{15}N along food chains: further evidence and the relation between $\delta^{15}N$ and animal age. Geochim Cosmochim Acta 48:1135–1140

Payo-Payo A, Ruiz B, Cardona L, Borrell A (2013) Effect of tissue decomposition on stable isotope signatures of striped dolphins *Stenella coeruleoalba* and loggerhead sea turtles *Caretta caretta*. Aquat Biol 18:141–147

Post DM (2002) Using stable isotopes to estimate trophic position: models, methods, and assumptions. Ecology 83:703–718

Post DM, Layman CA, Arrington DA, Takimoto G, Quattrochi J, Montaña CG (2007) Getting to the fat of the matter: models, methods and assumptions for dealing with lipids in stable isotope analyses. Oecologia 152:179–189

Prior B, Booth DT, Limpus CJ (2016) Investigating diet and

diet switching in green turtles (*Chelonia mydas*). Aust J Zool 63:365–375

R Development Core Team (2013) R: a language and environment for statistical computing. R Foundation for Statistical Computing, Vienna. www.R-project.org/

Revelles M, Cardona L, Aguilar A, Borrell A, Gloria F, San Félix M (2007) Stable C and N isotope concentration in several tissues of the loggerhead sea turtle *Caretta caretta* from the western Mediterranean and dietary implications. Sci Mar 71:87–93

Ricca MA, Miles AK, Anthony RG, Deng X, Hung SSO (2007) Effect of lipid extraction on analyses of stable carbon and stable nitrogen isotopes in coastal organisms of the Aleutian archipelago. Can J Zool 85:40–48

Ruiz-Cooley RI, Garcia KY, Hetherington ED (2011) Effects of lipid removal and preservatives on carbon and nitrogen isotope ratios on squid tissues: implications for ecological studies. J Exp Mar Biol Ecol 407:101–107

Ryan C, McHugh B, Trueman CN, Harrod C, Berrow SD, O'Connor I (2012) Accounting for the effects of lipids in stable isotope ($\delta^{13}C$ and $\delta^{15}N$ values) analysis of skin and blubber of balaenopterid whales. Rapid Commun Mass Spectrom 26:2745–2754

Sotiropoulos MA, Tonn WN, Wassenaar LI (2004) Effects of lipid extraction on stable carbon and nitrogen isotope analyses of fish tissues: potential consequences for food-web studies. Ecol Freshw Fish 13:155–160

Sweeting CJ, Polunin NVC, Jennings S (2006) Effects of chemical lipid extraction and arithmetic lipid correction on stable isotope ratios of fish tissues. Rapid Commun Mass Spectrom 20:595–601

Thomson JA, Heithaus MR, Burkholder DA, Vaudo JJ, Wirsing AJ, Dill LM (2012) Site specialists, diet generalists? Isotopic variation, site fidelity, and foraging by loggerhead turtles in Shark Bay, Western Australia. Mar Ecol Prog Ser 453:213–226

Todd SK, Holm B, Rosen DAS, Tollit DJ (2010) Stable isotope signal homogeneity and differences between and within pinniped muscle and skin. Mar Mamm Sci 26:176–185

Tomaszewicz CNT, Seminoff JA, Ramirez MD, Kurle CM (2015) Effects of demineralization on the stable isotope analysis of bone samples. Rapid Commun Mass Spectrom 29:1879–1888

Vanderklift MA, Ponsard S (2003) Sources of variation in consumer-diet ^{15}N enrichments: a meta-analysis. Oecologia 136:169–182

Wilson RM, Chanton JP, Balmer BC, Nowacek DP (2014) An evaluation of lipid extraction techniques for interpretation of carbon and nitrogen isotope values in bottlenose dolphin (*Tursiops truncatus*) skin tissue. Mar Mamm Sci 30:85–103

Yurkowski DJ, Hussey NE, Semeniuk C, Ferguson SH, Fisk AT (2015) Effects of lipid extraction and the utility of lipid normalization models on $\delta^{13}C$ and $\delta^{15}N$ values in Arctic marine mammal tissues. Polar Biol 38:131–143

16

Reproductive dynamics of the southern pink shrimp *Farfantepenaeus subtilis* in northeastern Brazil

Emanuell Felipe Silva[1,2,*], Nathalia Calazans[2], Leandro Nolé[3],
Thaís Castelo Branco[2], Roberta Soares[2], Maria Madalena Pessoa Guerra[4],
Flávia Lucena Frédou[3], Silvio Peixoto[2]

[1]Instituto Federal de Educação, Ciência e Tecnologia da Paraíba, Campus Cabedelo, Cabedelo, 58103-772 PB, Brazil

[2]Laboratório de Tecnologia em Aquicultura (LTA), Departamento de Pesca e Aquicultura,
Universidade Federal Rural de Pernambuco, 52171-900 Recife/PE, Brazil

[3]Laboratório de Estudos de Impactos Antrópicos na Biodiversidade Marinha e Estuarina (BIOIMPACT),
Departamento de Pesca e Aquicultura, Universidade Federal Rural de Pernambuco, 52171-900 Recife/PE, Brazil

[4]Laboratório de Andrologia, Departamento de Medicina Veterinária, Universidade Federal Rural de Pernambuco,
52171-900 Recife/PE, Brazil

ABSTRACT: This study focused on the reproductive dynamics of southern pink shrimp *Farfantepenaeus subtilis* populations off the coast of Pernambuco, northeastern Brazil. *F. subtilis* specimens were collected each month between August 2011 and July 2012 by an artisanal fishing vessel. A total of 1246 specimens were collected, of which the majority (56%) were females, and which were significantly larger than the males. Ovary maturation, based on histological and visual criteria, was classified into 4 stages — I: immature, II: maturing, III: mature, and IV: spent. Mature gonads were found in the females throughout the year, but at a higher proportion during the warmer months (October–March). A higher proportion of juveniles were observed between December and April. The mean total body length at first gonadal maturation in female pink shrimp was estimated to be 11.9 cm. These findings could be used to help guide the development of fishery management policies for *F. subtilis* in the region.

KEY WORDS: Reproduction · Maturation · Conservation · Sustainability

INTRODUCTION

The penaeid shrimps are a valuable crustacean fishery resource (Dall et al. 1990), comprising 42.2% of the total worldwide shrimp catch between 1970 and 2000 (FAO 2009). In Brazil, the catch of penaeids reached 38 373 t in 2010, with pink shrimp *Farfantepenaeus* spp. representing 17.9% of the total crustacean catch (MPA 2012). In the state of Pernambuco in northeastern Brazil, penaeids represent the second most important crustacean fishery resource after swimming crabs. In this region, artisanal shrimp fishing is typically carried out in small (8 to 12 m) wooden sailboats or motorboats (IBAMA 2005).

Reliable knowledge about the life cycle and biology of the southern pink shrimp *F. subtilis* is necessary in order to efficiently exploit this important fishery resource. A number of studies have focused on the reproductive biology of penaeid populations on the Brazilian coast (Peixoto et al. 2003, Santos et al. 2008, Almeida et al. 2012, Simões 2012, Heckler et al. 2013), with data on *F. subtilis* available from the

*Corresponding author: emanuellfelipe@yahoo.com.br

northern (Isaac et al. 1992, Cintra et al. 2004) and northeastern regions of Brazil (Coelho & Santos 1993, 1995). However, no studies have focused on the reproductive dynamics of *F. subtilis* combined with a histological analysis of the ovaries, which provides a more precise evaluation of gonadal maturity. Given the biological and economic importance of *F. subtilis* in northeastern Brazil, the goal of the present study was to describe the reproductive dynamics of this species off the coast of Pernambuco.

MATERIALS AND METHODS

Our study focused on the principal shrimping grounds off the coast of the northeastern Brazilian state of Pernambuco, in the municipality of Sirinhaém (Fig. 1) Specimens of *Farfantepenaeus subtilis* were collected monthly from August 2011 through July 2012 by a vessel of the local artisanal shrimping fleet. The shrimp were caught by trawling during the day in the full moon phase using double trawl nets (length: 10 m; mouth: 6.10 m; body mesh: 30 mm; tail mesh: 25 mm). Specimens were collected in 3 trawls of 2 h each; approximately 70 shrimp were selected at random from each trawl. Once selected, individuals were immediately stored on ice until analysis.

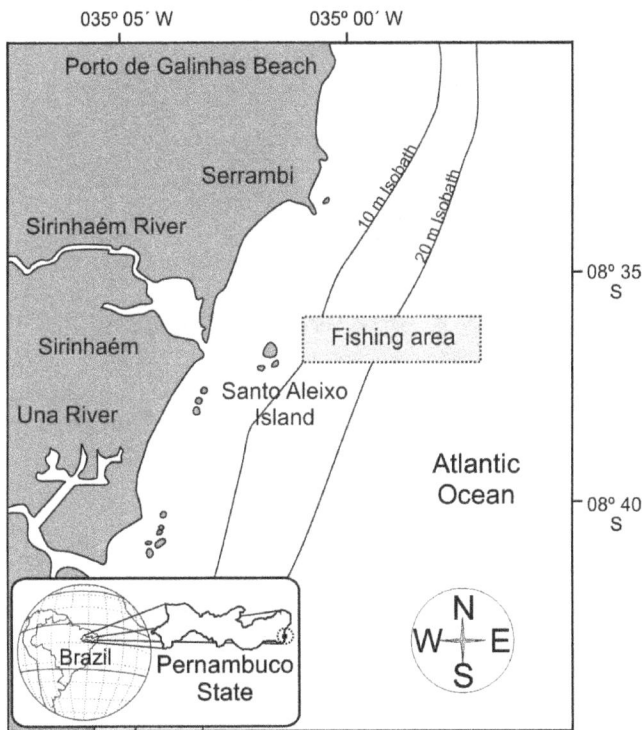

Fig. 1. Study area off the coast of Pernambuco, northeastern Brazil

The sex of each specimen was determined through the analysis of external traits (i.e. presence of a thelycum in females and the petasma in males). The length of the cephalothorax, along with the total length and total wet weight were recorded for each individual. The homogeneity of the sex ratio for the study period as a whole was tested using chi-squared analysis. Differences in mean (±SD) body length and weight between males and females were compared using a Student's t-test where the assumptions for a parametric test were satisfied.

Female gonadal development was evaluated based on the gonadosomatic index (GSI), which was calculated by dividing the weight of the dissected ovaries by that of the body, and the morphology and coloration of the gonads. The color of the fresh ovary was compared with a widely available chromatic scale (coated Pantone Matching System) to determine the most frequent color observed in the ovaries.

Females were selected according to their relative size and the shape of the ovary when observed through the exoskeleton using a flashlight. Samples of the median portion of the ovary were collected from 24 specimens for histological analysis. The tissue was fixed in Davidson solution for 24 h and then transferred to 70% ethanol before being set in paraffin at 58°C, sectioned (6 µm) and stained with hematoxylin-Eosin (Junqueira & Junqueira 1983, Bell & Lightner 1988).

Oocytes were classified according to their histological characteristics, based on descriptions available for other *Farfantepenaeus* species (Quintero & Gracia 1998, Peixoto et al. 2003). The proportion of each type of oocyte was calculated for each month and 30 oocytes representing each category (or the total number available, if less than 30) were measured using the software ImageTool v.2.0 for Windows (University of Texas Health Science Center San Antonio). Only oocytes with nuclei sectioned in the equatorial plane were selected for this analysis. Data were grouped into distinct maturation stages based on the most developed cell observed in each sample.

Total length (TL), cephalothorax length (CL), total weight (TW), weight of the ovary (WO), gonadosomatic index (GSI), frequency of each type of oocyte (FO), and oocyte diameter (OD) were compared among the different maturation stages of the ovaries using ANOVA, given the satisfaction of the assumptions for a parametric test. Tukey's HSD test was then applied to identify the maturation stages that were significantly different ($p < 0.05$) from one another.

In order to determine the breeding season of the species, the proportion of specimens in each stage of maturity was determined each month through the macro- and/or microscopic classification of the gonads. TL at first gonadal maturation of the females was determined by plotting the relative frequency of adults in each 0.5 cm size class. Specimens were considered to be adults when they presented well-developed or spent gonads. Body length classes were then plotted against the percentage of adult females, adjusted by the iterative non-linear least squares technique to obtain the value of TL_{50} using the logistic equation described by King (1995).

RESULTS

During the study period, a total of 1246 *Farfantepenaeus subtilis* specimens were collected, comprising 701 females and 545 males (1.28:1) ($\chi^2 = 0.00001$; p < 0.05), which represented 56 and 44% of the total number of specimens collected, respectively. In addi-

tion to being more abundant, females were significantly larger (in TL and CL) than males (p < 0.05), with mean (±SD) TLs of 11.42 ± 1.61 cm (range: 1.5 to 18.5 cm) and 10.5 ± 1.62 cm for females and males, respectively.

Based on histological characteristics and visual observations, ovarian development of *F. subtilis* was divided into 4 distinct stages. Stage I (immature) is characterized by a predominance of basophilic oocytes (BOs), with a large nucleus surrounded by a number of nucleoli (Fig. 2A, Table 1). In this stage the gonads are very fine and flaccid, which impairs their visualization through the exoskeleton, and the ovary is translucent (Pantone catalog [PC] category 607 PC; Fig. 2A). In Stage II (maturing) the vitellogenic oocytes (VOs) are distributed about the periphery of the ovarian lobes (Fig. 2B, Table 1). During this stage, the ovary can be visualized through the exoskeleton due to an increase in the volume and consistency of the gonad, which now fills the abdominal cavity and cov-

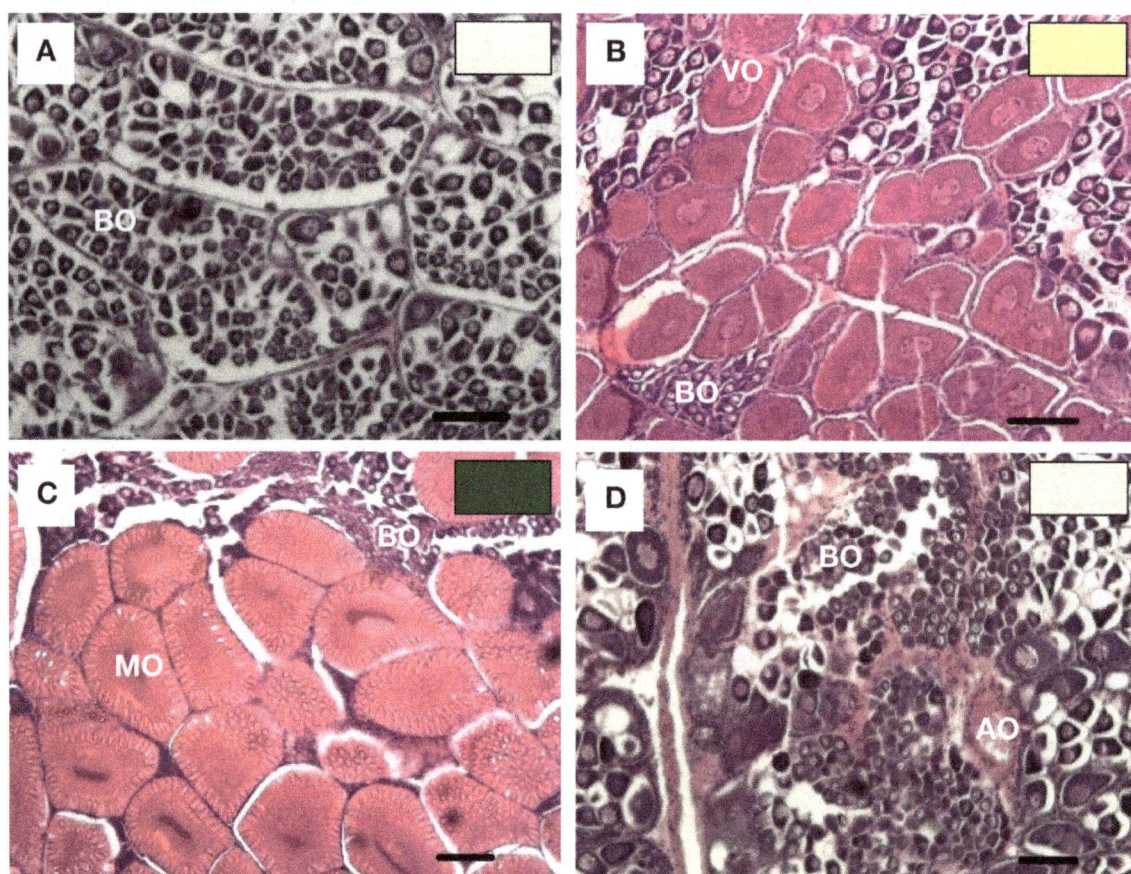

Fig. 2. Histological sections (10× magnification) showing colors representative of different development stages of the ovaries of the southern pink shrimp *Farfantepenaeus subtilis*. (A) Stage I (immature): basophilic oocytes (BO) and translucent ovary (Pantone catalog [PC]: 607 PC); (B) Stage II (maturing): vitellogenic oocytes (VO), ovary light yellow (386 PC); (C) Stage III (mature): mature oocytes (MO), dark green ovary (350 PC); (D) Stage IV (spent): atretic oocytes (AO), ovary same color as in Stage I. Scale bar: 100 μm

Table 1. Frequency and diameter (mean ± SD) of the basophilic (BO), vitellogenic (VO), mature (MO), and atretic (AO) oocytes observed in each stage of gonadal maturity (Stage I: immature; Stage II: maturing; Stage III: mature; Stage IV: spent) of female southern pink shrimp *Farfantepenaeus subtilis* specimens captured off Pernambuco, northeastern Brazil, between August 2011 and July 2012. Values in the same column marked with different superscript letters are significantly different (p < 0.05). NP: not present

	BO (%)	VO (%)	MO (%)	AO (%)	BO diameter (μm)	VO diameter (μm)	MO diameter (μm)
Stage I	100^a	0^a	0^a	0^a	31.08 ± 3.77^a	NP	NP
Stage II	73.56 ± 2.41^b	26.44 ± 2.41^b	0^a	0^a	34.96 ± 1.92^a	144.45 ± 12.98	NP
Stage III	66.00 ± 7.37^c	0^a	34.00 ± 7.37^b	0^a	33.51 ± 2.21^a	NP	219.09 ± 17.65
Stage IV	99.36 ± 0.11^a	0^a	0^a	0.64 ± 0.11^b	33.52 ± 3.83^a	NP	NP

ers part of the intestine in the cephalothorax region. The ovary is light yellow in color (category 386 PC; Fig. 2B). Stage III (mature) is characterized by the presence of mature oocytes (MOs) with peripheral cortical rods, which indicate the final stage of maturation (Fig. 2C, Table 1). The posterior lobes of the ovary occupy the entire abdominal cavity and the development of the lateral lobes can be observed in the cephalothorax. In this stage, the ovary is dark green (category 350 PC; Fig. 2C). Stage IV (spent) is characterized by the presence of atretic oocytes (AOs), which are mature oocytes in the process of being reabsorbed (Fig. 2D, Table 1). Macroscopically, the coloration of the ovary is equivalent to that of Stage I.

BOs were observed in all histological sections, irrespective of the stage of maturation (Fig. 2), although they were significantly more abundant in Stages I and IV. In Stages II and III, as the ovary developed a significant reduction was observed in the concentration of BOs (Table 1). No significant differences were found in the diameter of the BOs among developmental stages. The VO, MO, and AO were found exclusively in developmental stages II, III, and IV, respectively (Table 1, Fig. 2). No significant variation was found in body size (TL, CL or TW) among the different stages of gonadal maturity, although WO and the GSI were significantly higher in Stage III compared to all other stages (Table 2).

Overall, 56% of the female specimens collected during the study were immature, 13% were maturing, 10% were mature, and 21% had spent ovaries. The distribution of these frequencies over the year (grouped bimonthly) indicates that female southern pink shrimp with mature gonads were present in the population throughout the year, but were relatively

Table 2. Mean (±SD) total length (TL), cephalothorax length (CL), total weight (TW), ovary weight (OW), and gonadosomatic index (GSI) recorded in each stage of gonadal maturity in female southern pink shrimp *Farfantepenaeus subtilis* specimens captured off Pernambuco, northeastern Brazil, between August 2011 and July 2012. Values in the same column marked with different letters are significantly different (p < 0.05)

	TL (cm)	CL (cm)	TW (g)	OW (g)	GSI (%)
Stage I	10.65 ± 0.78^a	2.42 ± 0.28^a	10.52 ± 2.36^a	0.07 ± 0.04^a	0.66 ± 0.19^a
Stage II	12.53 ± 1.68^a	2.75 ± 0.26^a	14.55 ± 4.07^a	0.45 ± 0.18^a	2.96 ± 0.77^a
Stage III	12.93 ± 1.65^a	2.87 ± 0.46^a	18.45 ± 7.17^a	1.48 ± 0.95^b	6.60 ± 2.05^b
Stage IV	13.36 ± 1.87^a	2.87 ± 0.20^a	20.59 ± 10.20^a	0.31 ± 0.33^a	1.24 ± 0.86^a

more abundant between October and March (Fig. 3). The TL_{50} of the females was estimated to be 11.9 cm (Fig. 4). The smallest adult females had a TL of 9.0 cm; all those with TL > 13.5 cm were adults.

DISCUSSION

The sex ratio of *Farfantepenaeus subtilis* recorded in this study was 1.28:1 (females:males). Deviations in sex ratios are common in crustaceans (Wenner 1972), and tend to be related to differences in life cycle, migration, mortality, growth rates and behavior between males and females. Other factors such as molt, dispersal, and reproductive patterns may also be important (Botelho et al. 2001). A female bias appears to be characteristic of most penaeid populations (Coelho & Santos 1993, 1995), principally during the breeding season or in spawning grounds in the open sea, resulting in an increase in the capture of individuals of one sex (although a ratio close to 1:1 may indicate an area in which mating occurs). Collected specimens were also sexually dimorphic, with females attaining larger body size than males. This sexual size dimorphism is typical of the penaeids (Boschi 1969). Gab-Alla et al. (1990) suggested that the higher growth rate observed in females is related to reproductive processes: the larger size of the fe-

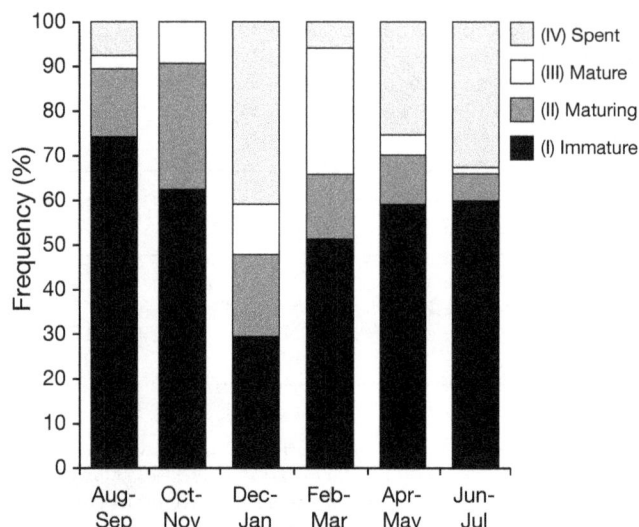

Fig. 3. Relative frequency of ovary developmental stages in southern pink shrimp *Farfantepenaeus subtilis* captured off Pernambuco, northeastern Brazil, between August 2011 and July 2012

males reflects the body space required for the development of ovaries and other reproductive structures (Hartnoll 1982).

The stages of ovarian development in a number of different penaeids have been defined based on the presence of basophilic, vitellogenic, mature, and atretic oocytes (Quintero & Gracia 1998, Peixoto et al. 2002, 2003, Dumont et al. 2007, Gonçalves et al. 2009). These criteria were also adopted for *F. subtilis* in the present study, to define 4 stages of ovarian development as immature, maturing, mature, and spent. BOs were observed in all developmental stages and did not differ significantly in diameter, although they were more abundant in the immature (100%) and spent (99.36%) stages. These similarities in the proportion of BOs hamper differentiation be-

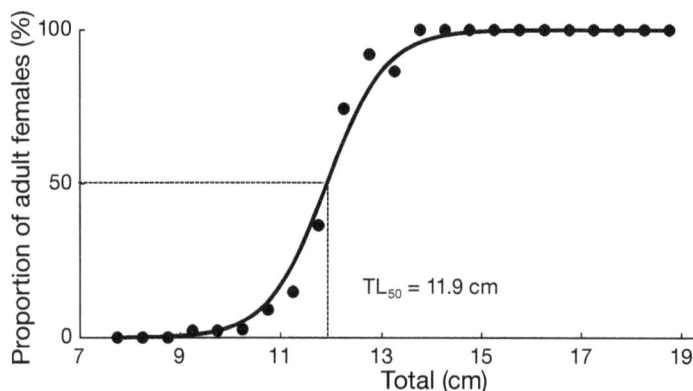

Fig. 4. Total length at first sexual maturity of female southern pink shrimp *Farfantepenaeus subtilis* captured off Pernambuco, northeastern Brazil, between August 2011 and July 2012

tween the immature and spent stages in penaeids (King 1948, Quintero & Gracia 1998, Peixoto et al. 2003), which is reinforced by the difficulties involved in macroscopic observations of the gonad, given that it cannot be visualized through the exoskeleton in either stage (Castille & Lawrence 1991). However, a number of studies have shown that the presence of AOs is an important characteristic for the distinction of immature and spent ovaries (King 1948, Martosubroto 1974, Quintero & Gracia 1998, Peixoto et al. 2003). Palacios et al. (1999) and Peixoto et al. (2005) concluded that the reabsorption of MOs (atresia) has a significant impact on reproductive performance, since this represents the presence of oocytes that were not liberated during the spawning process.

During ovarian development there is an increase in the size of the oocytes, which can be distinguished by their grainy cytoplasm, acidophilous reaction, and the presence of nucleoli arranged around the nucleus—that is, vitellogenic oocytes (Quintero & Gracia 1998, Peixoto et al. 2003). These modifications are accentuated when comparing the immature and maturing stages due to protein synthesis that occurs during this period (Quintero & Gracia 1998). Additionally, the color of the ovary during the maturation process in penaeids may range from yellow to green, depending on the degree of pigmentation (Dall et al. 1990, Browdy 1992, Quintero & Gracia 1998, Peixoto et al. 2003).

The presence of peripheral cortical rods in the cytoplasm of the MOs is commonly observed during oogenesis in other penaeids (Dall et al. 1990, Bell & Lightner 1988, Quintero & Gracia 1998, Ohtomi et al. 2003, Peixoto et al. 2003), and was also observed in the present study of *F. subtilis*. Clark et al. (1980) and Yano (1988) concluded that these rods exude their content after spawning, producing a gelatinous layer around the oocyte which probably guarantees the fixation of the sperm and the formation of the eclosion envelope.

Southern pink shrimp are known to reproduce continuously, with seasonal pulses in northern (Isaac et al. 1992, Cintra et al. 2004) and northeastern Brazil (Coelho & Santos 1993, 1995). A similar pattern was observed off Pernambuco in the present study—mature females were collected throughout the year, but a higher proportion were recorded during the hottest months (October–March), which is considered to be the reproductive peak. In the present study, 56% of the collected females were immature. This proportion could be related to the proximity of the collecting area to the estuary, which may have increased the probability of catching juveniles.

In *F. subtilis*, the TL_{50} at first sexual maturationwas estimated to be 11.9 cm in the present study. Coelho & Santos (1993) recorded a similar value (10.3 cm) from the Tamandaré region of Pernambuco, while Isaac et al. (1992) and Cintra et al. (2004) registered values of 11.0 and 12.65 cm, respectively, for populations from northern Brazil. Variations in the size of the shrimp at first maturity may be related to differences in the intensity of local fisheries, which may provoke precocious maturation (Sparre & Venema 1992, Fonteles-Filho 2011). Furthermore, Dall et al. (1990) suggested that size at first maturity was dependent on environmental parameters influenced by season, latitude and depth (Dall et al. 1990).

The findings presented here could be used to help guide the development of a fishery management policy for this species in the Pernambuco region, since this state has yet to implement legislation for the regulation of shrimp fisheries.

Acknowledgements. This study was supported by the Pernambuco State Foundation for Science and Technology (FACEPE), the Brazilian Federal Higher Education Training Program (CAPES), and the Brazilian National Science Council (CNPq). S.P., F.L.F. and M.M.P.G. hold CNPq research productivity fellowships.

LITERATURE CITED

➤ Almeida AC, Baeza JA, Fransozo V, Castilho AL, Fransozo A (2012) Reproductive biology and recruitment of *Xiphopenaeus kroyeri* in a marine protected area in the Western Atlantic: implications for resource management. Aquat Biol 17:57–69

Bell TA, Lightner DV (1988) A handbook of normal penaeid shrimp histology. World Aquaculture Society, Baton Rouge, LA

Boschi EE (1969) Estudio biológico pesquero del camarón *Artemesia longinaris* Bate de Mar del Plata. Bol Inst Biol Mar 18:1–47

Botelho ERRO, Santos MCF, Souza JRB (2001) Aspectos populacionais do Guaiamum, *Cardisoma guanhumi* Latreille, 1825, do estuário do Rio Una (Pernambuco – Brasil). Bol Técn Cient CEPENE 9:123–146

Browdy CL (1992) A review of the reproductive biology of *Penaeus* species: perspectives on controlled shrimp maturation systems for high quality nauplii production. In: Wyban J (ed) Proceedings of the special session on shrimp farming. World Aquaculture Society, Baton Rouge, LA, p 22–51

Castille FL, Lawrence AL (1991) Reproductive studies concerning natural shrimp populations: a description of changes in size and biochemical compositions of the gonads and digestive glands in penaeid shrimps. In: De Loach PF, Dougherty WJ, Davidson MA (eds) Frontiers of shrimp research, Vol 22. Elsevier, Amsterdam, p 17–32

Cintra IHA, Aragão JAN, Silva KCA (2004) Maturação gonadal do camarão-rosa, *Farfantepenaeus subtilis* (Pérez-Farfante, 1967), na região norte do Brasil. Bol

Técn Cient CEPNOR 4:21–29

➤ Clark WH Jr, Lynn JW, Persyo HO (1980) Morphology of the cortical reaction in the eggs of *Penaeus aztecus*. Biol Bull (Woods Hole) 158:175–186

Coelho PA, Santos MCF (1993) Época da reprodução do camarão rosa, *Penaeus subtilis* Pérez-Farfante, 1967 (Crustacea, Decapoda, Penaeidae), na região de Tamandaré, PE. Bol Técn Cient CEPENE 1:57–72

Coelho PA, Santos MCF (1995) Época da reprodução dos camarões *Penaeus schmitti* Burkenroad, 1936 e *Penaeus subtilis* Pérez-Farfante, 1967 (Crustacea, Decapoda, Penaeidae), na região da foz do Rio São Francisco. Bol Técn Cient CEPENE 3:122–140

Coleto ZF, Guerra MMP, Batista AM (2002) Avaliação do sêmen congelado de caprinos com drogas fluorescentes. Rev Bras Med Vet 24:101–104

Dall W, Hill BJ, Rothlisberg PC, Staples DJ (1990) The biology of the Penaeidae. Advances in marine biology 27. Academic Press, London

Dumont LFC, D'Incao F, Santos RA, Maluche S, Rodrigues LF (2007) Ovarian development of wild pink prawn (*Farfantepenaeus paulensis*) females in northern coast of Santa Catarina State, Brazil. Nauplius 15:65–71

FAO (2009) The state of world fisheries and aquaculture, 2008. FAO, Rome

Fonteles Filho AA (2011) Oceanografia, biologia e dinâmica populacional de recursos pesqueiros. Editora da Universidade Federal do Ceará, Fortaleza

➤ Gab-Alla AAFA, Hartnoll RG, Ghobashy AF, Mohammed SZ (1990) Biology of penaeid prawns in the Suez Canal lakes. Mar Biol 107:417–426

Gonçalves SM, Santos JL, Rodrigues ES (2009) Estágios de desenvolvimento gonadal de fêmeas do camarão-branco *Litopenaeus schmitti* (Burkenroad, 1936), capturadas na região marinha da Baixada Santista, São Paulo. Rev Cecil 1:96–100

Hartnoll RG (1982) Growth. In: Abele LG (ed) The biology of Crustacea, Vol 2: embryology, morphology, and genetics. Academic Press, New York, NY, p 111–196

➤ Heckler GS, Simões SM, Santos APF, Fransozo A, Costa RC (2013) Population dynamics of the seabob shrimp *Xiphopenaeus kroyeri* (Dendrobranchiata, Penaeidae) in south-eastern Brazil. Afr J Mar Sci 35:17–24

Holthuis LB (1980) FAO species catalogue, Vol 1: shrimps and prawns of the world. An annotated catalogue of species of interest to fisheries. FAO Fisheries Synopsis No.125. FAO, Rome

➤ Hossain MY, Ohtomi J (2008) Reproductive biology of the southern rough shrimp *Trachysalambria curvirostris* (Penaeidae) in Kagoshima Bay, southern Japan. J Crustac Biol 28:607–612

IBAMA (Instituto Brasileiro do Meio Ambiente e dos Recursos Naturais Renováveis) (2005) Boletim estatístico da pesca marítima e estuarina do nordeste do Brasil. IBAMA, Brasília

Isaac VJ, Neto J, Damasceno FG (1992) Camarão-rosa da costa norte: biologia, dinâmica e administração pesqueira. Série de Estudos de Pesca, Coleção Meio Ambiente, Brasília

Junqueira LCU, Junqueira LMMS (1983) Técnicas básicas de citologia e histologia. Livraria e Editora Santos, São Paulo

➤ Kevrekidis K, Thessalou-Legaki M (2013) Reproductive biology of the prawn *Melicertus kerathurus* (Decapoda: Penaeidae) in Thermaikos Gulf (N Aegean Sea). Helgol Mar Res 67:17–31

► King JE (1948) A study of the reproductive organs of the common marine shrimp, *Penaeus setiferus* (Linnaeus). Biol Bull (Woods Hole) 94:244–262

King M (1995) Fisheries biology, assessment and management. Blackwell Publishing, Oxford

Martosubroto P (1974) Fecundity of pink shrimp, *Penaeus duorarum* Burkenroad. Bull Mar Sci 24:606–627

MPA (Ministério da Pesca e Aquicultura) (2012) Boletim estatístico da pesca e aquicultura. Ministério da Pesca e Aquicultura, Brasília

► Ohtomi J, Tashiro T, Atsuchi S, Kohno N (2003) Comparison of spatiotemporal patterns in reproduction of the kuruma prawn *Marsupenaeus japonicus* between two regions having different geographic conditions in Kyushu, southern Japan. Fish Sci 69:505–519

► Palacios E, Rodriguez-Jaramillo C, Racotta IS (1999) Comparison of ovary histology between different-sized wild and pond-reared shrimp *Litopenaeus vannamei* (= *Penaeus vannamei*). Invertebr Reprod Dev 35:251–259

Peixoto S, Cavalli RO, D'Incao F, Wasielesky W, Aguado N (2002) Description of reproductive performance and ovarian histology of wild *Farfantepenaeus paulensis* from shallow waters in southern Brazil. Nauplius 10: 149–153

► Peixoto S, Cavalli RO, D'Incao F, Milach ÂM, Wasielesky W (2003) Ovarian maturation of wild *Farfantepenaeus paulensis* in relation to histological and visual changes. Aquacult Res 34:1255–1260

► Peixoto S, Coman G, Arnold S, Crocos P, Preston N (2005) Histological examination of final oocyte maturation and atresia in wild and domesticated *Penaeus monodon* (Fabricius) broodstock. Aquacult Res 36:666–673

Pérez-Farfante I (1969) Western Atlantic shrimps of the genus *Penaeus*. Fish Bull 67:461–591

► Quintero MES, Gracia A (1998) Stages of gonadal development in the spotted pink shrimp *Penaeus brasiliensis*. J Crustac Biol 18:680–685

Santos JL, Severino-Rodrigues E, Vaz-Dos-Santos AM (2008) Estrutura populacional do camarão-branco *Litopenaeus schmitti* nas regiões estuarina e marinha da baixada santista, São Paulo, Brasil. Bol Inst Pesca 34: 375–389

Simões SM (2012) Estrutura da comunidade e biologia reprodutiva dos camarões marinhos (Penaeidea e Caridea), no complexo baía-estuário de Santos e São Vicente/SP, Brasil. Tese de Doutorado, Universidade Estadual Paulista, Botucatu, São Paulo

Sparre P, Venema SC (1997) Introduction to tropical fish stock assessment. Part 1. Manual. FAO Fish Tech Pap No. 306/1. FAO, Rome

Vazzoler AEM (1996) Biologia da reprodução de peixes teleósteos: teoria e prática. Editora Eduem, Maringá

► Wenner AM (1972) Sex ratio as a function of size in marine Crustacea. Am Nat 106:321–350

► Yano I (1988) Oocyte development in the kuruma prawn *Peneaus japonicus*. Mar Biol 99:547–553

17

Thermal adaptations of embryos of six terrestrial hermit crab species

Katsuyuki Hamasaki[1,*], Takahiro Matsuda[1], Ken Takano[1], Mio Sugizaki[1],
Yu Murakami[1], Shigeki Dan[2], Shuichi Kitada[1]

[1]Graduate School of Marine Science and Technology, Tokyo University of Marine Science and Technology, Minato, Konan,
Tokyo 108-8477, Japan
[2]Research Centre for Marine Invertebrate, National Research Institute of Fisheries and Environment of Inland Sea,
Fisheries Research Agency, Momoshima, Onomichi, Hiroshima 722-0061, Japan

ABSTRACT: We evaluated the thermal adaptations of embryos of 6 terrestrial hermit crab species in the family Coenobitidae (genera *Birgus* and *Coenobita*): *B. latro*, *C. brevimanus*, *C. cavipes*, *C. purpureus*, *C. rugosus*, and *C. violascens*. Embryos of each species were cultured in vitro at 6 different temperatures (18 to 34°C) in artificial seawater to avoid air desiccation; the lower threshold temperatures for embryonic development were estimated using heat summation theory equations. Additionally, partial effective cumulative temperatures (> lower threshold temperature) until hatching were determined for ovigerous females of each species cultured in containers. The relationships between the embryonic growth index values (relative area of the embryonic body vs. total embryo surface) and effective cumulative temperatures were expressed using cubic equations. Lower threshold temperature was estimated to be 12.7 to 14.5°C. The effective cumulative temperature and egg incubation period estimates from the appearance of the embryonic body to hatching were higher in *B. latro* and *C. brevimanus*, followed by *C. rugosus*, *C. cavipes*, and *C. violascens*, and lower in *C. purpureus*, suggesting that *C. brevimanus* may retain an ancestral thermal adaptation trait for embryos, as in *B. latro*, which is considered the most ancestral species in the coenobitid phylogeny. Egg size varied among species but did not affect the thermal adaptations of embryos. The lower effective cumulative temperature and shorter egg incubation period may be advantageous to producing broods during the shorter summer breeding season in *C. purpureus*, which has the northern-most geographical distribution.

KEY WORDS: Coconut crab · Land hermit crab · Embryonic development · Lower threshold temperature · Effective cumulative temperature

INTRODUCTION

Terrestrial hermit crabs in the family Coenobitidae diverged from a marine ancestor between 84 and 39 million years ago (Bracken-Grissom et al. 2013). They comprise land hermit crabs of the genus *Coenobita*, with 16 species, as well as the coconut crab *Birgus latro* (Linnaeus, 1767), which is the only species in the genus *Birgus* (McLaughlin et al. 2010). Land hermit crabs carry gastropod shells; however, the shell-carrying behavior of *B. latro* appears only

during the juvenile stage (Harms 1938, Reese 1968, Kadiri-Jan & Chauvet 1998, Hamasaki et al. 2014b).

Terrestrial hermit crabs live in a potentially desiccating environment except during the larval phase. Females extrude their eggs terrestrially, which are subsequently attached to the setae of the pleopods on the left ventral side of the abdomen (de Wilde 1973, Greenaway 2003, Sato & Yoseda 2008, 2009, Drew et al. 2010). The *B. latro* egg mass is afforded little physical protection from the environment by the mother's abdomen and is therefore susceptible to

*Corresponding author: hamak@kaiyodai.ac.jp

desiccating conditions. To minimize dehydration of the egg mass, ovigerous *B. latro* females require a high-humidity shelter (Schiller et al. 1991). Additionally, *B. latro* females maintain hydration of their eggs by using branchial water reserves and body fluids while grooming the egg mass with their fifth pair of pereiopods (Schiller et al. 1991). In contrast, the egg mass of *Coenobita* spp. land hermit crabs is well protected against physical damage and desiccation by the female's shell (de Wilde 1973). Land hermit crabs carry water within their shells (de Wilde 1973, Greenaway 2003), and ovigerous females often moisten their eggs using shell water but never permanently bathe them (de Wilde 1973). Coenobitid females return to the sea to hatch the embryos (Hartnoll 1988, Schiller et al. 1991, Nakasone 2001). Newly hatched larvae develop through planktonic zoeal stages to megalopae in the sea (Hamasaki et al. 2015b), similar to marine hermit crabs. After settlement, coenobitid megalopae recognize and acquire gastropod shells and then migrate onto land (Reese 1968, Reese & Kinzie 1968, Harvey 1992, Brodie 1999, Hamasaki et al. 2011, 2015a).

Coenobitid crabs mainly occur in subtropical and tropical coastal regions and have been divided into 2 groups based on their geographical occurrence patterns: (1) widely distributed species in the Indo-West Pacific, and (2) relatively narrower distributed species in particular regions, such as the western Atlantic, West Africa, west coast of North America, Northern Australia, and the northwestern Pacific (Hartnoll 1988, Nakasone 1988, Harvey 1992, McLaughlin et al. 2007, Wang et al. 2007, Hamasaki et al. 2015b).

Coenobita purpureus Stimpson, 1858 has a limited distribution and mainly occurs in oceanic islands north of 24° N in the Ryukyu Archipelago, Izu Islands, and the Ogasawara (Bonin) Islands, Japan. This crab is also found on the Pacific coasts of mainland Japan (<35° N) (Hamasaki et al. 2016). Additionally, several widely distributed coenobitids, including *Coenobita brevimanus* Dana, 1852, *C. cavipes* Stimpson, 1858, *C. rugosus* H. Milne-Edwards, 1837, *C. violascens* Heller, 1862, and *B. latro*, commonly occur in the southern oceanic islands, Japan (Nakasone 1988, 2001, Hamasaki et al. 2016). Hamasaki et al. (2016) examined the phylogenetic relationships between *C. purpureus* and its widely distributed congeners based on partial 16S mitochondrial rDNA sequences. Their phylogenetic tree demonstrated that *C. purpureus* clustered with *C. rugosus*. They also hypothesized that ancestral *Coenobita* species may have expanded their distribution into the northwestern Pacific region (>24° N) and evolved into *C. purpureus* in the Ryukyu region land masses during the Pliocene.

Temperature is the most important environmental factor affecting biological processes of ectothermic organisms, including behavior, physiology, growth, and survival of all life history stages. Therefore, ectothermic organisms must adapt to the thermal conditions of their habitats (Stillman & Somero 2000, Hall & Thatje 2009), and thus temperature may have acted as a selective force when *C. purpureus* diverged in the Ryukyu Archipelago region. We hypothesize that the thermal adaptations of *C. purpureus* may differ from those of other widely distributed coenobitid species. Our objective in this study was to test this hypothesis and discuss thermal adaptations of coenobitid crabs in the context of evolutionary and ecological traits. We investigated thermal adaptations of embryos in 6 coenobitid crabs (*B. latro*, *C. brevimanus*, *C. cavipes*, *C. purpureus*, *C. rugosus*, and *C. violascens*) by estimating the lower threshold temperatures (LTT, °C) at which embryonic development ceases and the sum of daily temperatures above LTT, i.e. effective cumulative temperatures (ECT, degree-days [°D]), required for embryonic development to hatching. LTT and ECT have been estimated in order to evaluate thermal adaptations of ectothermic organisms including insects (Honěk 1996, Kiritani 2012) and decapod crustaceans (Hamasaki 1996, 2002, 2003, Hamasaki et al. 2009). To obtain the biological data for estimating these 2 life history parameters, we conducted 2 laboratory experiments. In Expt 1, we incubated embryos of each species at different temperatures to estimate the LTT values. In Expt 2, we cultured ovigerous females to hatch their embryos and examined the relationships between growth of the embryos and ECT values. Additionally, we tested whether or not egg size affected the thermal adaptation traits of the embryos of these coenobitids, because it has been suggested that larger eggs show slower developmental rates in closely related decapod crustaceans (Wear 1974), and interspecific variation in egg size is known for coenobitid species (Nakasone 2001).

MATERIALS AND METHODS

Culture of test animals

We captured ovigerous females of *Birgus latro*, *Coenobita brevimanus*, *C. cavipes*, *C. purpureus*, *C. rugosus*, and *C. violascens* during late June to early July of 2005 to 2015 on Hatomajima Island (24° 28′ N, 123° 49′ E), Ishigakijima Island (24° 23–31′ N, 124° 07–18′ E), and/or Miyakojima Island (24° 43–50′ N,

125° 15–21′ E), Okinawa Prefecture, Japan, and cultured them until their embryos hatched into larvae that were used in experiments. In 2005, ovigerous females of *B. latro* were cultured in a laboratory (29 to 30°C) at Yaeyama Station, Seikai National Fisheries Research Institute, Fisheries Research Agency on Ishigakijima Island, and from 2006 to 2015, ovigerous females of all species were cultured in a laboratory (27 to 28°C) at the Tokyo University of Marine Science and Technology (TUMSAT), Tokyo, Japan. Ovigerous females were maintained in tanks equipped with simulated land and sea areas (34 ppt salinity), according to the methods of Hamasaki et al. (2009) and Hamasaki (2011). The photoperiod of the culture room was 13–14 h light:11–10 h dark. Air temperatures were recorded by data loggers every 20 to 60 min during the culture period.

All *Coenobita* spp. in Japan are recognized as a Natural Monument Animal to promote their conservation. Therefore, *Coenobita* spp. were collected and cultured with permission of the Agency for Cultural Affairs, Ministry of Education, Culture, Sports, Science, and Technology of Japan. In addition, *B. latro* is listed as 'vulnerable' in the Red Data Book by the Ministry of the Environment of Japan. Therefore, these crabs were returned to their natural habitat after the culture experiments were completed.

Expt 1: LTT for embryonic development

In general, to investigate the effects of temperature on the embryonic development of decapod crustacean species, individual females are cultured in containers at different temperatures from spawning to hatching (e.g. Wear 1974, Hamasaki et al. 2003, Stevens et al. 2008). However, this methodology requires a large number of test animals. For our experiments, the number of females of each coenobitid species available was limited because they are a protected and/or endangered species, and we have not yet developed the culture methodology to spawn coenobitid females in the laboratory; therefore, we adopted an *in vitro* incubation methodology for embryos separated from ovigerous females. In decapod crustacean aquaculture, artificial incubation of embryos separated from the mother is considered beneficial because the potential negative effects of the mother (e.g. egg loss and death of the mother) can be alleviated (Balasundaram & Pandian 1981, Nakata et al. 2004, Zeng 2007). Several methodologies for *in vitro* incubation of embryos have been employed (e.g. Balasundaram & Pandian 1981, Zeng

2007). We used the simple method developed by Nakata et al. (2004) for artificial incubation of embryos of Japanese crayfish *Cambaroides japonicus* De Haan, 1841. Crayfish embryos were successfully reared individually in wells of cell-culture microplates with sterile water (0.125 to 10 ml) without aeration or water exchange. Nakata et al. (2004) stated that an advantage of this method is that it avoids negative effects caused by the presence of other eggs, such as a decline in water quality caused by dead eggs.

Two different ovigerous females (Females 1 and 2) of each species were used for the embryo incubation experiments conducted during 2012 to 2014 at TUMSAT. In 2012, 2 females each of *B. latro* and *C. rugosus*; in 2013, 1 female of *C. cavipes* and 2 females of *C. purpureus*; and in 2014, 1 female of *C. cavipes* and 2 females each of *C. brevimanus* and *C. violascens* were used. To avoid desiccation of the embryos and to maintain the same humid conditions for the *in vitro* incubation, we cultured the embryos by immersing them in artificial seawater. The concentration of the artificial seawater was regulated at 2 salinity levels (27 and 34 ppt), considering the osmolality of early stage *B. latro* embryos (807 mOsm kg^{-1} H$_2$O ≈ 27 ppt salinity) (Schiller et al. 1991). Early stage embryos were carefully removed from ovigerous females of each species using small forceps before eye pigmentation had appeared on the eye placodes, and were placed in a plastic dish with seawater (27 ppt). Individual embryos were separated carefully with small dissecting needles (see Fig. S1 in the Supplement at www.int-res.com/articles/suppl/b025p083_supp.pdf for photographs of embryos from females of each species) and rinsed with the designated salinity seawater in another dish. One embryo was stocked in each of 10 plastic tubes (volume: 2.8 ml) containing 2 ml of the designated salinity seawater. A total of 20 tubes for the 2 salinity levels were set in each of 6 incubation chambers (14 h light:10 h dark cycle) (MT1-201, Tokyo Rikakikai) controlled at 6 different temperatures between 18 and 34°C (120 embryos for each female). This temperature range was adopted considering the LTT estimate for *B. latro* zoeal development (Hamasaki et al. 2009) and the high temperature that coenobitids can experience in their natural habitats in the Ryukyu Archipelago. The seawater used for culture was not renewed during the incubation period, and evaporation was prevented by placing caps on the tubes. Incubation temperatures were recorded every 30 min with data loggers placed in the incubation chambers; the mean (±SD) incubation temperatures during the culture period are shown in Table S1 in the Supplement for each female.

To estimate the LTT for embryonic development, the number of days required to reach the designated developmental stage should be determined for embryos incubated at different temperatures. In the present study, we counted the number of days from the onset of incubation until eye pigmentation appeared on the eye placodes because of its easy detection by daily observation of embryos under a stereomicroscope. Embryos that appeared cytologically normal were considered survivors, and dead embryos became cloudy. Additionally, to compare the development of *in vitro* and *in situ* embryos, we attempted to determine the number of days required until eye pigmentation appeared on embryos of *C. brevimanus*, *C. cavipes*, and *C. violascens* ovigerous females cultured in 2014.

The relationship between mean temperature (*T*) and the number of days (*D*) required until eye pigmentation appeared on embryos incubated at the respective salinity levels was expressed by the following equation for each female: $D = a/(T - b)$. In this equation, the parameter *a* varies within species; it is larger when younger embryos are used for incubation. The parameter *b* is the so-called 'LTT' for embryonic development which is to be estimated in the present study. The parameters were estimated using a non-linear ordinary least-squares method and evaluated with *t*-tests.

Our experimental design—incubating embryos from 2 different females at different salinity levels—allows us to statistically test the interspecific variation of the LTT estimates. We applied a linear mixed-effects model (Everitt & Hothorn 2009, Zuur et al. 2009) to examine differences in the LTT among the 6 coenobitids using estimates for embryos incubated at 2 different salinities. In this analysis, species was the categorical explanatory variable, and female identity was included as a random effect to account for a potential correlation of repeated measurements (2 salinity levels) within females. Statistical significance of the explanatory variable was evaluated using the 'lme' and 'anova' functions in the 'nlme' package (https://cran.r-project.org/web/packages/nlme/nlme.pdf) in R v.3.1.0 (R Core Team 2014).

Expt 2: ECT for embryonic development

If coenobitid females had been available for controlled spawning in the laboratory, we could have evaluated the ECT for the entire embryonic period. However, we used wild ovigerous females which were cultured to obtain newly hatched larvae in the laboratory. Therefore, we developed an alternative methodology to estimate the ECT for partial embryonic development using ovigerous female coenobitids based on the method of Hamasaki (1996). In his study, the embryonic growth of laboratory-spawned ovigerous females of the swimming crab *Portunus trituberculatus* (Miers, 1876) was examined daily from the onset of appearance of embryonic body to hatching, and development was quantified by an embryonic growth index (EGI) measured as the relative area of the embryonic body versus the total embryo surface in a lateral view of the egg. ECT values were calculated as the sum of the daily effective temperature above LTT (°D) for partial embryonic development from the day of appearance of embryonic body to the day when hatching occurred in the designated females. Hamasaki (1996) showed that the ECT required for partial embryonic development to hatching decreased with increasing EGI values. When the relationship between the EGI and ECT values can be formulated, the intercept of the equation is interpreted as the mean ECT required for partial embryonic development, starting when the embryonic body appeared until hatching. Additionally, the equation between the EGI and ECT values allows us to predict the hatching day of embryos.

B. latro females are highly fecund and, depending on body size, produce ~50 000 to 250 000 eggs brood[−1] (Sato & Yoseda 2008). Moreover, it is easy to sample eggs from ovigerous females because the egg mass is not protected by a gastropod shell. Therefore, we measured the EGI daily and calculated the partial ECT from the time of appearance of embryonic body to hatching in 4 different ovigerous females of *B. latro* in 2005; the number of individual estimates of partial ECT based on 1 observation of EGI and known hatch time for each female ranged from 21 to 24 (total: 91) (see Table S2 in the Supplement). An additional 1 to 16 estimates (total: 89) of the partial ECT were obtained for 20 different *B. latro* females during 2006 to 2014. In contrast to *B. latro*, females of *Coenobita* spp. land hermit crabs are less fecund, and it is difficult to sample eggs from the egg mass because females must be partially pulled out from their shell. To reduce stress to the *Coenobita* ovigerous females from egg sampling, we collected only a small number of estimates of the partial ECT (1 to 8) for each female. The dates of measurements of EGI were arbitrarily determined for each female so as to cover the entire period from the appearance of embryonic body to hatching in the species. The total number of partial ECT estimates ranged from 33 to 82 and were based on 18 to 36 different females of each *Coenobita* species (Table S2).

When we observed embryos, a small cluster of ~10 eggs was removed from the outer margin of the egg mass, placed on a glass slide in seawater (34 ppt) under a stereomicroscope, and lateral views of 5 eggs were photographed with a digital camera (Nikon Digital Sight). Lengths of the major axis (L) and minor axis (S) at the outer egg membrane, and the areas of the embryonic body (EA) and yolk mass (YA) were determined from the digital photographs using an image analyzing system (NIS-Elements software; Nikon). Egg volume was calculated as $\pi LS^2/6$. The EGI was calculated as: $EA/(EA + YA) \times 10^2$. In the present study, the date of egg extrusion was not known but the date of hatching was determined for each female in Expt 2; therefore, the EGI values and egg sizes can be presented in relation to the number of days before embryo hatching.

Hatching occurred at night in all species. The partial ECT from the day when the EGI value was measured to the day when hatching occurred in each female was calculated as the sum of the daily effective temperature above the LTT for embryonic development (°D). The relationships between EGI and ECT were empirically fitted with the following cubic equation: $ECT = aEGI^3 + bEGI^2 + cEGI + d$; where a, b, c, and d are parameters, determined by a linear mixed-effects model using the 'lme' function and evaluated with t-tests in R. In this analysis, female identity was included as a random effect to account for a potential correlation of repeated measurements within females.

Effect of egg size on thermal adaptations of embryos

Intraspecific variation of offspring size occurs in many decapod crustacean species (e.g. Gardner 1997, Moland et al. 2010, Guay et al. 2011), including *B. latro* (Sato & Suzuki 2010) in which larger females produce larger offspring. Although intraspecific variation of offspring size dependent on maternal size has not been reported for other coenobitid species, possible maternal size effects should be considered as a confounding factor in interspecific comparisons of egg size. However, we were not able to conduct such a statistical test because we did not measure female body size in the present study. Alternatively, to examine interspecific variation in egg size while accounting the potential effects of female identity on egg size, we applied a nested ANOVA, in which species was the independent factor and female identity was nested within each of the 6 species.

To evaluate differences in egg diameters and volumes at the same embryo stage among the 6 coenobitids, measurements of eggs within 1 d of hatching, determined for 4 to 8 different females of each species were summarized. Interspecific variation in egg size (volume) was statistically compared using the 'aov' function. Finally, we used the Pearson's product-moment correlation (r) analysis, whose significance was determined by t-test, to evaluate the relationship between mean egg volume of each species and 2 measures of thermal adaptation of embryos, i.e. LTT from Expt 1 and ECT as represented by parameter d in the cubic equation from Expt 2.

RESULTS

Expt 1: LTT for embryonic development

Embryos of all of the coenobitid crab species developed in artificial seawater, although embryo survival tended to be low at lower and higher temperatures in *Coenobita cavipes* Female 1 (see Fig. S2 in the Supplement at www.int-res.com/articles/suppl/b025 p083_supp.pdf). Temperature strongly affected the rate of embryonic development; the number of days from the onset of incubation until eye pigmentation appeared decreased exponentially with increasing incubation temperature (Fig. 1). Among the female broods, the number of days until eye pigmentation appeared on embryos attached to the female pleopods was determined for *C. brevimanus* Female 2 (3 d), *C. cavipes* Female 2 (8 d), and *C. violascens* Females 1 (5 d) and 2 (6 d). These developmental times coincided with those of embryos incubated *in vitro* at the same temperature (~28°C) (*C. brevimanus* Female 2: 3–4 d; *C. cavipes* Female 2: 7–8 d; *C. violascens* Females 1 and 2: 5–6 d and 6–7 d, respectively).

The heat summation theory equation seemed to be a good fit to the relationship between mean temperature and the number of days until eye pigmentation appeared, as shown by the regression curves in Fig. 1 which were drawn using the parameter estimates of equations in Table S3 in the Supplement. The LTT for embryonic development (i.e. parameter b in the equation) was estimated to vary from 12.4 to 15.0°C (Table S3 in the Supplement). LTT values are summarized by box plots in Fig. 2 using estimates for the 27 and 34 ppt seawater incubations from 2 females of each species. The LTT was significantly different among species ($F = 6.018$, df = 5, 6, p = 0.0247). The means ± SD of the LTT estimates for 2 females were: 13.5 ± 0.3, 13.6 ± 0.5, 12.7 ± 0.3, 14.5 ± 0.6, 13.5 ± 0.6,

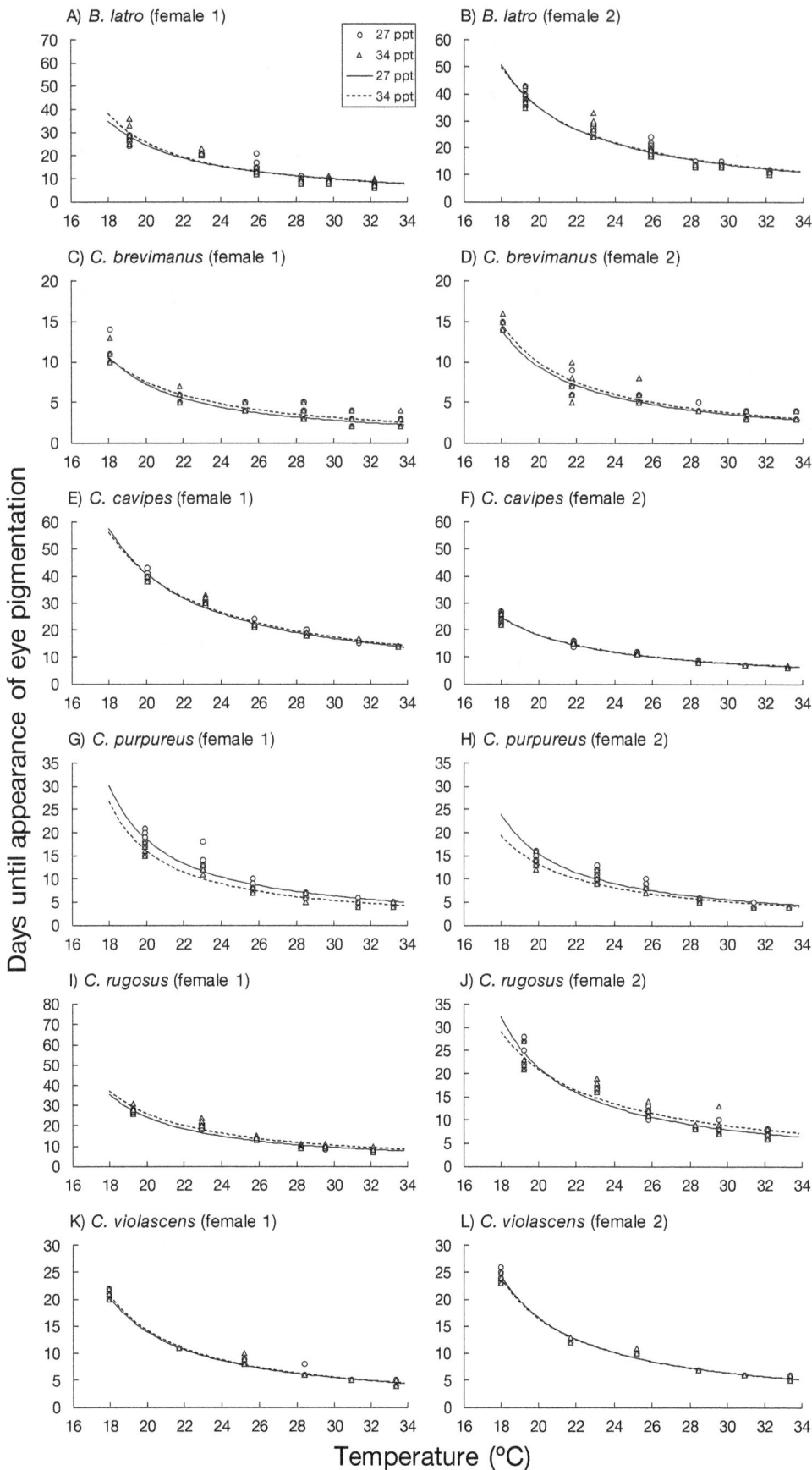

Fig. 1. Relationships between mean incubation temperature and the number of days until eye pigmentation appeared on embryos of the 6 coenobitids incubated *in vitro* in seawater at 2 salinity levels and 6 different temperatures. Experiments were conducted with 2 different females of each species: (A,B) *Birgus latro*, (C,D) *Coenobita brevimanus*, (E,F) *C. cavipes*, (G,H) *C. purpureus*, (I,J) *C. rugosus*, and (K,L) *C. violascens*. Regression curves were drawn for each salinity based on the heat summation theory equations applied to the relationship between temperature and incubation period shown in Table S3 in the Supplement at www.int-res.com/articles/suppl/b025p083_supp.pdf

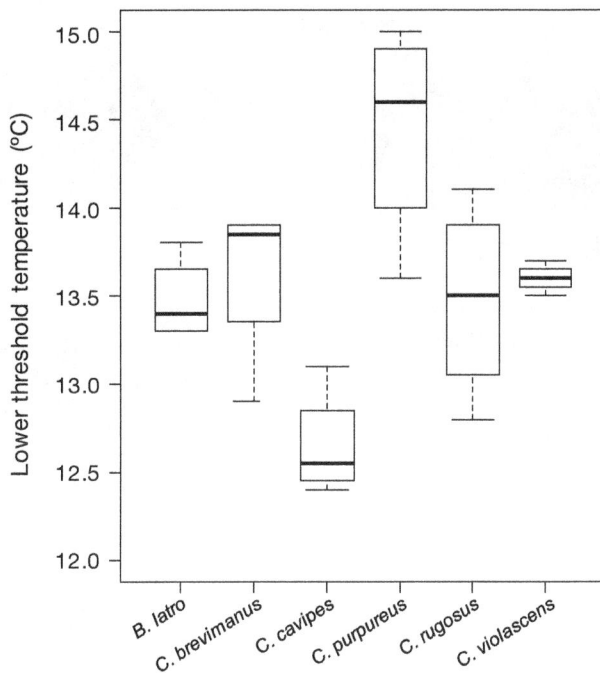

Fig. 2. Lower threshold temperatures for embryonic development in the coenobitids *Birgus latro*, *Coenobita brevimanus*, *C. cavipes*, *C. purpureus*, *C. rugosus*, and *C. violascens* using estimates for 2 different females from each species incubated in 27 and 34 ppt seawater (see Fig. 1)

and 13.6 ± 0.1°C for *Birgus latro*, *Coenobita brevimanus*, *C. cavipes*, *C. purpureus*, *C. rugosus*, and *C. violascens*, respectively, and they were slightly lower in *C. cavipes* and higher in *C. purpureus*.

Expt 2: ECT for embryonic development

Changes in the external morphology of the embryos in lateral view from the day when the embryonic body appeared to the day of hatching are represented in photographs of a *B. latro* female cultured in 2005 shown in Fig. 3. Additionally, changes in the EGI values and egg diameters of 4 *B. latro* females observed daily during incubation in 2005 are shown in Fig. 4, and those for all females are shown in Figs. S3 & S4 in the Supplement. The germinal disk formed as a yolk-free portion (Fig. 3A); that yolk-free portion (i.e. the embryonic body) then increased, while the rudimentary appendages, including antennules, antennae, mandibles, maxillules, maxillae, and mandibles as well as the eye placodes and pleon, developed (Fig. 3B–H). Pigmentation first appeared on the eye placodes (Fig. 3I) when the EGI value was around 25%. The eye-pigmented area then increased with decreasing yolk mass while red/brownish pigmenta-

tion developed on the body (Fig. 3J–X). The rate of increase in the EGI value was slow until it reached about 15%, at which point EGI values increased fairly linearly in all species (Fig. 4 & Fig. S3 in the Supplement). The S length of the eggs changed little during embryonic development, whereas L length clearly increased in all species (Fig. 4 & Fig. S4).

The partial ECT values calculated from the day of measurement until the day of hatching are plotted against the EGI values of each species in Fig. 5. The cubic equations seemed to be a good fit to the relationship between the EGI and ECT values as shown by the regression curves in Fig. 5, which were drawn using the parameter estimates of equations in Table 1. The estimates of parameter d in these equations are the mean ECTs required for partial embryonic development starting when the embryonic body appeared until hatching. The egg incubation periods from the appearance of the embryonic body to hatching were calculated under air temperatures of 24 to 29°C during the main reproductive season of coenobitid species as: partial ECT from the appearance of the embryonic body to hatching / (mean T – LTT). As shown in Fig. S5 in the Supplement, *B. latro*: 23 to 35 d; *C. brevimanus*: 24 to 38 d; *C. cavipes*: 18 to 27 d; *C. purpureus*: 15 to 25 d; *C. rugosus*: 20 to 32 d; and *C. violascens*: 17 to 26 d. The partial ECT and egg incubation period estimates from the appearance of the embryonic body to hatching were higher in *B. latro* and *C. brevimanus*, followed by *C. rugosus*, *C. cavipes*, and *C. violascens*, and lower in *C. purpureus* (Fig. 6 & Fig. S5).

Effect of egg size on thermal adaptations of embryos

The dimensions (lengths of L and S) and volume of eggs within 1 d of hatching for each species are summarized in Table 2. The values of egg size (volume) are also summarized by box plots for each female in Fig. S6 in the Supplement. Egg volume varied significantly among species ($F = 313.177$, df = 5, p < 0.001) and among females (nested in each species) ($F = 26.934$, df = 27, p < 0.001), showing that the factor species accounted for 65% and females for 30% of the total variance (Table S4 in the Supplement). Egg size tended to be larger in *B. latro* and *C. brevimanus*, followed by *C. purpureus*, *C. rugosus*, and *C. violascens*, but was smaller in *C. cavipes*. Mean egg size did not significantly affect the LTT or ECT values for embryonic development in the 6 coenobitid species (LTT: r = 0.374, t = 0.808, df = 4, p = 0.465; ECT: r = 0.501, t = 1.159, df = 4, p = 0.311).

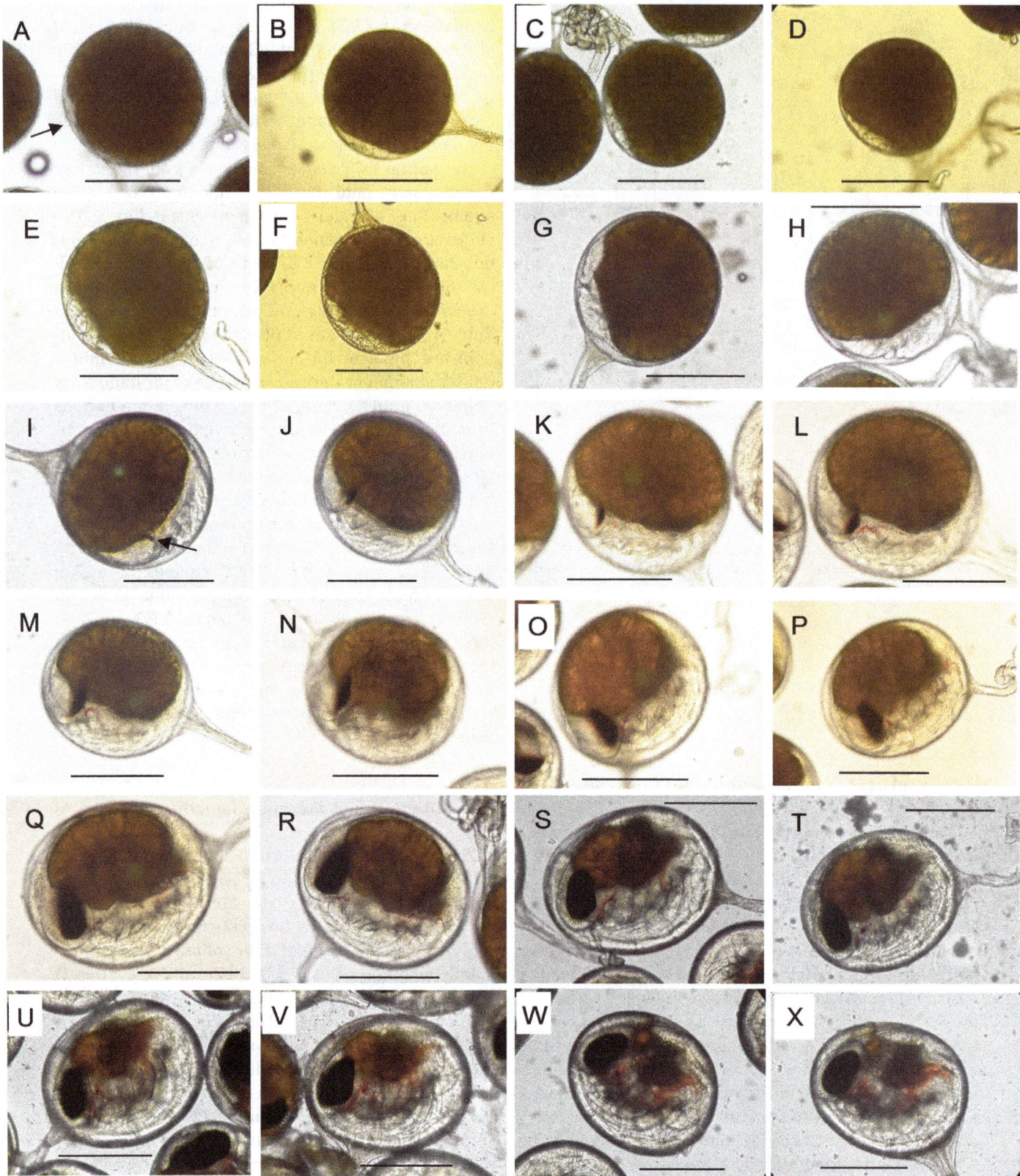

Fig. 3. (A–X) Embryonic development of *Birgus latro* observed daily for 24 d before hatching. Hatching occurred on the night of the last observation of embryos (panel X). Arrows in (A) and (I) show the embryonic body and pigmented eye placode, respectively. Scale bars = 0.5 mm

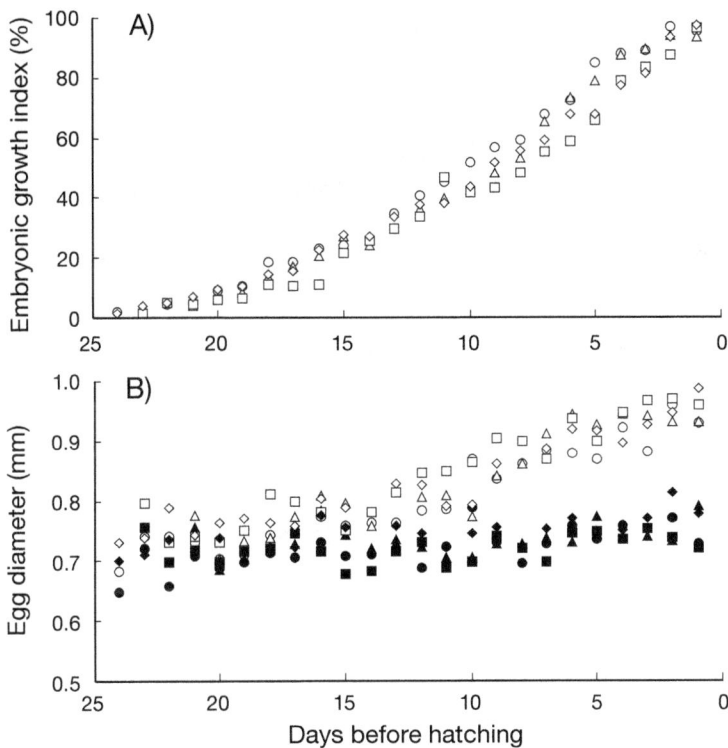

Fig. 4. Changes in (A) the embryonic growth index (EGI) and (B) egg diameter before hatching in 4 *Birgus latro* broods. The EGI represents the relative area of the embryonic body vs. the total embryo surface measured in lateral view of the egg. Egg diameter is shown as major axis (white symbols) and minor axis (black symbols) lengths. Different symbols indicate data from 4 different females

DISCUSSION

In the present study, the eggs of 6 terrestrial coenobitid species were incubated in seawater of different salinities (27 and 34 ppt) to avoid air desiccation during *in vitro* incubation, and they successfully developed. In a previous experiment, we successfully cultured *Birgus latro* embryos *in vitro* (34 ppt seawater) until hatching occurred, and the larvae were reared to metamorphosis into megalopae (Hamasaki et al. unpubl. data). These results indicate that coenobitid embryos can breathe in air and seawater, which has been demonstrated for some subtidal and semi-terrestrial brachyuran crabs by measuring embryo oxygen uptake in air and seawater (Cannicci et al. 2011, Simoni et al. 2011, 2013). We confirmed for *Coenobita brevimanus*, *C. cavipes*, and *C. violascens* that the developmental rates of the embryos were similar during the *in vitro* (with seawater) and *in situ* (with mothers) incubations under identical temperature conditions as in the present study, suggesting that coenobitid embryos may have similar physiological performance in air and seawater.

Changes in external morphology, including size increases in eggs during embryonic development of the 6 coenobitid species observed in the present study, were similar to those documented for many other decapod crustacean species (e.g. Nagao et al. 1999, Yamaguchi 2001, García-Guerrero & Hendrickx 2005, Stevens 2006), including *B. latro* in Vanuatu, South Pacific (Schiller et al. 1991). The increase in egg volume is attributed in part to slow but steady osmotic uptake of water (Davis 1968). Additionally, the egg volume increases sharply because of rapid embryonic growth after the metanaupliar stage in some decapod crustaceans (Furota 1996, Nagao et al. 1999, Hamasaki et al. 2003), and the EGI is about 15% when eggs of *Portunus trituberculatus* initiate a large increase in volume (Hamasaki et al. 2003), which is similar to the embryonic developmental profiles of the coenobitid crabs observed in the present study.

The present study demonstrated that temperature largely affected the rate of embryonic development in 6 coenobitid species, and the duration of embryo incubation decreased exponentially with increasing temperature. While this has been shown in many marine decapod crustaceans (e.g. Wear 1974, Hamasaki 2002, 2003, Hamasaki et al. 2003), this may be the first report demonstrating temperature-dependent embryonic development in terrestrial decapod crustaceans. In this study, we incubated embryos in seawater without aeration or water exchange. Oxygen is less soluble in warmer than in colder water and it could be depleted more rapidly in warmer water due to the higher metabolic rate of the embryo. This may affect embryonic development negatively at higher incubation temperatures. However, apparent negative effects, such as prolonged developmental periods of embryos, were not observed at higher temperatures in any species tested. Therefore, we believe that the culture methodology had little influence on interspecific comparisons of thermal adaptations of embryos in these 6 coenobitid species.

Nakasone (2001) documented that the breeding season based on the chronology of larval release and the occurrence of ovigerous females was similar in *C. cavipes* and *C. purpureus*, extending from late May to late August, but the season was somewhat longer in *C. rugosus* (late May to November) on Okinawajima Island and/or Kudakajima Island (26° 8–10' N, 127° 45–54' E), Okinawa Prefecture, Japan. Sato &

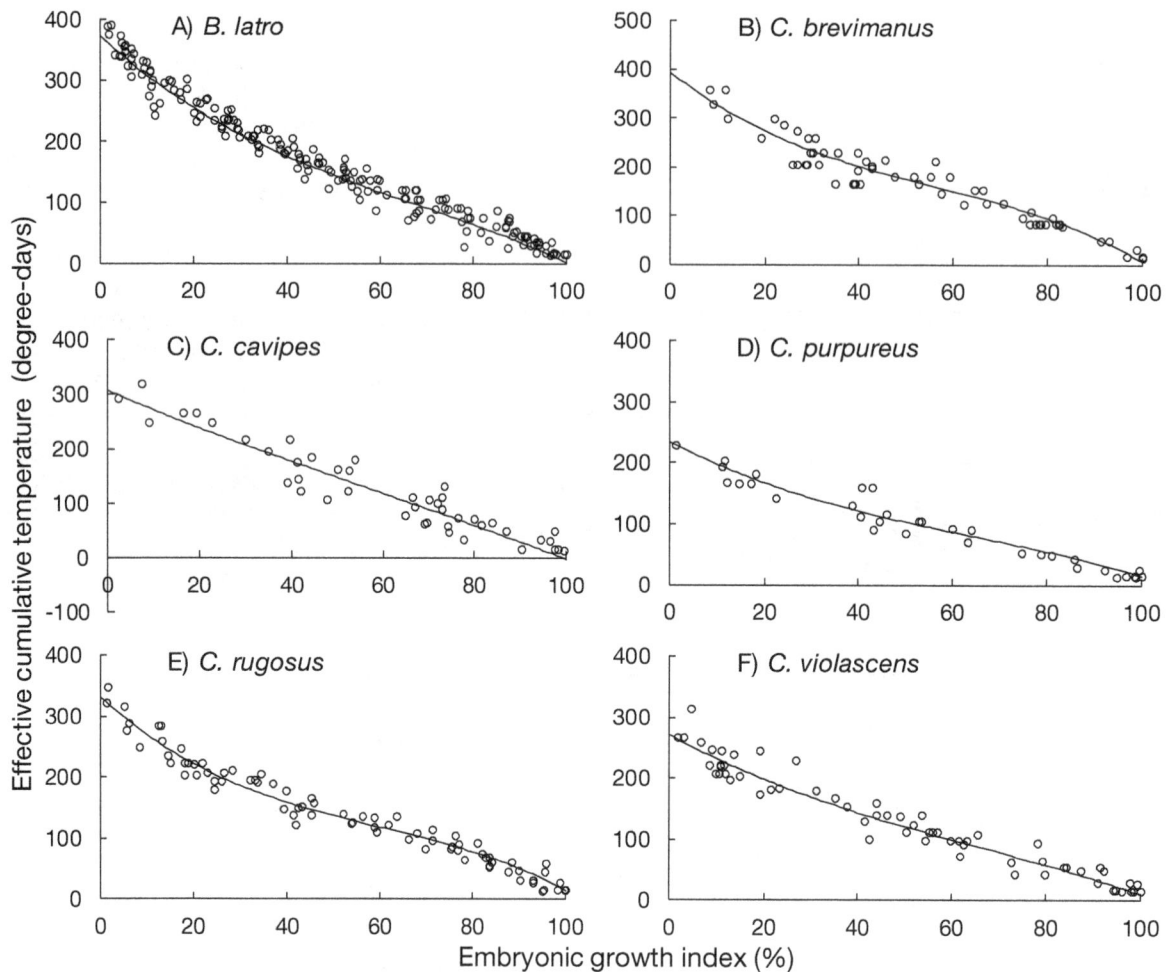

Fig. 5. Relationships between the embryonic growth index (EGI) and partial effective cumulative temperatures (ECT) from the day when EGI values were measured to the day of hatching in 6 coenobitids: (A) *Birgus latro*, (B) *Coenobita brevimanus*, (C) *C. cavipes*, (D) *C. purpureus*, (E) *C. rugosus*, and (F) *C. violascens*. The partial ECT values were calculated as the sum of the daily effective temperature for embryonic development (> lower threshold temperature). Regression curves were drawn from the cubic equations applied to the relationships between the EGI and ECT values as shown in Table 1

Yoseda (2008) reported that ovigerous *B. latro* females were found from early June to late August on Hatomajima Island (24° 28′ N, 123° 49′ E), Okinawa Prefecture. Monthly (10 d interval) mean temperatures at Naha (26° 12′ N, 127° 41′ E) on Okinawajima Island, the capital of Okinawa Prefecture, fluctuate from 24.5°C (late May) to 29.0°C (mid-July) and 28.6°C (late August) to 21.1°C (late November) and those on Ishigakijima Island (24° 20′ N, 124° 9′ E), near Hatomajima Island, range from 26.1°C (early June) to 29.4°C (late August) during the breeding seasons of coenobitids (www.data.jma.go.jp). Therefore, the incubation periods of coenobitid crab embryos vary with these temperature changes during the breeding season.

The heat summation theory equations fit the relationship between mean incubation temperature and the number of days the embryos were incubated until eye pigmentation appeared in the 6 coenobitid species. LTT estimates for embryonic development were slightly lower in *C. cavipes* (12.7°C) but higher in *C. purpureus* (14.5°C), compared with those of *B. latro*, *C. brevimanus*, *C. rugosus*, and *C. violascens* (13.5 to 13.6°C). Hamasaki et al. (2009) estimated a LTT of ~18°C for *B. latro* zoeal development. Thus, the LTT for development was lower for embryos than for zoeal larvae. This is a common finding because the embryonic developmental period precedes that of larvae during the reproductive season. It has been suggested that the LTT represents the degree of lower-temperature adaptation in ectothermic organisms. Honěk (1996) examined variation of the LTT estimates for 355 insect species in relation to their geographic origins, demonstrating that the LTT esti-

Table 1. Parameter estimates for the cubic equations describing the relationship between degree of embryonic development expressed by the embryonic growth index (EGI, %) and the effective cumulative temperature required until hatching (ECT, °D) in 6 coenobitid species in the genera *Birgus* and *Coenobita*: ECT = $aEGI^3 + bEGI^2 + cEGI + d$, where a, b, c, and d are parameters. N: number of observations; SE: standard error

Species	N	Parameter a Estimate (SE)	Parameter a t-value	Parameter a p	Parameter b Estimate (SE)	Parameter b t-value	Parameter b p	Parameter c Estimate (SE)	Parameter c t-value	Parameter c p	Parameter d Estimate (SE)	Parameter d t-value	Parameter d p
B. latro	180	-3.181×10^{-4} (4.517×10^{-5})	−7.042	<0.001	6.514×10^{-2} (6.847×10^{-3})	9.514	<0.001	−7.036 (0.292)	−24.111	<0.001	372.4 (4.6)	81.138	<0.001
C. brevimanus	62	-5.588×10^{-4} (1.112×10^{-4})	−5.025	<0.001	9.376×10^{-2} (1.902×10^{-2})	4.930	<0.001	−7.654 (0.998)	−7.672	<0.001	392.8 (15.9)	24.742	<0.001
C. cavipes	44	-6.266×10^{-5} (8.016×10^{-5})	−0.782	0.434	1.179×10^{-2} (1.349×10^{-2})	0.874	0.382	−3.638 (0.675)	−5.389	<0.001	307.1 (11.3)	27.266	<0.001
C. purpureus	33	-2.074×10^{-4} (9.994×10^{-5})	−2.076	0.038	3.988×10^{-2} (1.661×10^{-2})	2.401	0.016	−4.105 (0.766)	−5.361	<0.001	234.7 (9.8)	24.022	<0.001
C. rugosus	82	-4.822×10^{-4} (5.230×10^{-5})	−9.219	<0.001	8.693×10^{-2} (8.236×10^{-3})	10.555	<0.001	−7.032 (0.371)	−18.948	<0.001	331.5 (5.1)	64.516	<0.001
C. violascens	62	-1.723×10^{-4} (1.173×10^{-4})	−1.469	0.142	3.359×10^{-2} (1.849×10^{-3})	1.816	0.069	−4.242 (0.800)	−5.304	<0.001	270.8 (8.6)	31.568	<0.001

mates were high with little geographic variation in the tropics, and tended to decrease with increasing geographical latitude, although the scatter of data was large in subtropical to temperate regions. Though little is known about the LTT for embryonic development in decapod crustaceans, Hamasaki (2003) compared LTT estimates for embryos of 2 closely related mud crabs, *Scylla serrata* (Forsskål, 1775) (Hamasaki 2003) and *Scylla paramamosain* Estampador, 1949 (Hamasaki 2002). The results showed that LTT was higher in *S. serrata* (15.7°C), which mainly occurs in subtropical and tropical areas, than in *S. paramamosain* (14.0°C), which occupies the northern limit of the mud crab species distribution, extending into warm temperate areas. They suggested that *S. paramamosain* may be adapted to the lower-temperature region. However, the LTT for embryonic development in the present study did not appear to be related to the geographical distributions of the coenobitid species because the highest LTT estimate was determined for *C. purpureus*, which has the most northern geographical distribution, extending to the main island of Japan.

We also successfully estimated partial ECT from the appearance of the embryonic body to hatching in the 6 coenobitids; the estimates were higher in *B. latro* and *C. brevimanus*, followed by *C. rugosus*, *C. cavipes*, and *C. violascens*, and lower in *C. purpureus*. Wear (1974) suggested that interspecific variation in egg size affects embryonic development; larger egg size slows the developmental rate in closely related decapod species. Although significant interspecific differences were found in the egg sizes of the 6 coenobitids, as previously shown for *C. cavipes*, *C. purpureus* and *C. rugosus* by Nakasone (2001), egg size did not affect embryonic development, such as the LTT and ECT values. Hamasaki et al. (2016) examined the phylogenetic relationships among 7 coenobitid species based on 16S rDNA sequences. They found that *B. latro* branched off from the outer diogenids (sister marine hermit crabs) first, followed by *C. brevimanus*, and that the other species comprised 3 clusters: (1) *C. cavipes* and *C. perlatus* H. Milne-Edwards, 1837, (2) *C. violascens*, and (3) *C. rugosus* and *C. purpureus*. Bracken-Grissom et al. (2013) analyzed 2 mitochondrial and 3 nuclear markers as well as 156 morphological characters for 19 families, 77 genera, and 137 species, including *B. latro* and 3 *Coenobita* species, to reconstruct the evolutionary history of Anomura, and demonstrated that *B. latro* first evolved from sister diogenid species. Additionally, Hamasaki et al. (2014a) demonstrated that the larval morphology of *C. brevimanus* is more similar to that of *B. latro*

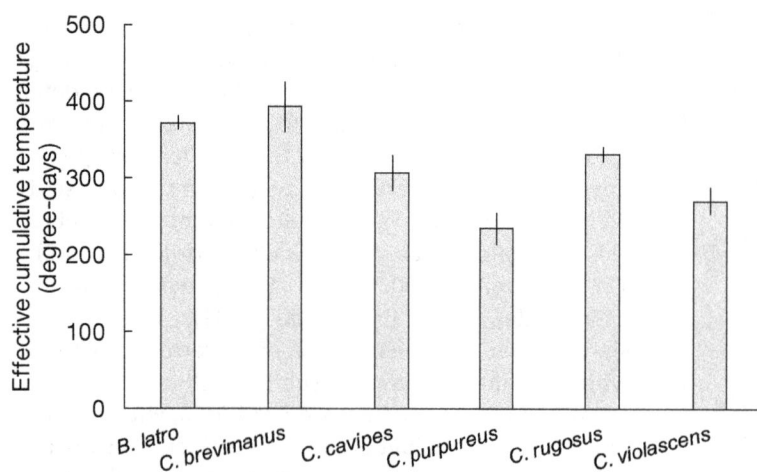

Fig. 6. Estimates of effective cumulative temperature (ECT) required from the appearance of the embryonic body to hatching in 6 coenobitids (*Birgus latro, Coenobita brevimanus, C. cavipes, C. purpureus, C. rugosus,* and *C. violascens*). Estimates are intercept values of the cubic equations applied to the relationship between embryonic growth index (EGI) and ECT shown in Table 1. Vertical bars: 95% CI of the estimates

Table 2. Mean dimensions of eggs within 1 d of hatching in 6 coenobitid species in the genera *Birgus* and *Coenobita*. N: number of different females examined; SD: standard deviation

Species	N	Major axis (mm)	SD	Minor axis (mm)	SD	Volume (mm³)	SD
B. latro	8	0.991	0.050	0.788	0.047	0.320	0.058
C. brevimanus	4	0.919	0.028	0.743	0.017	0.266	0.018
C. cavipes	4	0.783	0.032	0.591	0.038	0.144	0.023
C. purpureus	4	0.890	0.038	0.687	0.028	0.221	0.028
C. rugosus	7	0.877	0.052	0.678	0.047	0.213	0.039
C. violascens	6	0.927	0.060	0.693	0.036	0.235	0.038

rather than other *Coenobita* spp. Thus, *C. brevimanus* is thought to have retained the ancestral traits of *B. latro*, which is considered the most ancestral species in the coenobitid phylogeny. Therefore, the ECT and egg incubation period estimates may be similar in *B. latro* and *C. brevimanus*. *C. purpureus* and other coenobitid crabs are abundant in the northern (>27° 19′–26′ N) and southern (<27° 01′–04′ N) Ryukyu Archipelago regions, respectively (Kagoshima Prefectural Board of Education 1987, Okinawa Prefectural Board of Education 1987, 2006). Hamasaki et al. (2016) suggested that an ancestral *C. purpureus* population may have been isolated in the Ryukyu region land masses during the Pliocene, considering the divergence time between *C. purpureus* and *C. rugosus* and the paleogeography of the Ryukyu Archipelago region. It has been inferred that *B. latro* females produce only one brood per reproductive season (Schiller

et al. 1991, Sato & Yoseda 2008). Meanwhile, Nakasone (2001) reported that some *C. cavipes, C. purpureus,* and *C. rugosus* females produce at least 2 broods during the breeding season because ovigerous females with brooded embryos have mature oocytes simultaneously; moreover, some *C. rugosus* females, a species with a protracted breeding season, may produce a third brood. Therefore, it can be assumed that a reduction in the ECT values and incubation periods for *C. cavipes, C. purpureus,* and *C. rugosus* embryonic development may be an adaptive trait to produce more than one brood by accelerating the embryonic development rate during the reproductive season. A higher ECT value and longer egg incubation periods of *C. rugosus* among these 3 species may also relate to its protracted breeding season. In particular, a lower ECT values and shorter egg incubation periods in *C. purpureus* may be advantageous to produce broods during the shorter summer that occurs in the northern Ryukyu Archipelago region (see Fig. S7 in the Supplement for fluctuations in monthly [10 d interval] mean temperatures at 3 locations from the southern, middle, and northern Ryukyu Archipelago regions).

Little is known about the egg incubation periods of coenobitid crabs. Schiller et al. (1991) attempted to culture *B. latro* females with freshly extruded eggs in enclosures to monitor egg development, but all animals aborted their eggs. Alternatively, Schiller et al. (1991) estimated that the egg incubation period is 25 to 45 d (majority: 27 to 29 d) using radio-tracking and mark–recapture techniques. It has also been estimated that wild *Coenobita clypeatus* (Fabricius, 1787) incubate their eggs for 3 to 4 wk until hatching (de Wilde 1973). The cubic equations used to apply the relationships between the EGI and ECT values for the 6 coenobitid species provided the egg incubation periods from the appearance of the embryonic body to hatching as shown in Fig. S5 in the Supplement. These equations can also be used to estimate the hatching periods by measuring the EGI values and using the *in situ* temperature data from natural and laboratory settings. Further studies that examine the duration of the egg cleavage stage would be required to estimate the entire egg incubation period from egg laying to hatching in coenobitid crabs.

Our results suggest that coenobitid embryos have adapted to thermal conditions within an evolutionary constraint and/or in terms of an ecological trait. The reproductive biology and ecology of other species, such as *C. brevimanus* and *C. violascens*, should be investigated to explain the interspecific variations of the thermal adaptations in coenobitid crabs. Additionally, the methodological approach employed in the present study, i.e. estimates of LTT and ECT required for embryonic development based on *in vitro* and *in situ* culture experiments, would also be effective to examine thermal adaptations of the embryos of threatened crustacean species, of which available numbers of test animals are limited.

Acknowledgements. We thank Okinawa Prefectural Board of Education for permission to culture the land hermit crabs. We are grateful to Dr. Bernard Sainte-Marie and 3 anonymous reviewers for their constructive comments and suggestions, which improved the manuscript substantially. This study was supported, in part, by Grants-in-Aid for Scientific Research C20580198 and B24310171 from the Ministry of Education, Culture, Sports, Science, and Technology of Japan.

LITERATURE CITED

➤ Balasundaram C, Pandian TJ (1981) In vitro culture of *Macrobrachium* eggs. Hydrobiologia 77:203–208

➤ Bracken-Grissom HD, Cannon ME, Cabezas P, Feldmann RM and others (2013) A comprehensive and integrative reconstruction of evolutionary history for Anomura (Crustacea: Decapoda). BMC Evol Biol 13:128

➤ Brodie RJ (1999) Ontogeny of shell-related behaviors and transition to land in the terrestrial hermit crab *Coenobita compressus* H. Milne Edwards. J Exp Mar Biol Ecol 241: 67–80

➤ Cannicci S, Simoni R, Giomi F (2011) Role of the embryo in crab territorialisation: an ontogenetic approach. Mar Ecol Prog Ser 430:121–131

Davis CC (1968) Mechanisms of hatching in aquatic invertebrate eggs. Oceanogr Mar Biol Annu Rev 6:325–376

de Wilde PAWJ (1973) On the ecology of *Coenobita clypeatus* in Curaçao with reference to reproduction, water economy and osmoregulation in terrestrial hermit crabs. Stud Fauna Curaçao Other Caribbean Isl. 144:1–138

➤ Drew MM, Harzsch S, Stensmyr M, Erland S, Hansson BS (2010) A review of the biology and ecology of the robber crab, *Birgus latro* (Linnaeus, 1767) (Anomura: Coenobitidae). Zool Anz 249:45–67

Everitt BS, Hothorn T (2009) A handbook of statistical analyses using R, 2nd edn. CRC Press, Boca Raton, FL

➤ Furota T (1996) Life cycle studies on the introduced spider crab *Pyromaia tuberculate* (Lockington) (Brachyura: Majidae). I. Egg and larval stages. J Crustac Biol 16:71–76

➤ García-Guerrero MU, Hendrickx ME (2005) Embryology of decapod crustaceans, II: Gross embryonic development of *Petrolisthes robsonae* Glassell, 1945 and *Petrolisthes armatus* (Gibbes, 1850) (Decapoda, Anomura, Porcellanidae). Crustaceana 78:1089–1097

➤ Gardner C (1997) Effect of size on reproductive output of giant crabs *Pseudocarcinus gigas* (Lamarck): Oziidae. Mar Freshw Res 48:581–587

Greenaway P (2003) Terrestrial adaptations in the Anomura (Crustacea: Decapoda). Mem Mus Vic 60:13–26

➤ Guay C, Sainte-Marie B, Brêthes JC (2011) Strong maternal effects and extreme heterogeneity of progeny development in the caridean shrimp *Sclerocrangon boreas* (Crangonidae). Mar Biol 158:2835–2845

➤ Hall S, Thatje S (2009) Global bottlenecks in the distribution of marine Crustacea: temperature constraints in the family Lithodidae. J Biogeogr 36:2125–2135

Hamasaki K (1996) Study on the reproduction and development of the swimming crab, *Portunus trituberculatus*. Spec Sci Rep Japan Sea-Farming Assoc 8:1–124 (in Japanese with English Abstract)

Hamasaki K (2002) Effects of temperature on the survival, spawning and egg incubation period of overwintering mud crab broodstock, *Scylla paramamosain* (Brachyura: Portunidae). Aquacult Sci 50:301–308

➤ Hamasaki K (2003) Effects of temperature on the egg incubation period, survival and developmental period of larvae of the mud crab *Scylla serrata* (Forskål) (Brachyura: Portunidae) reared in the laboratory. Aquaculture 219: 561–572

Hamasaki K (2011) Early life history of coconut crabs inferred from culture experiments. Cancer 20:73–77 (in Japanese)

Hamasaki K, Fukunaga K, Maruyama K (2003) Egg development and incubation period of the swimming crab *Portunus trituberculatus* (Decapoda: Portunidae) reared in the laboratory. Crustac Res 32:45–54

➤ Hamasaki K, Sugizaki M, Dan S, Kitada S (2009) Effect of temperature on survival and developmental period of coconut crab (*Birgus latro*) larvae reared in the laboratory. Aquaculture 292:259–263

➤ Hamasaki K, Sugizaki M, Sugimoto A, Murakami Y, Kitada S (2011) Emigration behaviour during sea-to-land transition of the coconut crab *Birgus latro*: effects of gastropod shells, substrata, shelters and humidity. J Exp Mar Biol Ecol 403:81–89

➤ Hamasaki K, Ishiyama N, Yamashita S, Kitada S (2014a) Survival and growth of juveniles of the coconut crab *Birgus latro* under laboratory conditions: implications for mass production of juveniles. J Crustac Biol 34:309–318

➤ Hamasaki K, Kato S, Hatta S, Murakami Y, Dan S, Kitada S (2014b) Larval development and emigration behaviour during sea-to-land transition of the land hermit crab *Coenobita brevimanus* Dana, 1852 (Crustacea: Decapoda: Anomura: Coenobitidae) under laboratory conditions. J Nat Hist 48:1061–1084

➤ Hamasaki K, Hatta S, Ishikawa T, Yamashita S, Dan S, Kitada S (2015a) Emigration behaviour and moulting during the sea-to-land transition of terrestrial hermit crabs under laboratory conditions. Invertebr Biol 134: 318–331

➤ Hamasaki K, Kato S, Murakami Y, Dan S, Kitada S (2015b) Larval growth, development and duration in terrestrial hermit crabs. Sex Early Dev Aquat Org 1:93–107

➤ Hamasaki K, Iizuka C, Sanda T, Imai H, Kitada S (2016) Phylogeny and phylogeography of the land hermit crab *Coenobita purpureus* (Decapoda: Anomura: Coenobitidae) in the Northwestern Pacific region. Mar Ecol 37, doi:10.1111/maec. 12369

Harms JW (1938) Lebenslauf und Stammesgeschichte des

Birgus latro L., von der Weihnachtsinsel. Jena Zeitschr Naturwiss 71:1–34

Hartnoll RG (1988) Evolution, systematic, and geographical distribution. In: Burggren WW, McMahon BR (eds) Biology of the land crabs. Cambridge University Press, Cambridge, p 6–54

▶ Harvey AW (1992) Abbreviated larval development in the Australian terrestrial hermit crab *Coenobita variabilis* McCulloch (Anomura: Coenobitidae). J Crustac Biol 12:196–209

Honěk A (1996) Geographical variation in thermal requirements for insect development. Eur J Entomol 93:303–312

Kadiri-Jan T, Chauvet C (1998) Distribution of the juvenile coconut crab, *Birgus latro* (L.), on the island of Lifou, New Caledonia. Ecoscience 5:275–278

Kagoshima Prefectural Board of Education (1987) A report of the emergence survey on the distribution and ecology of land hermit crabs in Kagoshima Prefecture. Cultural Assets Division, Kagoshima Prefectural Board of Education, Kagoshima (in Japanese)

Kiritani K (2012) The low development threshold temperature and the thermal constant in insects and mites in Japan, 2nd edn. Bull Natl Inst Agro Environ Sci 31:1–74 (in Japanese with English Abstract)

McLaughlin PA, Rahayu DL, Komai T, Chan TY (2007) A catalog of the hermit crabs (Paguroidea) of Taiwan. National Taiwan Ocean University, Keelung

McLaughlin PA, Komai T, Lemaitre R, Rahayu DL (2010) Annotated checklist of anomuran decapod crustaceans of the world (exclusive of the Kiwaoidea and families Chirostylidae and Galatheidae of the Galatheoidea) Part I – Lithodoidea, Lomisoidea and Paguroidea. Raffles Bull Zool Suppl 23:5–107

▶ Moland E, Olsen EM, Stenseth NC (2010) Maternal influences on offspring size variation and viability in wild European lobster *Homarus gammarus*. Mar Ecol Prog Ser 400:165–173

▶ Nagao J, Munehara H, Shimazaki K (1999) Embryonic development of the hair crab *Erimacrus isenbeckii*. J Crustac Biol 19:77–83

Nakasone Y (1988) Land hermit crabs from the Ryukyus, Japan, with a description of a new species from the Philippines (Crustacea, Decapoda, Coenobitidae). Zoolog Sci 5:165–178

▶ Nakasone Y (2001) Reproductive biology of three land hermit crabs (Decapoda: Anomura: Coenobitidae) in Okinawa, Japan. Pac Sci 55:157–169

▶ Nakata K, Matsubara H, Goshima S (2004) Artificial incubation of Japanese crayfish (*Cambaroides japonicus*) eggs by using a simple, easy method with a microplate. Aquaculture 230:273–279

Okinawa Prefectural Board of Education (1987) AMAN: a report on the distribution and ecology of land hermit crabs in Okinawa Prefecture. Ryokurindo-Shoten, Ginowan (in Japanese)

Okinawa Prefectural Board of Education (2006) A report on the distribution and ecology of land hermit crabs in Okinawa Prefecture – II. Cultural Assets Division, Okinawa Prefectural Board of Education, Naha (in Japanese)

R Core Team (2014) R: a language and environment for statistical computing. R Foundation for Statistical Computing, Vienna

▶ Reese ES (1968) Shell use: an adaptation for emigration from the sea by the coconut crab. Science 161:385–386

Reese ES, Kinzie RA III (1968) The larval development of the coconut or robber crab *Birgus latro* (L.) in the laboratory (Anomura, Paguridea). Crustaceana Suppl 2:117–144

▶ Sato T, Suzuki N (2010) Female size as a determinant of larval size, weight, and survival period in the coconut crab, *Birgus latro*. J Crustac Biol 30:624–628

▶ Sato T, Yoseda K (2008) Reproductive season and female maturity size of coconut crab *Birgus latro* on Hatoma Island, southern Japan. Fish Sci 74:1277–1282

▶ Sato T, Yoseda K (2009) Egg extrusion site of coconut crab *Birgus latro*: direct observation of terrestrial egg extrusion. Mar Biodivers Rec 2:e37

Schiller CD, Fielder R, Brown IW, Obed A (1991) Reproduction, early life-history and recruitment. In: Brown IW, Fielder DR (eds) The coconut crab: aspects of *Birgus latro* biology and ecology in Vanuatu. ACIAR Monograph, No. 8. Australian Centre for International Agricultural Research, Canberra, p 13–35

▶ Simoni R, Cannicci S, Anger K, Pörtner HO, Giomi F (2011) Do amphibious crabs have amphibious eggs? A case study of *Armases miersii*. J Exp Mar Biol Ecol 409: 107–113

▶ Simoni R, Giomi F, Spigoli D, Pörtner HO, Cannicci S (2013) Adaptations to semi-terrestrial life in embryos of East African mangrove crabs: a comparative approach. Mar Biol 160:2483–2492

▶ Stevens BG (2006) Embryo development and morphometry in the blue king crab *Paralithodes platypus* studied by using image and cluster analysis. J Shellfish Res 25: 569–576

▶ Stevens BG, Swiney KM, Buck L (2008) Thermal effects on embryonic development and hatching for blue king crab *Paralithodes platypus* (Brandt, 1850) held in the laboratory, and a method for predicting dates of hatching. J Shellfish Res 27:1255–1263

▶ Stillman JH, Somero GN (2000) A comparative analysis of the upper thermal tolerance limits of eastern Pacific porcelain crabs, genus *Petrolisthes*: influences of latitude, vertical zonation, acclimation, and phylogeny. Physiol Biochem Zool 73:200–208

▶ Wang FL, Hsieh HL, Chen CP (2007) Larval growth of the coconut crab *Birgus latro* with a discussion on the development mode of terrestrial hermit crabs. J Crustac Biol 27:616–625

▶ Wear RG (1974) Inculcation in British decapod Crustacea, and the effects of temperature on the rate and success of embryonic development. J Mar Biol Assoc UK 54: 745–762

▶ Yamaguchi T (2001) Incubation of eggs and embryonic development of the fiddler crab, *Uca lactea* (Decapoda, Brachyura, Ocypodidae). Crustaceana 74:449–458

▶ Zeng C (2007) Induced out-of-season spawning of the mud crab, *Scylla paramamosain* (Estampador) and effects of temperature on embryo development. Aquacult Res 38: 1478–1485

Zuur AF, Ieno EN, Walker NJ, Saveliev AA, Smith GM (2009) Mixed effects models and extensions in ecology with R. Springer, New York, NY

18

Micro CTD data logger reveals short-term excursions of Japanese sea bass from seawater to freshwater

Naoyuki Miyata[1,*], Tomohiko Mori[1], Masaaki Kagehira[2], Nobuyuki Miyazaki[3], Michihiko Suzuki[4], Katsufumi Sato[1]

[1]Atmosphere and Ocean Research Institute, The University of Tokyo, 5-1-5 Kashiwanoha, Kashiwa, Chiba, 277-8564, Japan
[2]Oita Prefecture Southern Region Bureau Rural Community Promotion Department, Saiki, Oita, 876-0813, Japan
[3]Japan Marine Science Foundation, Taito, Tokyo, 110-0008, Japan
[4]Little Leonardo, Bunkyo, Tokyo, 113-0021, Japan

ABSTRACT: We conducted calibration tests and field deployments of a newly developed micro CTD data logger on Japanese sea bass *Lateolabrax japonicus* to measure fine-scale movement and reconstruct micro-salinity profiles of estuarine habitat. In June, July and November 2013 and 2014, Japanese sea bass were caught in the Ohno (33° 12′ N, 131° 36′ E) and Oita (33° 16′ N, 131° 41′ E) Rivers in Oita Prefecture, Japan. In seawater tanks, the measurement drift for salinity was negligible over a 2 wk period (n = 8). Moreover, in the estuary, the values of standard seawater measured at pre- and post-deployment of sensors on fish were consistent to the first decimal place; therefore, sensor drift was negligible (n = 4, duration: 53–129 h). *In situ* measurements of depth, temperature, and salinity (conductivity) successfully revealed the microenvironment, and showed that Japanese sea bass often ascended from deep seawater to shallow freshwater at night. In all fish, over 99% of these excursions were completed within 10 min. The maximum duration of these excursions was about 6 h. This might be an effective strategy for temporal utilization of hypo-osmotic environments for foraging without physiological and energetic costs. This is the first report to clearly validate the accuracy of salinity measurement without sensor drift in free-ranging fish. The device and methods presented here can be applied to other euryhaline fish.

KEY WORDS: Bio-logging · Estuary · Euryhaline fish · Osmoregulation · Salinity

INTRODUCTION

Estuaries rank among the most biologically productive ecosystems on Earth (Pinet 2011). Hence, estuarine fishes are typically abundant but represent relatively few species, largely because of the dynamic nature of their environment (Pinet 2011). One of the most variable characteristics of estuarine water is salinity, which is controlled by river inflow, tides, rainfall etc. Thus, salinity is an important variable in investigations of the behavioral ecology of estuarine fish. Salinity may act directly or indirectly in controlling the distribution of estuarine-dependent fish (Bulger et al. 1993, Lankford & Targett 1994). Spatial behaviors may also be affected by the distribution of predators and prey within estuaries, both of which are affected by salinity (Boesch 1977, Weinstein et al. 1980, Currin et al. 1984, Lankford & Targett 1994). Water salinity directly affects physiological processes such as osmoregulation, respiration, metabolism, and growth (Plaut 1999, Bœuf & Payan 2001, Tseng & Hwang 2008, Takei et al. 2014). Physiologically, estuarine-dependent fish are expected to be able to cope with the large salinity changes that occur in estuaries. It is widely accepted that catadromous, anadromous, amphidromous, and other estuarine species are euryhaline organisms (Zydlewski & Wilkie 2012). Members of the diadromous group have the ability to switch their osmoregulatory mechanisms as they migrate from freshwater to seawater or vice versa (Zydlewski & Wilkie 2012); however, these mechanisms come with physiological and energetic costs. Detailed salinity information is required to reveal the habitat use and the reaction to salinity variation of estuarine fish on tidal and daily time scales.

In recent years, telemetry and bio-logging devices have produced large amounts of data on marine animals (Fedak 2004) at fine temporal and spatial resolutions. While depth and temperature sensors are relatively stable, methods used to measure conductivity in order to calculate salinity have presented great challenges (Fedak 2004). In several research projects, data loggers deployed on marine mammals were successfully used to collect salinity data measurements (Lydersen et al. 2002, Hooker & Boyd 2003, Biuw et al. 2007, Boehme et al. 2008a,b, Charrassin et al. 2008, Roquet et al. 2009, Meredith et al. 2011). However, the large size of the loggers (545 ± 5 g in air; see Boehme et al. 2009 for detailed specifications) means they are only suitable for use with relatively large organisms. One small conductivity logger using a 2-AC bipolar sensor is commercially available (DST CTD, Star-Oddi) and has been used to investigate the migration patterns of Atlantic salmon *Salmo salar* L. and sea trout *Salmo trutta* L. (Sturlaugsson & Sigmar 1997). Bipolar sensors can be affected by erosion and fouling because the metal electrodes are directly exposed to the water. Furthermore, measurements are affected by the proximity of other objects (near-field effect), since the sensor measures conductivity by generating an electromagnetic field around the electrodes. To date, there have been no studies that test the accuracy of salinity measurements using 2-AC bipolar sensors, although these may be affected by drift over time and the near-field effect.

In this study, we conducted calibration tests and initial field deployments using a new micro-salinity data logger (ORI400-DTC, Little Leonardo) on Japanese sea bass *Lateolabrax japonicus* to measure fine-scale sea bass movement and reconstruct the micro-salinity of their estuarine habitat. Japanese sea bass are euryhaline marine fish inhabiting coastal areas and estuaries in Japan, occasionally appearing in freshwater (Ochiai & Tanaka 1998). The migration patterns of larval and juvenile Japanese sea bass have previously been investigated and amphidromous characteristics have been found in some popu-

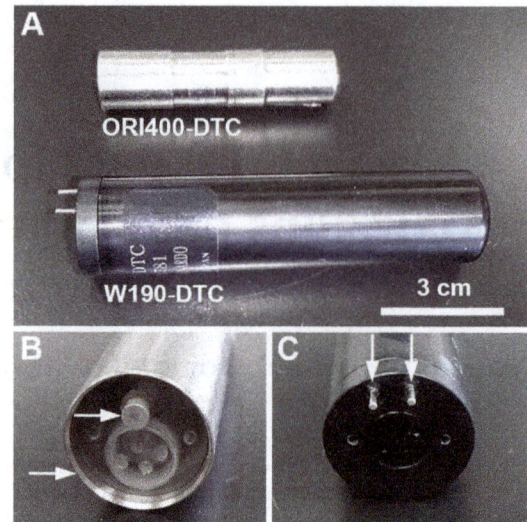

Fig. 1. Two types of salinity data logger and design of conductivity sensor. (A) Comparison of sizes. (B) Sensor design of ORI400-DTC, which measures the current between the titanium pole and cylinder body. (C) Sensor design of the W190-DTC, which uses 2 stainless steel poles. This is used as control for comparison with the improved model (ORI400-DTC). Arrows indicate electrodes

lations (Tanaka 1997), with fish immigrating to freshwater at the larva–juvenile transformation phase or early juvenile stage. There are few reports of adult fish migration; however, catches have been reported in the upper reaches of inland rivers (Shoji et al. 2002). According to local anglers, some populations of adult fish seasonally travel up rivers, following baitfish.

MATERIALS AND METHODS

Data loggers

We used 2 types of data loggers to record depth, temperature, and conductivity experienced by Japanese sea bass. (ORI400-DTC and W190-DTC, Little Leonardo: 14 mm diameter, 58 mm length, 19 g in air and 19 mm diameter, 90 mm length, 45g in air, respectively; Fig. 1). The 2-AC bipolar sensor was applied to these data loggers. ORI400-DTC measures conductivity between the cylindrical body and a pole made of titanium, and W190-DTC measures conductivity between 2 stainless steel poles. We used an older model (W190-DTC) as a control to compare measurement drift of salinity with that of the improved model, ORI400-DTC. Therefore, only the specifications of ORI400-DTC are shown (Table 1). Measured conductivity was converted to salinity using software (Salin-

Table 1. Specifications of the ORI400-DTC data logger. Time constant represents the time required to reach 63.2% of its final value following a sudden water condition change from one stable condition to another. Values in parentheses are the ranges in which accuracy was guaranteed

	Depth (m)	Temperature (°C)	Salinity
Range	0–400	−10 to 50	0–40
Resolution	0.2	0.1	0.5
Accuracy	±2 (0–100)	±0.5	±1.5 (0–35)
	±2% (100–400)		
Time constant (s)	<1	<15	–

ity Logger Tools, Little Leonardo), according to the Practical Salinity Scale of 1978 (UNESCO 1981). The duration of recording ranged from 40 to 300 d, depending on the sampling interval of the data logger (1–60 s).

Calibration experiments

The day before the calibration experiments took place, the data loggers were activated and the surfaces of the electrodes were wiped with 99% ethanol and then brushed with a cotton swab soaked in flux (SUSSOL-F, Hakko) to remove any oxide coating. The data loggers were then soaked in seawater for over 24 h at room temperature to make the oxide film stable underwater.

Accuracy tests of the ORI400-DTC were conducted in plastic (polyethylene) buckets containing water of various salinities (0, 7.0, 13.9, 21.0, 27.6, and 34.8). The salinity of water from each bucket was measured using the ORI400-DTC and a conventional CTD profiler (Compact-CTD Lite, JFE Advantech, accuracy ± 0.03). Using the ORI-400DTC, the seawater in the buckets was tested for a near-field effect at various distances (0. 0.5, 1.0, 2.0, 3.0, 4.0, and 5.0 cm) from the polyethylene. According to Hooker & Boyd (2003), polyethylene negatively affected salinity measurements more than wood or nylon carpet. Different sensor angles were also tested. To check measurement drift, the data loggers, which had been prepared as described above, were soaked in aerated seawater at 15°C. The salinity of standard seawater, (with a constant salinity of 34) stored in sealed containers, was measured every other day over 2 wk. Drift was represented by the difference between measurements made using the data loggers (ORI400-DTC and W190-DTC) and those using the conventional CTD profiler. After each test, measurements of data loggers were uploaded to a PC, and the data were calibrated using the software.

Field experiments

The field study procedures were approved by the ethics committee at the University of Tokyo (Institutional Animal Care and Use Committee Protocol P12-14, P13-11).

In the field experiments, only the new smaller data logger (ORI400-DTC) was tested. Before the data loggers were attached to the fish, we measured standard seawater (salinity: 34.0, measured using a conventional CTD) as a reference for salinity calibration. In June, July and November 2013 and 2014, a total of 9 Japanese sea bass *Lateolabrax japonicus* were caught by lure fishing in the Ohno and Oita Rivers in Oita Prefecture (33° 12′ N, 131° 36′ E and 33° 16′ N, 131° 41′ E, respectively), Japan. The fish were first anaesthetized using 1/5000 2-phenoxyethanol. Using a needle, 2 small holes were made in the dorsal musculature in front of their dorsal fin. Two plastic cable ties were then passed through the holes to fix a plastic mesh in place. A tag consisting of the data logger (ORI400-DTC), the VHF transmitter (MM130B, Advanced Telemetry Systems; 16 mm diameter, 60 mm length, and 20 g in air), a time-release mechanism (RT-4, Little Leonardo; 16 mm diameter, 25 mm length, and 16 g in air), and a float were attached to the plastic mesh. After recovery from anesthesia, fish were released in the same location in which they were caught. Tags automatically detached from each fish using a time-scheduled release system. After detachment, the floating tags were located using the VHF signals they transmitted and were retrieved by a boat.

To check the validity of the salinity and temperature profiles estimated from the fish-borne data loggers, we performed a series of relatively short tagging experiments in 2013 (deployment duration: 1–2 h; n = 5 fish, total length: 480–825 mm, body mass: 0.83–3.97 kg). After the fish had been released, the water column was measured by conventional CTD at the point of release. To investigate fish behavior and habitat use, we conducted longer deployments in November 2014 (deployment duration: 53–158 h; n = 4 fish, total length: 488–775 mm, body mass: 0.88–4.94 kg). After retrieval of the data loggers, we again measured the same standard seawater used for the pre-deployment salinity measurement to assess instrument drift, which can be caused by sensor erosion and fouling in the natural environment. Exposure to air was kept to under 1 h to preserve sensor surface conditions.

Data analysis

To visualize the thermal and salinity structure surrounding the tagged fish, (1) a temperature-at-depth and salinity-at-depth matrix with 1 h time bins was constructed by calculating the average temperature and salinity at each depth value within each time bin (fish profiles), and (2) the isotherm and isohaline were constructed based on the resulting temperature-depth and salinity-depth field (Watson & Merriam 1992). In the short tagging experiment, the data for the first hour of fish profiles were compared with those of the CTD profiler. In the longer experiment, original depth data of where the fish swam were averaged with 1-h time bins, and the result was plotted in the salinity structure. Temperature and salinity data obtained by fish-borne data loggers were analyzed using Igor Pro (Wave metrics). To estimate the impact of ambient water salinity on fish physiology, we used a threshold salinity of 12, which has roughly the same osmolality as the body fluid of teleost fish, to distinguish between different behavioral responses.

RESULTS

Calibration of data loggers

Salinity measurements taken by the ORI400-DTC in all 6 water tanks (salinity: 0, 7.0, 13.9, 21.0, 27.6, and 34.8) were accurate to within 1.5 of measurements by conventional CTD profilers, and the precision of the data loggers was ± 0.3. As expected, the near-field effect on the conductivity sensor (a titanium pole) was pronounced when the obstacle (polyethylene) was immediately adjacent to the titanium pole of the loggers, and the effects were negligible when the distance between obstacle and titanium pole was >1 cm (Fig. 2A,B). The angle of the titanium pole relative to the obstacle also affected salinity measurements (Fig. 2C). These results suggest that when the data loggers (ORI400-DTC) are attached to animals, the titanium pole should be kept at sufficient distance (we suggest at least 1 cm) from the fish's body. If other instruments are deployed on the same fish, and these instruments have to be fixed to each other, salinity calibration must be done in a fixed setting without any obstacles within a 1 cm radius of the titanium pole. In our field study, we used a VHF transmitter, float, and a time-release mechanism to retrieve the data loggers; we fixed them to each other and performed calibration under the same conditions (Fig. 3). Accuracy and precision

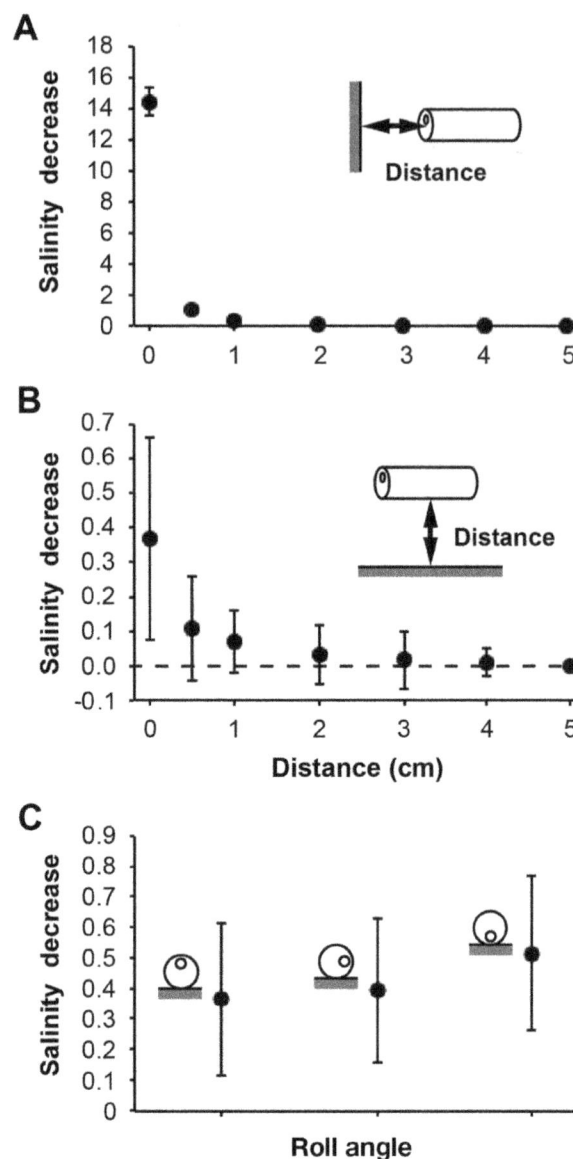

Fig. 2. Near-field effect of the obstacle (polyethylene) on the conductivity sensor (ORI400-DTC). (A) Longitudinal distance between the logger and the obstacle, (B) distance in dorsoventral axis direction, and (C) roll angle of sensor relative to obstacle. Data are means ± SD (n = 8)

did not differ from the values mentioned at the beginning of this paragraph.

Fig. 4 shows the drift of salinity measurements recorded by 2 types of data loggers soaked in aerated seawater for 2 wk. The measurement drift for the ORI400-DTC salinity readings, which measured the current between the titanium pole and cylinder body, was ≤1 throughout the experiment. This stability was better than that for W190-DTC, which used 2 stainless steel poles. In both types, significant variation of salinity values occurred within the first 2 d. These

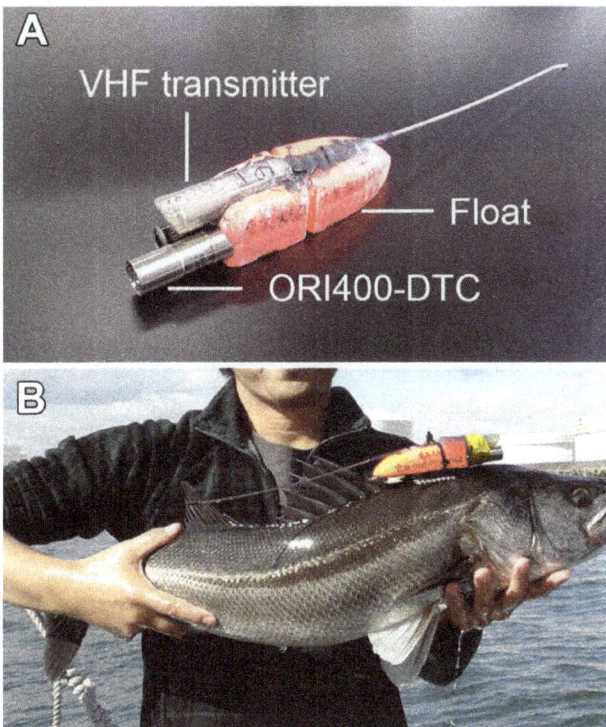

Fig. 3. (A) A tag consisting of the data logger (ORI400-DTC), VHF transmitter and float. (B) The tag was attached in front of the dorsal fin of Japanese fish bass.

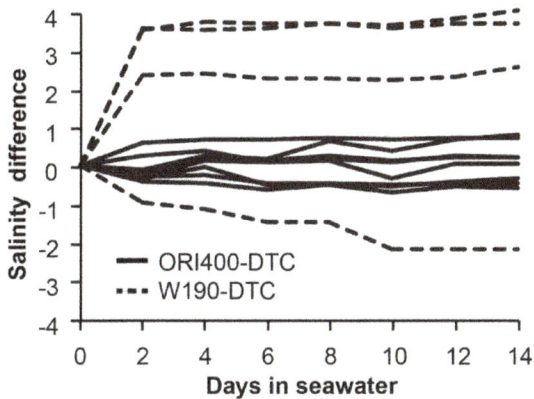

Fig. 4. Salinity measurement drift in seawater for the 2 data logger types ORI400-DTC (n = 8) and W190-DTC (n = 4). The data loggers were soaked in aerated seawater for 14 d at 15°C after calibration, and measured in standard seawater every second day

results indicate that the ORI400-DTC can provide reliable salinity readings over several weeks. They also indicate that the reliability of the measurements may have been improved by the pre-conditioning procedure in which the data loggers were soaked in seawater (we recommend >3 d) before being attached to the fish.

Field experiments

When the fish remained in a stable environment (n = 4; Fig. 5A,C), fish profiles coincided well with vertical profiles measured by a conventional CTD profiler lowered manually into the water column, and measurement error was negligible even in the non-averaged original data. On the other hand, when fish experienced a wide range of salinity levels around the mouth of the river (n = 1, Fig. 5B), original data from fish-borne loggers varied widely. However, averaged fish profiles coincided well with CTD profiles.

For longer measurements made in natural habitats (n = 4, 53–158 h), salinity values of standard seawater measured pre- and post-deployment were consistent to the first decimal place (Table 2). This result indicates that the erosion and fouling effects on the conductivity sensor were negligible in this experiment. Three fish were caught in a freshwater area and the other fish were caught in a seawater area (Fig. 6A). All fish experienced a wide range of salinity that was in the same range when swimming from seawater to freshwater; however, mean salinity ranged from 13.9 to 32.9 (Fig. 6B). All of the mean values were above one-third of the salinity of seawater.

When we analyzed salinity data with a resolution of 1 h, we found that fish often ascended to shallow water with low salinity at night (Fig. 7). When we examined data at a more detailed time scale (sampling interval: 1 s), the duration of each freshwater (FW) excursion, in which fish were in lower osmotic water than body fluid (defined salinity <12), was found to be <1 h in 3 of 4 fish and <6 h in the other fish (Fig. 8). In total, over 99 % of FW excursions were finished within 10 min (fish1: 100.0 %, fish2: 99.7 %, fish3: 99.4 %, fish4: 99.0 %).

DISCUSSION

Micro CTD data logger

The results of this study indicate that a newly developed CTD data logger (ORI400-DTC) can successfully measure micro variations in the salinity levels of the habitat surrounding a mid-sized predator, the Japanese sea bass. This is the first report which clearly validates the accuracy of salinity measurement without sensor drift in free-ranging fish, as the measurement drifts were negligible over several days. As is the case with drifter buoys, salinity (conductivity) sensor fouling is a problem with animal-borne data loggers, although maybe less of a prob-

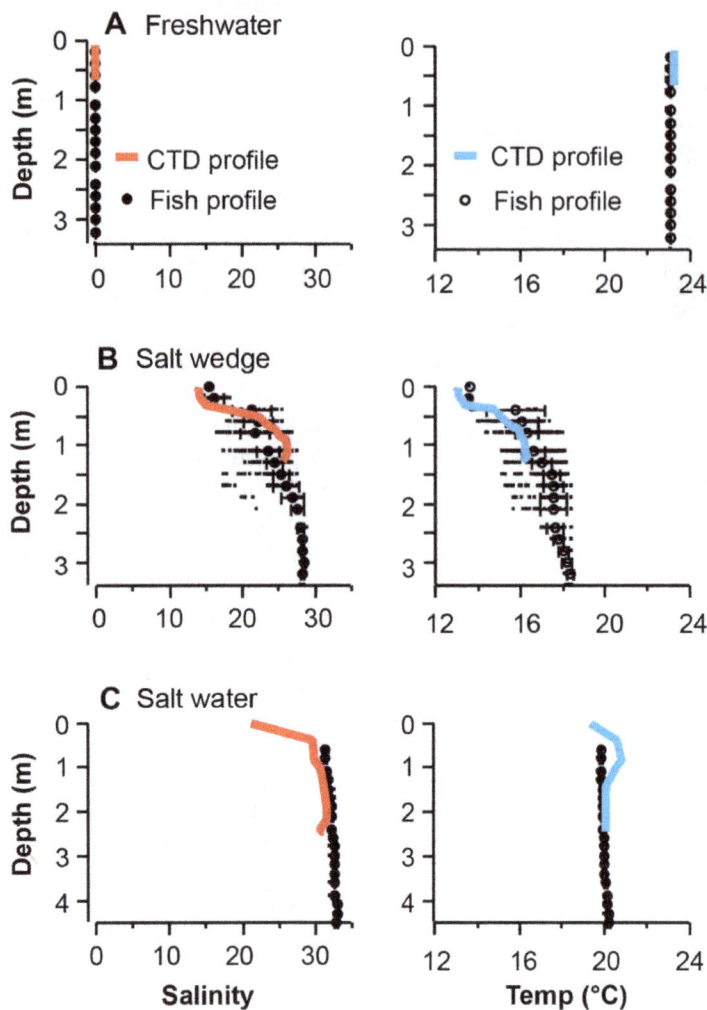

Fig. 5. Comparison of vertical profiles of salinity and temperature measured by a conventional CTD and a fish-borne data logger (n = 3). Red and blue lines represent vertical profiles of salinity and temperature, respectively, measured by a conventional CTD profiler in (A) a freshwater area, (B) a salt wedge area; in which salinity change is >10 between surface and bottom and (C) a seawater area. Filled and open circles represent averaged data calculated from 1 h data sets from fish-borne data loggers (fish profiles), error bars represent SD, and dots indicate original 1 h data

Table 2. Salinity measurement drift in standard seawater, as measured by CTD data logger (ORI400-DTC), between pre- and post-deployment on fish. Values are means ± SEM

ID	Salinity of standard seawater		
	Pre-deployment	Post-deployment	Deployment duration (h)
Fish1	34.0 ± 0.0	34.0 ± 0.0	129
Fish2	34.0 ± 0.1	34.0 ± 0.1	109
Fish3	34.0 ± 0.0	34.0 ± 0.0	158
Fish4	34.0 ± 0.0	34.0 ± 0.0	53

lem because of the animals' continuous movement through the water (Fedak 2004). This problem is greater when using a 2-AC bipolar sensor than an inductive conductivity sensor because the metal electrodes are directly exposed to water. We resolved this problem by using sensor electrodes made of stable titanium and by pre-conditioning the sensor electrodes.

Fine-scale and direct measurement of salinity surrounding aquatic organisms is considerably more difficult using the other conventional methods. Recently, many studies of anadromous and catadromous fish migration in estuarine habitats have been made easier by measuring the strontium (Sr) concentration in the otoliths, because Sr is generally recognized as a proxy of water salinity (Elsdon & Gillanders 2003, Gillanders 2005, Reis-Santos et al. 2013). However, this method has 2 limitations. Firstly, use of otoliths requires killing the individual. Secondly, in practice, the assay and interpretation of this chemical record is less than clear-cut, in part due to physiological filters between the ambient water and the otolith (Kalish 1989). Therefore, it is also unlikely that otolith chemistry would reveal short-term excursions into freshwater such as those observed in the present study. The data logger used in the present study directly measures ambient water at a fine scale. Therefore, data loggers can play a significant role in revealing the habitat use and the reaction to micro-salinity habitat variation of estuarine fish at tidal and daily time scales.

This new data logger and the methods presented here for Japanese sea bass can also be applied to other fish inhabiting estuaries. However, 2 issues should be considered when applying this method to fish research: (1) the near-field effect and (2) measurement error caused by the slow response time of the temperature sensor. The near-field effect was reduced to a negligible level by fixing instruments to each other and creating a space around the conductivity sensor. Nevertheless, the proximity of a sensor on a fish to any inorganic and organic materials, such as rocks, seaweeds and sea grasses occurring in the fish's habitat, can result in temporary errors. When reconstructing the environment from information from an animal-borne data logger, the sensor response time affects the resulting profiles. Previous studies have thus attempted to estimate the true temperature from meas-

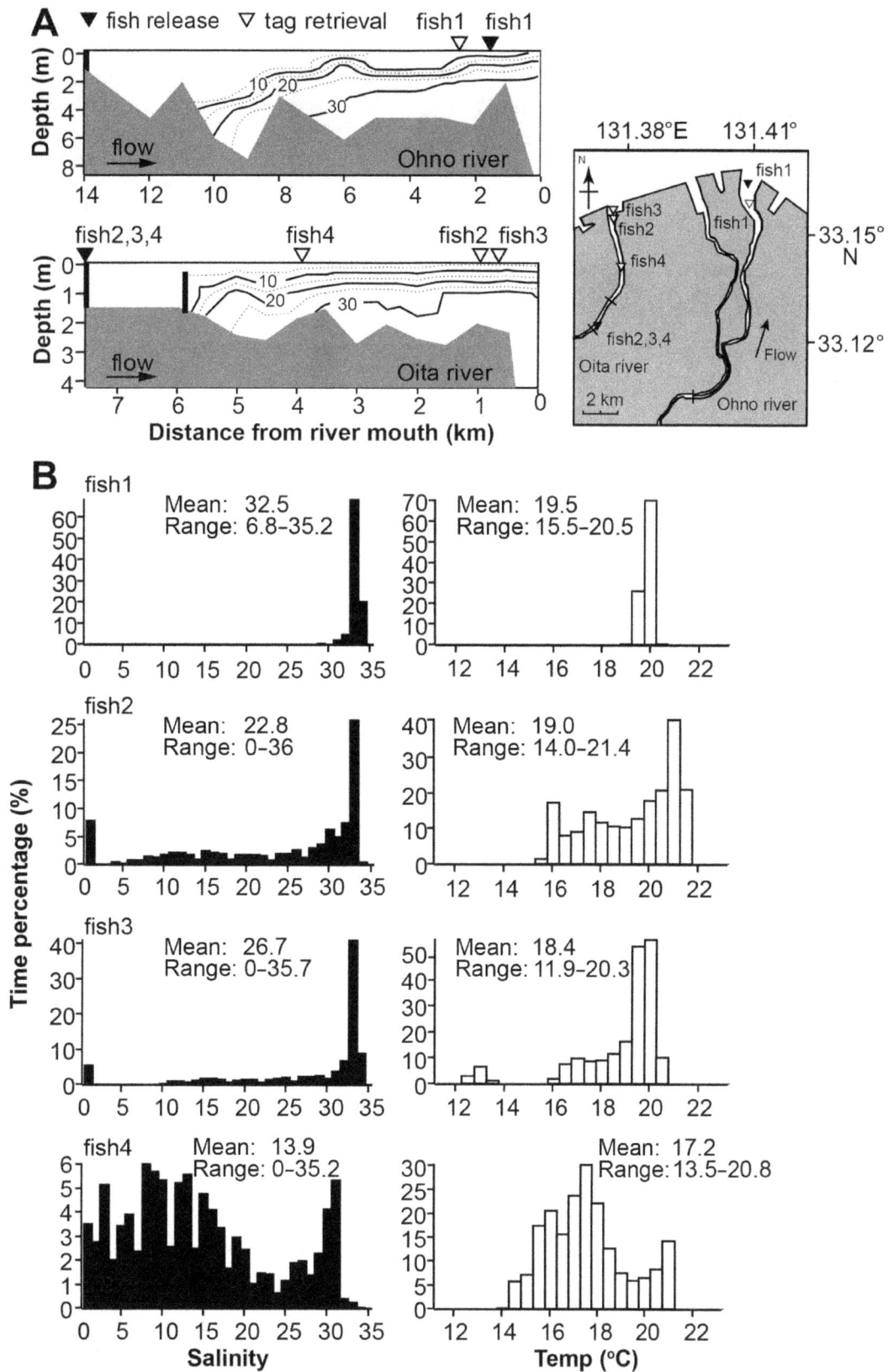

Fig. 6. Salinity at the study site and the water conditions that the fish experienced. (A) Contours represent salinity profiles measured by conventional CTD at the 2 study sites when the tide level was highest at spring tide. Black bars indicate ground-sill. (B) Percentage of time spent by fish at a certain salinity and temperature as measured by fish-borne data loggers

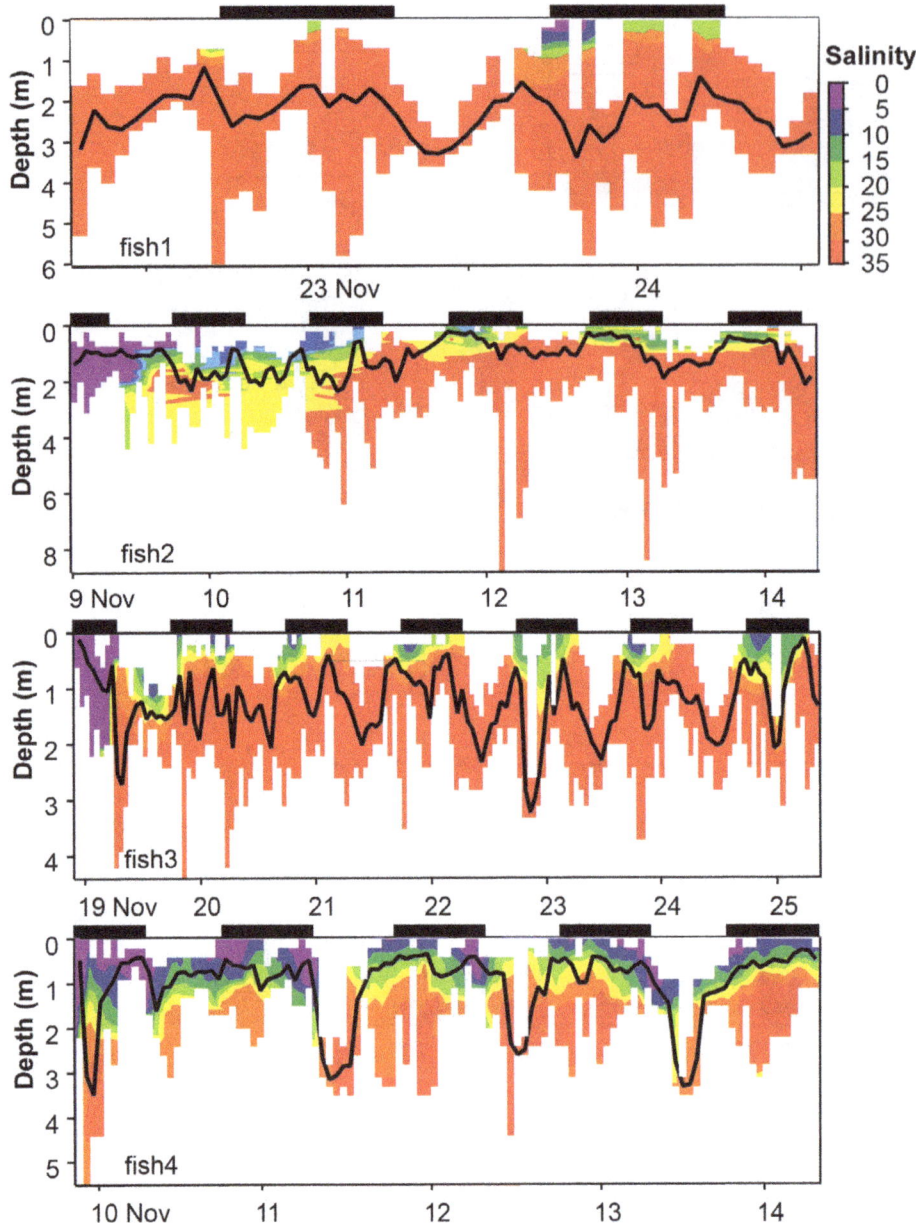

Fig. 7. Isohalines converted from the time series 'fish profiles' for the swimming activity of each fish. The salinity range is shown in different colors (see scale bar). The extent of the color bars depicts the range of the collected data for each time bin. Black lines represent the mean depth for each 1 h time bin. Black horizontal bars indicate night

resolve any inconsistencies in the main features of the water column (i.e. the depths and strength of thermo- and halocline) using the data logger. However, it should be noted that the accuracy of vertical profiles could be affected by fish swimming speed in the vertical direction and the amount of time fish remained at the same depth.

Behavioral ecology of fish

All fish showed a clear diel pattern; they often ascended to shallow, low-salinity water at night. Generally, there are 3 possible explanations for excursions from seawater to freshwater: predator avoidance, removal of ectoparasites, and foraging. Japanese sea bass are considered to be the largest predators at our study sites; therefore, it is unlikely that their purpose was predator avoidance. Heavily infested sea trout have been shown to return to freshwater prematurely (Birkeland 1996). Sea lice-infected Japanese sea bass have been observed in Tokyo Bay (T. Mori pers. obs.); however, no sea lice-infected fish were found in our study site in Oita Prefecture. Moreover, these short-term excursions were considered ineffective as a means of removing sea lice because sea lice can survive in freshwater for 1–2 wk (Finstad et al. 1995). A reasonable explanation for the night excursions seems to be that the

ured data (Daunt et al. 2003, Charrassin et al. 2004, Takahashi et al. 2008). The ORI400-DTC data logger also has a considerable time constant (15 s) on the temperature sensor (Table 1), which affects the salinity calculation. In this study, data averaging reduced the differences between profiles measured by data logger and conventional CTD. Despite a slight discrepancy in time and location, the 2 measurements yielded quite similar profiles, thus permitting us to

sea bass were foraging on post-spawning Ayu fish *Plecoglossus altivelis* which spawn in October and November, when this research took place. The Ayu fish is one of the main prey species of Japanese sea bass (Ochiai & Tanaka 1998) and has often been found in the stomach contents of sea bass in this season (N. Miyata pers. obs.). Ayu fish mainly spawn between 16:00 and 20:00 h in freshwater and die shortly after reproducing (Ochiai & Tanaka 1998). Capturing

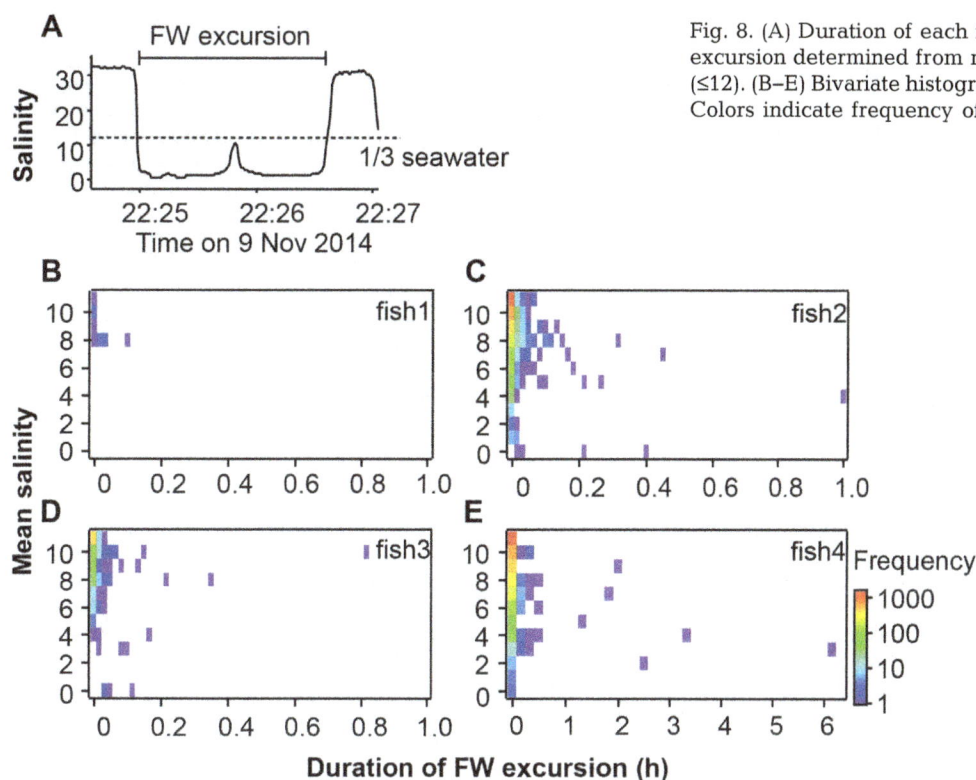

Fig. 8. (A) Duration of each freshwater (FW) excursion determined from raw salinity data (≤12). (B–E) Bivariate histogram for each fish. Colors indicate frequency of FW excursions

floating Ayu fish dying after spawning would be an efficient and energy-saving strategy for foraging sea bass.

Physiologically, teleost fish are known to maintain their osmotic concentration at about one-quarter to one-third of that in seawater (Schmidt-Nielsen 1997). Hence, Japanese sea bass in our field study were, on average, in a hyper-osmotic environment. This indicates a higher probability that their body-fluid osmoregulatory system would acclimate to seawater, even though 3 out of 4 fish were caught in a freshwater river (Fig. 6A). Their short-term excursions from seawater to freshwater might be an effective strategy for temporal utilization of hypo-osmotic environments without physiological and energetic cost; however, further insight into this aspect is a matter for future research.

In this study, we have demonstrated that the newly developed micro CTD data logger can be used to collect fine-scale salinity information from estuarine-dependent fish with negligible measurement drift. Applying this method in future studies will give better insights into the habitat use, foraging strategies and osmoregulatory mechanisms of estuary-dependent species. This basic information about behavioral ecology of euryhaline fish is important to help understand estuarine ecosystems. Top-down forcing and trophic cascades often have pronounced effects on

the abundance and species composition of autotrophs, leading to regime shifts and alternative states of ecosystems (Terborgh & Estes 2010). Euryhaline fish can affect freshwater, estuarine and marine ecosystems through predation; hence, knowledge of their basic ecology is critical for the conservation of ecosystems around estuaries, which are some of the environments most vulnerable to anthropogenic change.

Acknowledgements. This study project was financially supported by the Bio-logging Science project, the University of Tokyo (UTBLS), a Sasakawa Scientific Research Grant (to T.M.) and the Japan Society for the Promotion of Science (24241001 and 25660152 to K.S.). We thank K. Utsumi and the staff of Oita Agriculture and Forestry Fisheries Study Instruction Center for their assistance. We also thank K. Kiyosue and T. Harada for catching experimental fish.

LITERATURE CITED

➤ Birkeland K (1996) Consequences of premature return by sea trout (*Salmo trutta*) infested with the salmon louse (*Lepeophtheirus salmonis* Krøyer): migration, growth, and mortality. Can J Fish Aquat Sci 53:2808–2813
➤ Biuw M, Boehme L, Guinet C, Hindell M and others (2007) Variations in behavior and condition of a Southern Ocean top predator in relation to *in situ* oceanographic conditions. Proc Natl Acad Sci USA 104:13705–13710
➤ Boehme L, Meredith MP, Thorpe SE, Biuw M, Fedak M (2008a) Antarctic circumpolar current frontal system in the South Atlantic: monitoring using merge Argo and

animal-borne sensor data. J Geophys Res 113:C09012, doi:10.1029/2007JC004647

▶ Boehme L, Thorpe SE, Biuw M, Fedak M, Meredith MP (2008b) Monitoring Drake Passage with elephant seals: frontal structures and snapshots of transport. Limnol Oceanogr 53:2350–2360

▶ Boehme L, Lovell P, Biuw M, Roquet F and others (2009) Technical note: Animal-borne CTD-satellite relay data loggers for real-time oceanographic data collection. Ocean Sci 5:685–695

Boesch DF (1977) A new look at the zonation of benthos along the estuarine gradient. In: Coull BC (ed) Ecology of marine benthos. University of South Carolina Press, Columbia, SC, p 245–266

Bœuf G, Payan P (2001) How should salinity influence fish growth? Comp Biochem Physiol C Toxicol Pharmacol 130:411–423

▶ Bulger AJ, Hayden BP, Monaco ME, Nelson DM, McCormick-Ray MG (1993) Biologically-based estuarine salinity zones derived from a multivariate analysis. Estuaries 16:311–322

▶ Charrassin JB, Park YH, Le Maho Y, Bost CA (2004) Fine resolution 3D temperature fields off Kerguelen from instrumented penguins. Deep-Sea Res I 51:2091–2103

▶ Charrassin JB, Hindell M, Rintoul SR, Roquet F and others (2008) Southern Ocean frontal structure and sea-ice formation rates revealed by elephant seals. Proc Natl Acad Sci USA 105:11634–11639

▶ Currin BM, Reed JP, Miller JM (1984) Growth, production, food consumption, and mortality of juvenile spot and croaker: a comparison of tidal and nontidal nursery areas. Estuaries 7:451–459

▶ Daunt F, Peters G, Scott B, Grémillet D, Wanless S (2003) Rapid-response recorders reveal interplay between marine physics and seabird behaviour. Mar Ecol Prog Ser 255:283–288

▶ Elsdon TS, Gillanders BM (2003) Reconstructing migratory patterns of fish based on environmental influences on otolith chemistry. Rev Fish Biol Fish 13:217–235

Fedak M (2004) Marine animals as platforms for oceanographic sampling: a ' win/win ' situation for biology and operational oceanography. Mem Natl Inst Polar Res Spec Issue 58:133–147

▶ Finstad B, Bjørn PA, Nilsen ST (1995) Survival of salmon lice, Lepeophtheirus salmonis Krøyer, on Arctic charr, Salvelinus alpinus (L.), in fresh water. Aquacult Res 26: 791–795

▶ Gillanders BM (2005) Otolith chemistry to determine movements of diadromous and freshwater fish. Aquat Living Resour 18:291–300

▶ Hooker SK, Boyd IL (2003) Salinity sensors on seals: use of marine predators to carry CTD data loggers. Deep-Sea Res I 50:927–939

▶ Kalish JM (1989) Otolith microchemistry: validation of the effects of physiology, age and environment on otolith composition. J Exp Mar Biol Ecol 132:151–178

▶ Lankford TE, Targett TE (1994) Suitability of estuarine nursery zones for juvenile weakfish (Cynoscion regalis): effects of temperature and salinity on feeding, growth and survival. Mar Biol 119:611–620

▶ Lydersen C, Nost OA, Lovell P, McConnell BJ and others (2002) Salinity and temperature structure of a freezing Arctic fjord—monitored by white whales (Delphinapterus

leucas). Geophys Res Lett 29:34-1–34-4, doi:10.1029/2002GL015462

▶ Meredith MP, Nicholls KW, Renfrew IA, Boehme L, Biuw M, Fedak M (2011) Seasonal evolution of the upper-ocean adjacent to the South Orkney Islands, Southern Ocean: results from a 'lazy biological mooring'. Deep-Sea Res II 58:1569–1579

Ochiai A, Tanaka M (1998) Ichthyology (new version), Part 2. Koseishakoseikaku, Tokyo

Pinet PR (2011) Invitation to oceanography, 6th edn. Jones & Bartlett, Burlington, MA

▶ Plaut I (1999) Effects of salinity acclimation on oxygen consumption in the freshwater blenny, Salaria fluviatilis, and the marine blenny, S. pavo. Mar Freshw Res 50:655–659

▶ Reis-Santos P, Tanner SE, Elsdon TS, Cabral HN, Gillanders BM (2013) Effects of temperature, salinity and water composition on otolith elemental incorporation of Dicentrarchus labrax. J Exp Mar Biol Ecol 446:245–252

▶ Roquet F, Park YH, Guinet C, Bailleul F, Charrassin JB (2009) Observations of the fawn trough current over the Kerguelen plateau from instrumented elephant seals. J Mar Syst 78:377–393

Schmidt-Nielsen K (1997) Animal physiology: adaptation and environment, 5th edn. Cambridge University Press, Cambridge

Shoji N, Sato K, Ozaki M (2002) Distribution and utilization of the stock. In: Tanaka M, Kinoshita I (eds) Temperate bass and biodiversity—new perspective for fisheries biology. Koseisha Koseikaku, Shinjuku, p 9–20

Sturlaugsson J, Sigmar G (1997) Tracking of Atlantic salmon (Salmo salar L.) and sea trout (Salmo trutta L.) with Icelandic data storage tags. In: Boehlert GW (ed) Application of acoustic tags and archival tags to assess estuarine, nearshore, and offshore habitat utilization and movement by salmonids. NOAA-TM NM. National Marine Fisheries Service, San Diego, CA, p 52–54

▶ Takahashi A, Matsumoto K, Hunt GL, Shultz MT and others (2008) Thick-billed murres use different diving behaviors in mixed and stratified waters. Deep-Sea Res II 55: 1837–1845

▶ Takei Y, Hiroi J, Takahashi H, Sakamoto T (2014) Diverse mechanisms for body fluid regulation in teleost fishes. Am J Physiol Regul Integr Comp Physiol 307:R778–R792

Tanaka M (1997) Inshore migration of coastal marine fish in relation to metamorphosis. Kaiyo Monthly 29:199–204

Terborgh J, Estes JA (eds) (2010) Trophic cascades: predators, prey, and the changing dynamics of nature. Island Press, Wachington, DC

▶ Tseng YC, Hwang PP (2008) Some insights into energy metabolism for osmoregulation in fish. Comp Biochem Physiol C Toxicol Pharmacol 148:419–429

UNESCO (1981) Tenth report of the joint panel on oceanographic tables and standards. UNESCO Tech Pap Mar Sci 36:1–25

Watson DF, Merriam D (1992) Contouring: a guide to the analysis and display of spatial data. Pergamon Press, Oxford

▶ Weinstein MP, Weiss SL, Walters MF (1980) Multiple determinants of community structure in shallow marsh habitats, Cape Fear River estuary, North Carolina, USA. Mar Biol 58:227–243

▶ Zydlewski J, Wilkie MP (2012) Freshwater to seawater transitions in migratory fishes. Fish Physiol 32:253–326

Permissions

All chapters in this book were first published in AB, by Inter-Research; hereby published with permission under the Creative Commons Attribution License or equivalent. Every chapter published in this book has been scrutinized by our experts. Their significance has been extensively debated. The topics covered herein carry significant findings which will fuel the growth of the discipline. They may even be implemented as practical applications or may be referred to as a beginning point for another development.

The contributors of this book come from diverse backgrounds, making this book a truly international effort. This book will bring forth new frontiers with its revolutionizing research information and detailed analysis of the nascent developments around the world.

We would like to thank all the contributing authors for lending their expertise to make the book truly unique. They have played a crucial role in the development of this book. Without their invaluable contributions this book wouldn't have been possible. They have made vital efforts to compile up to date information on the varied aspects of this subject to make this book a valuable addition to the collection of many professionals and students.

This book was conceptualized with the vision of imparting up-to-date information and advanced data in this field. To ensure the same, a matchless editorial board was set up. Every individual on the board went through rigorous rounds of assessment to prove their worth. After which they invested a large part of their time researching and compiling the most relevant data for our readers.

The editorial board has been involved in producing this book since its inception. They have spent rigorous hours researching and exploring the diverse topics which have resulted in the successful publishing of this book. They have passed on their knowledge of decades through this book. To expedite this challenging task, the publisher supported the team at every step. A small team of assistant editors was also appointed to further simplify the editing procedure and attain best results for the readers.

Apart from the editorial board, the designing team has also invested a significant amount of their time in understanding the subject and creating the most relevant covers. They scrutinized every image to scout for the most suitable representation of the subject and create an appropriate cover for the book.

The publishing team has been an ardent support to the editorial, designing and production team. Their endless efforts to recruit the best for this project, has resulted in the accomplishment of this book. They are a veteran in the field of academics and their pool of knowledge is as vast as their experience in printing. Their expertise and guidance has proved useful at every step. Their uncompromising quality standards have made this book an exceptional effort. Their encouragement from time to time has been an inspiration for everyone.

The publisher and the editorial board hope that this book will prove to be a valuable piece of knowledge for researchers, students, practitioners and scholars across the globe.

List of Contributors

David M. Stafford
Moss Landing Marine Laboratories, Moss Landing, California 95039, USA

Susan M. Sogard
Fisheries Ecology Division, Southwest Fisheries Science Center, National Marine Fisheries Service, NOAA, Santa Cruz, California 95060, USA

Steven A. Berkeley
University of California Santa Cruz, Santa Cruz, California 95060, USA

Juan B. Ortiz-Delgado and Carmen Sarasquete
Instituto de Ciencias Marinas de Andalucía-ICMAN/CSIC, Campus Universitario Río San Pedro, Apdo. Oficial, 11510, Puerto Real, Cádiz, Spain

Ignacio Fernández
IRTA, Centre de Sant Carles de la Ràpita (IRTA-SCR), Unitat de Cultius Experimentals, Crta. del Poble Nou s/n, 43540 Sant Carles de la Ràpita, Spain
Centre of Marine Sciences (CCMAR), University of Algarve, Campus de Gambelas, 8005-139 Faro, Portugal

Enric Gisbert
IRTA, Centre de Sant Carles de la Ràpita (IRTA-SCR), Unitat de Cultius Experimentals, Crta. del Poble Nou s/n, 43540 Sant Carles de la Ràpita, Spain

N. Takai, Y. Kozuka, T. Tanabe, Y. Sagara, M. Ichihashi, S. Nakai and M. Suzuki, N. Mano, S. Itoi, K. Asahina, T. Kojima and H. Sugita
Department of Marine Science and Resources, College of Bioresource Sciences, Nihon University, Fujisawa 252-0880, Japan

Charles A. Gray
NSW Primary Industries, Cronulla Fisheries Research Centre, Cronulla, NSW 2230, Australia
WildFish Research, Grays Point, Sydney, NSW 2232, Australia
University of New South Wales, Randwick, NSW 2052, Australia

Lachlan M. Barnes
NSW Primary Industries, Cronulla Fisheries Research Centre, Cronulla, NSW 2230, Australia
Cardno, St Leonards, NSW 2065, Australia

Dylan E. van der Meulen
NSW Primary Industries, Cronulla Fisheries Research Centre, Cronulla, NSW 2230, Australia
Batemans Bay Fisheries Centre, NSW 2536, Australia

Benjamin W. Kendall
NSW Primary Industries, Cronulla Fisheries Research Centre, Cronulla, NSW 2230, Australia
Seglaregatan, Gothenburg 41457, Sweden

Faith A. Ochwada-Doyle
NSW Primary Industries, Cronulla Fisheries Research Centre, Cronulla, NSW 2230, Australia
Sydney Institute of Marine Science, Mosman, NSW 2088, Australia

William D. Robbins
NSW Primary Industries, Cronulla Fisheries Research Centre, Cronulla, NSW 2230, Australia
8Wildlife Marine, Sorrento, Perth, WA 6020, Australia

Marie-Anne Blanchet and Rolf A. Ims
Norwegian Polar Institute, Framsentret, 9296 Tromsø, Norway
Department of Arctic and Marine Biology, UiT-Arctic University of Norway, 9037 Tromsø, Norway

Christian Lydersen, Andrew D. Lowther and Kit M. Kovacs
Norwegian Polar Institute, Framsentret, 9296 Tromsø, Norway

Sean Bignami
Division of Marine Biology and Fisheries, Rosenstiel School of Marine and Atmospheric Science, University of Miami, Miami, FL 33149, USA
Concordia University Irvine, Irvine, CA 92612, USA

Su Sponaugle
Division of Marine Biology and Fisheries, Rosenstiel School of Marine and Atmospheric Science, University of Miami, Miami, FL 33149, USA
Department of Integrative Biology, Oregon State University, Corvallis, OR 97331, USA
Hatfield Marine Science Center, Oregon State University, Newport, OR 97365, USA

Robert K. Cowen
Division of Marine Biology and Fisheries, Rosenstiel School of Marine and Atmospheric Science, University of Miami, Miami, FL 33149, USA

Hatfield Marine Science Center, Oregon State University, Newport, OR 97365, USA

D. M. Moore and I. D. McCarthy
School of Ocean Sciences, College of Natural Sciences, Bangor University, Menai Bridge, Anglesey LL59 5AB, UK

Jeremy J. Kiszka
Littoral Environnement et Sociétés (LIENSs), UMR 7266 CNRS-Université de la Rochelle, 2 rue Olympe de Gouges, 17000 La Rochelle, France
Marine Sciences Program, Department of Biological Sciences, Florida International University, 3000 NE 151st Street, North Miami, Florida 33181, USA

Kevin Charlot, Paco Bustamante, Benoit Simon-Bouhet and Florence Caurant
Littoral Environnement et Sociétés (LIENSs), UMR 7266 CNRS-Université de la Rochelle, 2 rue Olympe de Gouges, 17000 La Rochelle, France

Nigel E. Hussey
Great Lakes Institute for Environmental Research, University of Windsor, 401 Sunset Avenue, Ontario N9B 3P4, Canada

Michael R. Heithaus
Marine Sciences Program, Department of Biological Sciences, Florida International University, 3000 NE 151st Street, North Miami, Florida 33181, USA

Frances Humber
Blue Ventures, Level 2 Annex, Omnibus Business Centre, 39-41 North Road, London N7 9DP, UK
Centre for Ecology and Conservation, College of Life and Environmental Sciences, University of Exeter, Penryn TR10 9FE, UK

Maritza Sepúlveda and Macarena Santos-Carvallo
Centro de Investigación y Gestión de los Recursos Naturales (CIGREN), Instituto de Biología, Facultad de Ciencias, Universidad de Valparaíso, Gran Bretaña 1111, Playa Ancha, Valparaíso 2360102, Chile
Centro de Investigación Eutropia, Ahumada 131 Oficina 912, Santiago 8320238, Chile

Danai Olea
Centro de Investigación y Gestión de los Recursos Naturales (CIGREN), Instituto de Biología, Facultad de Ciencias, Universidad de Valparaíso, Gran Bretaña 1111

Pablo Carrasco and Renato A. Quiñones
Programa de Investigación Marina de Excelencia (PIMEX), Facultad de Ciencias Naturales y Oceanográficas, Casilla 160-C, Universidad de Concepción, Concepción 4070205, Chile

Interdisciplinary Center for Aquaculture Research (INCAR-FONDAP), Universidad de Concepción, O'Higgins 1695, Concepción 4070007, Chile

Jorge Castillo
Instituto de Fomento Pesquero, Casilla 8v, Valparaíso, Almte. Manuel Blanco Encalada 839, Valparaíso 2361827, Chile

Hernán J. Sacristán and Laura S. López Greco
Biology of Reproduction and Growth in Crustaceans, Department of Biodiversity and Experimental Biology, FCE y N, University of Buenos Aires, Cdad. Univ. C1428EGA, Buenos Aires, Argentina
Instituto de Biodiversidad y Biologia Experimental y Aplicada (IBBEA), CONICET-UBA, Argentina

Héctor Nolasco-Soria
Centro de Investigaciones Biológicas del Noroeste, S. C. La Paz, Baja California Sur 23090, México

Emma Samson, Jacob W. Brownscombe and Steven J. Cooke
Fish Ecology and Conservation Physiology Laboratory, Ottawa-Carleton Institute for Biology, Carleton University, 1125 Colonel By Dr., Ottawa, ON K1S 5B6, Canada

Kazutaka Takahashi
Department of Aquatic Bioscience, Graduate School of Agricultural and Life Sciences, The University of Tokyo 1-1-1, Yayoi,Bunkyo-ku, Tokyo 113-8657, Japan

Norio Nagao
Institute of Bioscience, Universiti Putra Malaysia, 43400 Serdang, Selangor Darul Ehsan, Malaysia

Satoru Taguchi
Faculty of Engineering, Soka University, 1-236 Tangi-cho, Hachioji, Tokyo 192-8577, Japan

Bianka Grunow, Tina Kirchhoff and Steffen Harzsch
Cytology and Evolutionary Biology, Zoological Institute and Museum, Ernst-Moritz-Arndt-University of Greifswald, Soldmannstrasse 23, 17487 Greifswald, Germany

Tabea Lange
Cytology and Evolutionary Biology, Zoological Institute and Museum, Ernst-Moritz-Arndt-University of Greifswald, Soldmannstrasse 23, 17487 Greifswald, Germany
Deutsches Meeresmuseum, Katharinenberg 14–20, 18439 Stralsund, Germany

Timo Moritz
Deutsches Meeresmuseum, Katharinenberg 14–20, 18439 Stralsund, Germany

Katherina Brokordt, Hernán Pérez, Catalina Herrera and Alvaro Gallardo
Centro de Estudios Avanzados en Zonas Áridas (CEAZA), Universidad Católica del Norte, Larrondo 1281, Coquimbo, Chile

Thaisa F. Bergamo and Margareth Copertino
Laboratório de Ecologia Vegetal Costeira, Instituto de Oceanografia, Universidade Federal do Rio Grande - FURG, CP 474, Rio Grande, RS 96203-900, Brazil
Programa de Pós-Graduação em Oceanografia Biológica, Instituto de Oceanografia, Universidade Federal do Rio Grande - FURG, CP 474, Rio Grande, RS 96203-900, Brazil

Silvina Botta
Programa de Pós-Graduação em Oceanografia Biológica, Instituto de Oceanografia, Universidade Federal do Rio Grande - FURG, CP 474, Rio Grande, RS 96203-900, Brazil
Laboratório de Ecologia e Conservação da Megafauna Marinha, Universidade Federal do Rio Grande - FURG, CP 474, Rio Grande, RS 96203-900, Brazil

Emanuell Felipe Silva
Instituto Federal de Educação, Ciência e Tecnologia da Paraíba, Campus Cabedelo, Cabedelo, 58103-772 PB, Brazil

Nathalia Calazans, Roberta Soares, Thaís Castelo Branco and Silvio Peixoto
Laboratório de Tecnologia em Aquicultura (LTA), Departamento de Pesca e Aquicultura, Universidade Federal Rural de Pernambuco, 52171-900 Recife/PE, Brazil

Leandro Nolé and Flávia Lucena Frédou
Laboratório de Estudos de Impactos Antrópicos na Biodiversidade Marinha e Estuarina (BIOIMPACT), Departamento de Pesca e Aquicultura, Universidade Federal Rural de Pernambuco, 52171-900 Recife/PE, Brazil

Maria Madalena Pessoa Guerra
Laboratório de Andrologia, Departamento de Medicina Veterinária, Universidade Federal Rural de Pernambuco, 52171-900 Recife/PE, Brazil

Katsuyuki Hamasaki, Takahiro Matsuda, Ken Takano, Mio Sugizaki, Yu Murakami and Shuichi Kitada
Graduate School of Marine Science and Technology, Tokyo University of Marine Science and Technology, Minato, Konan, Tokyo 108-8477, Japan

Shigeki Dan
Research Centre for Marine Invertebrate, National Research Institute of Fisheries and Environment of Inland Sea, Fisheries Research Agency, Momoshima, Onomichi, Hiroshima 722-0061, Japan

Naoyuki Miyata, Tomohiko Mori and Katsufumi Sato
Atmosphere and Ocean Research Institute, The University of Tokyo, 5-1-5 Kashiwanoha, Kashiwa, Chiba, 277-8564, Japan

Masaaki Kagehira
Oita Prefecture Southern Region Bureau Rural Community Promotion Department, Saiki, Oita, 876-0813, Japan

Nobuyuki Miyazaki
Japan Marine Science Foundation, Taito, Tokyo, 110-0008, Japan

Michihiko Suzuki
Little Leonardo, Bunkyo, Tokyo, 113-0021, Japan

Index

www.ingramcontent.com/pod-product-compliance
Lightning Source LLC
Chambersburg PA
CBHW080658200326
41458CB00013B/4901